PROGRAMMABLE CONTROLLERS USING THE ALLEN-BRADLEY SLC-500 FAMILY

Second Edition

David A. Geller

Upper Saddle River, New Jersey
Columbus, Ohio

Assistant Vice President and Publisher: Charles E. Stewart, Jr.
Assistant Editor: Mayda Bosco
Production Editor: Alexandrina Benedicto Wolf
Design Coordinator: Diane Ernsberger
Cover Designer: Keith Van Norman
Cover art: Digital Vision
Production Manager: Matt Ottenweller
Marketing Manager: Ben Leonard

This book was set in Times Roman by Integra. It was printed and bound by Phoenix Color Corp. The cover was printed by Phoenix Color Corp.

Pearson Education Ltd.
Pearson Education Singapore Pte. Ltd.
Pearson Education Canada, Ltd.
Pearson Education—Japan
Pearson Education Australia Pty. Limited
Pearson Education North Asia Ltd.
Pearson Educación de Mexico, S.A. de C.V.
Pearson Education Malaysia Pte. Ltd.

10 9 8 7 6 5 4 3 2 1
ISBN 0-13-113052-8

PREFACE

The programmable logic controller (PLC) is now the control computer of choice in virtually all new industrial control design projects. It has replaced relay logic in almost every control system in current use by being retrofitted into older automation equipment. This complete takeover of the controls market is based on the PLC's ease of programming, reliability, and ability to provide real-time operating data. The computing power of PLCs has steadily increased to the point where the top-of-the-line controllers now have as much, or more, computing power than the standard desktop PC.

Programmable controllers can be used to control a variety of industrial machines and processes. These include control applications for automation machinery, process control systems, and HVAC control and monitoring in large buildings. Because PLC systems are now so widely used in industry, everyone from electrical engineers to maintenance technicians needs to have a basic understanding of PLC interfacing and programming.

Due to many years of corporate downsizing, most manufacturing engineering groups and maintenance groups are now staffed at minimum levels. This means that engineering and maintenance employees must now cover a much broader range of technology in their organizations. The days of being a mechanical or computer specialist are gone, and employees working in these manufacturing disciplines are expected to be able to handle any problem that arises. Because of this broadening of responsibility, it is now almost mandatory that all engineering and maintenance employees have some background in PLC systems and control systems in general.

ORGANIZATION

This book presents a comprehensive study of the use of PLC systems in the design of automation equipment.

The book begins with a chapter on binary mathematics because understanding binary math is essential to understanding memory addressing, math functions, and I/O address allocations used in PLC systems.

Chapters 2 and 3 cover the typical input and output devices and modules used in control systems. Examples are provided that demonstrate how sensing devices are interfaced to the PLC control system.

Chapters 4 and 5 cover the basic concepts of ladder logic, state diagrams, and flow charts, and how these elements are used to develop ladder programs to control a machine or process. Also covered in detail are the PLC's internal memory layout and I/O scan process.

Chapter 6 covers the equipment design process from a project management point of view. This design process has evolved to a team-based design approach that is now used by most companies.

Chapter 7 covers the software used in programming the PLC. This includes addressing formats, ladder instruction formats, and their proper uses. Also included are directions for setting up a project from scratch and defining the system's I/O configuration. A sample real-world design problem is presented and a control program is generated to solve the design problem.

Chapter 8 covers the basic mechanical building blocks of machine control. The control and interface of air cylinders and air valves are explained along with both rotary and linear indexer devices.

Chapters 9 through 11 cover timer and counter instructions with example programs for a variety of real-world control problems. Math instructions, compare instructions, and data handling instructions are also covered in this section.

Chapter 12 is devoted to understanding the design problems encountered in controlling manually activated machines. A sample program for a machine of this type is developed and written.

Chapter 13 deals with the design problems encountered in controlling continuously running machines. A sample program for a machine of this type is developed and written. Assembly line data tracking and word shift registers are also included in this chapter.

Chapters 14 and 15 cover ways to manipulate and track assembly line data. The use of sequencer-driven programs is also covered in detail.

Chapter 16 introduces the design of diagnostic programs for monitoring the performance of automation systems.

Chapter 17 covers HMI displays and their layout, programming, and interface to the control PLC.

CHANGES TO THE SECOND EDITION

In response to student requests, a chapter was added to cover the different input devices and sensor devices commonly used in most control system designs today. Because input devices and sensor devices account for more than 75% of the I/O used in most control designs, it was felt that a chapter covering the interfacing and proper use of these devices was required.

Each chapter now begins with a list of learning objectives. These objectives will help the student determine if he or she has mastered the concepts presented in each chapter.

Hands-on lab projects have been included at the end of each chapter beginning with Chapter 7. The lab projects are designed to reinforce the material introduced in each chapter. Working through the lab project allows the student to experiment and become comfortable with the new information and concepts presented in each chapter.

Chapter 13 has a new section that explains how to set up the proper timing sequences for continuously running machines. A sample program was added to demonstrate the basic concepts of cycle timers and methods for stopping the auto cycle only at the end of each assembly operation.

Chapter 15, which covers sequencer programming, now contains a sample program for demonstrating how to set up and load an SQO/SQC sequencer pair function.

The book now comes with a student CD that contains copies of all the sample programs covered in the book. As the student is working through the chapter material he or she can download the sample program and examine how it functions on the lab kit. The sample program can then be modified to gain a greater understanding of the concepts covered in the chapter.

An Instructor's Manual containing PowerPoint slides of figures, completed lab projects, and all the sample programs found in the book is available online.

STUDENT INFORMATION

The reader needs to have a basic understanding of electrical systems and machine concepts. Each chapter starts with a list of learning objectives that will be covered in the chapter. Most chapters end with a series of review questions that will test the reader's level of understanding. These review questions should help readers evaluate their level of comprehension of the material presented in the chapter. In addition, beginning with Chapter 7, all chapters end with lab projects that will provide the student with practical hands-on experience on the material covered in the chapter.

ACKNOWLEDGEMENTS

I would like to thank all my previous students for their insightful questions and patience in helping me develop this course material. A special thanks to the many technical staff and support staff members at Allen-Bradley who provided pictures and other technical information that made this text possible. I also thank the reviewers for this edition, Sam Guccione of Eastern Illinois University and Richard Windley of ECPI College of Technology, for their invaluable feedback.

David Geller

CONTENTS

To my loving wife, Diane, for her patience, help, and understanding during the rewrite of this text. And to my many mentors who, over the years, helped me transition from the world of theory to the real world of manufacturing, where projects must be completed on time, be under budget, and function 24-7.

I

Introduction to Programmable Controllers

Prior to the invention of the microprocessor chip in the 1970s, all machine control was accomplished by the use of relays. Control systems were designed with discrete wiring between all the logic elements of the control system (relays, switches, push buttons). To design control systems of this type, the design engineer had to draw electrical wiring diagrams to indicate how all of the components were connected. An example of such a drawing is shown in Figure I–1. As you can see from the example, this type of electrical drawing has the main power buses running vertically and the logic connections running horizontally between them. This layout resembles a ladder and for this reason became known as a *ladder diagram*.

Notice the numbers running down the outside of the power rails in Figure I–1. The numbers to the left are rung numbers. Each connection across the power rails is considered a single rung even if it has more than one branch. Notice that rungs 1, 2, and 7 have multiple start branches. The numbers on the right side of the power rail indicate in which rungs the control relay contacts are used. In this way an electrical engineer can find his or her way around the circuit. This example is only 7 rungs long; normal drawings for large machines often contain hundreds of rungs.

You should also notice that each wire used in the drawing has a number. This was necessary for troubleshooting large machine designs that may have had several hundred wire numbers. As each wire was connected, this number was attached to each termination point.

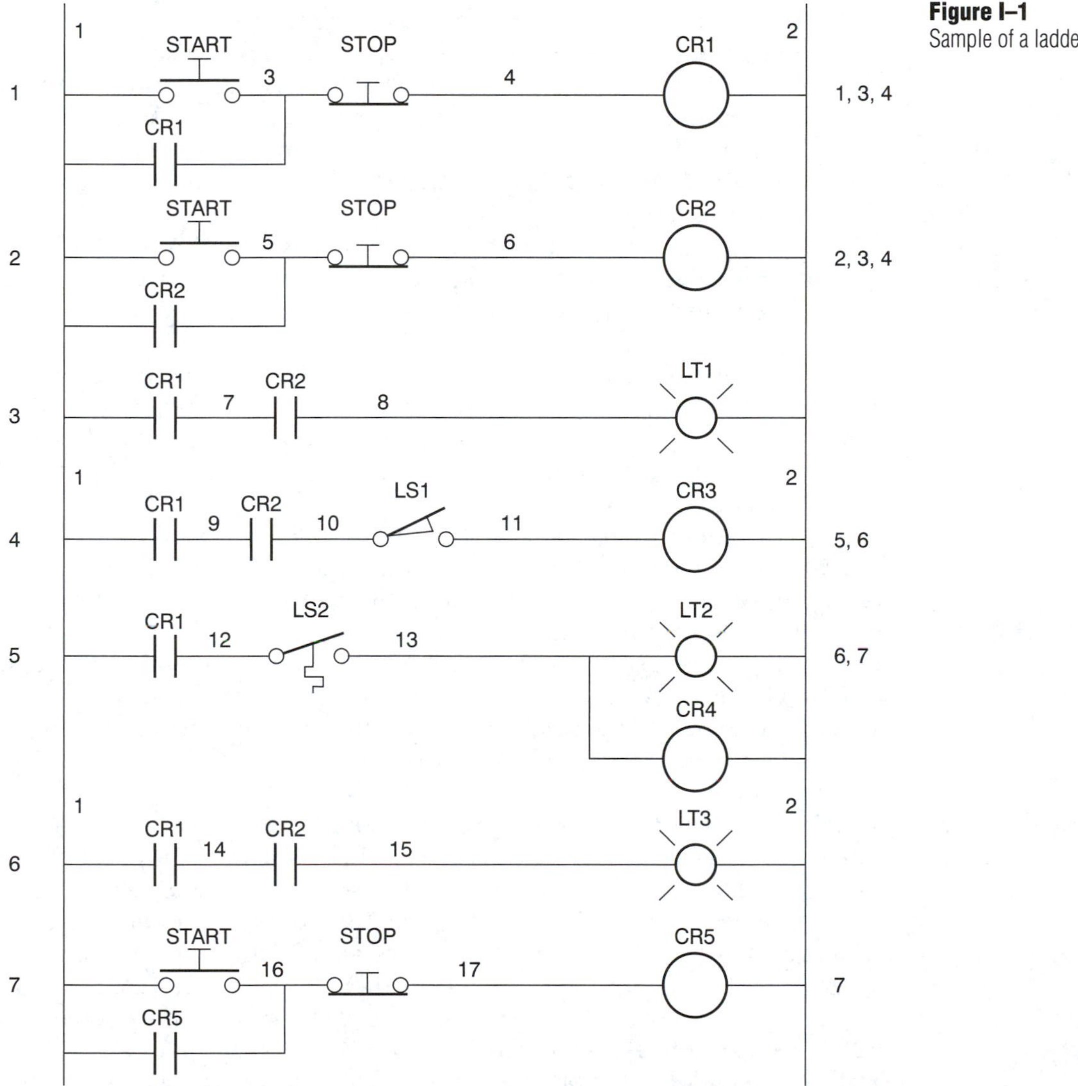

Figure I–1
Sample of a ladder diagram

Although this control system concept was used for many years, it has some major design and maintenance problems, some of which follow:

1. Each control relay had, at most, four sets of contacts. If more than four were needed, a second parallel relay had to be added.
2. Limit switches connected in series with relay contacts (rung 4) required long runs of wire on each contact.
3. Adding additional logic elements to the ladder was a very time-consuming process. To add a limit switch to rung 4 between LS1 and CR3 required the engineer to find both ends of wire number 11 and remove it. Then a new wire number 11 had to be added from the LS1 switch to one contact of the new switch and then a new wire number (18) added from CR3 to the other contact point of the new switch.
4. Debugging wiring mistakes in a control system with several hundred wire numbers was a long process. Keeping the documentation up to date was also a major problem.

The maintenance of such a system was also a major problem. Consider a control system with several hundred switches, relays, and contacts. Even a very good relay has a life span of 1 or 2 million cycles. In some high-speed machines this might be only 6 months of use. As relays start to degrade, they seldom fail all at once. Instead, they usually start to miss contact closures once in a while. This contact failure may exist on only one contact out of the four in each relay. This intermittent failure mode gave many a gray hair to the people in the maintenance department. When a machine was malfunctioning, the first action most maintenance men tried was changing relays. A second major problem with this type of control system was wiring documentation. Often changes were made to a piece of equipment, but the wiring drawings were never updated. To troubleshoot a control system without a good set of wiring drawings was almost impossible.

With the introduction of the programmable controller, many of these problems were eliminated. As you will learn in Chapter 2, all input/output (I/O) connections to a programmable controller are simplified because they all have one side connected to a common bus and the other connected to an I/O terminal. All logic functions are handled in software. Logic changes no longer require wiring changes unless additional I/O devices are needed. Because all outputs are transistor or triac driven, the old problem of mechanical wear on relays has been eliminated.

Documentation of the program is much easier to keep updated because it can be printed out on demand each time a change is made. Troubleshooting a large system is also easier because the programmer can monitor the logic functions and I/O status from the programming terminal. All in all, the programmable controller has simplified the design of control systems and greatly improved their reliability. In the following chapters you will learn how to design electrical control systems around programmable controllers as well as create programs for controlling these systems when used for automation machine control.

In the following chapters and labs we will be focusing on the Allen-Bradley SLC-500 family of PLC controllers. It is the author's belief that learning one PLC controller family well is a good idea before attempting to learn about the entire spectrum of available controllers. We will make references to other families, such as the MicroLogix 1000, 1200, and 1500 and the Control Logix 5000 family of controllers, but the instructional focus and all the lab work will be on the SLC-500 family.

In preparing this text to instruct students in the proper use of programmable controllers, we also designed and built a teaching kit to be used in the lab portion of the course. This kit was built into one metal carrying case to protect the components and to make storage easy (Figure I–2). Teaching PLC control systems and programming without interfacing to real moving components is a waste of time. To give the student a feel for programming real-time systems, we have provided a motion panel mounted in the lid of our PLC case (Figure I–3). Mounted into the lid of this case are five air

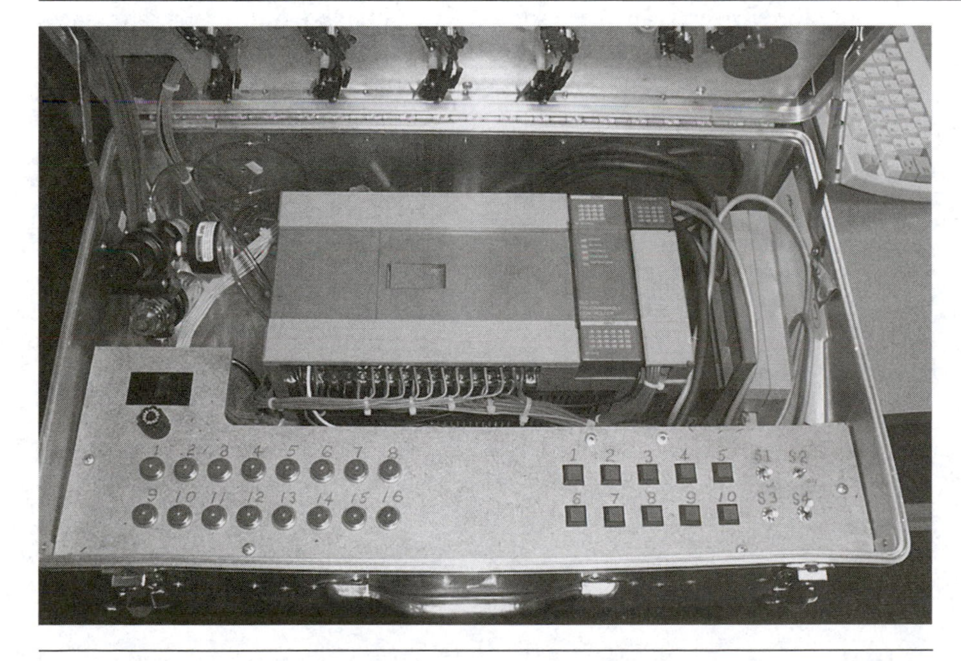

Figure I–2
PLC controller kit

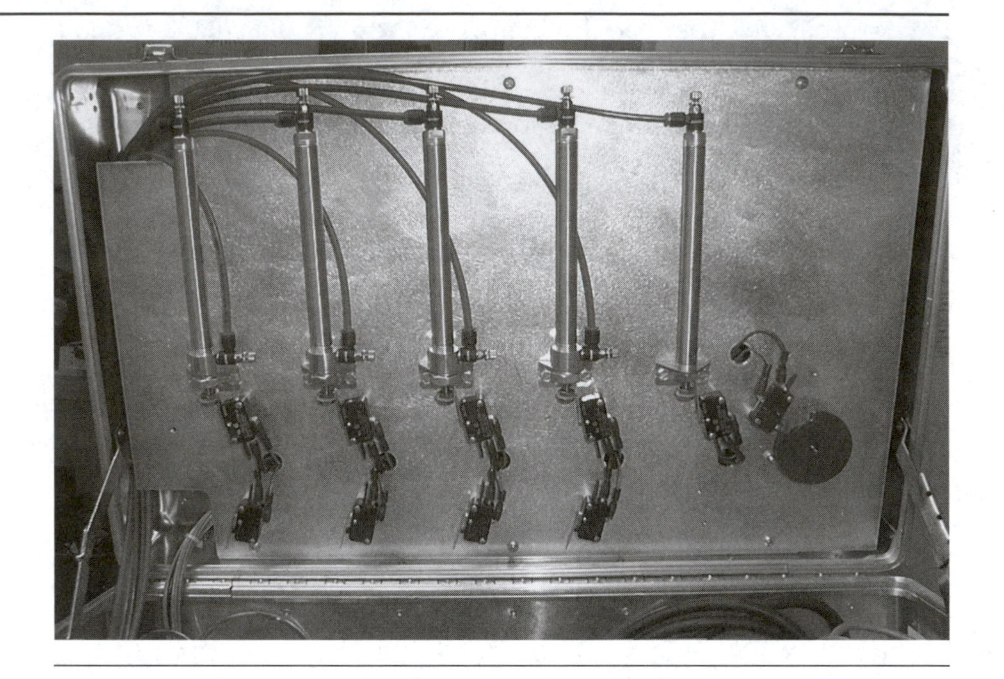

Figure I–3
Motion panel for PLC controller kit

cylinders, five air valves, an air regulator, nine proximity switches, and a timing motor with cam and cam switch. We will make numerous references to this kit in the following text. All the labs and program examples are designed to be able to be run on this simulator kit. If this book is to be used as a teaching text, I recommend you build a similar kit. Electrical schematics are shown in Appendix A.

Allen-Bradley has introduced the MicroLogix 1200 and 1500 families since the first publication of this book. The teaching kits we use are SLC500 based, but one could use the MicroLogix 1200 family of PLCs in any new teaching kit construction. All of the labs and programs in this book run on the 1200 family with only slight I/O address modifications. The 1762-L40BWA would be a good replacement for the 1747-L40E we used. For more information contact dgeller@oakton.edu.

CHAPTER 1

Binary Mathematics

LEARNING OBJECTIVES

After completing this chapter the reader should understand:

1. Binary number systems.
2. Conversion between binary, hex, and decimal number systems.
3. Logical functions such as AND, OR, and XOR.
4. Binary addition and subtraction.
5. Dotted decimal notation.

To design and program PLCs, the user needs to have a basic understanding of binary mathematics. Basic binary functions included in this text are ADD, SUBTRACT, AND, OR, XOR, and COMPLEMENT. We will also cover the hex and dotted decimal numbering systems.

The student should be well acquainted with the decimal number system we use every day. This system is a column-weighted system. Each column of numbers is valued at 10 times the column to its right. In the decimal number system we use 10 different symbols to express the column value (0 through 9). As an example the number 3526 shown in Figure 1–1 can be broken down by column as shown.

In the binary number system we also have a column-weighted system. The binary number system is based on powers of two. Each column value is therefore two times the value of the column to its right (see Figure 1–2). The binary number system uses only two symbols: 0 and 1. Binary numbers are used in computers because digital systems have only two states. The off state is represented by 0 volts and is a binary 0, and the on state is represented by 5 volts and is a binary 1. The first four columns of a binary count progression are shown in Figure 1–3.

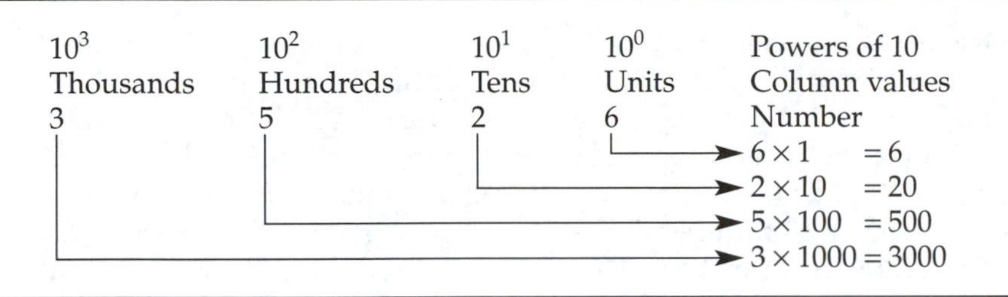

Figure 1–1
Powers of ten in decimal numbers

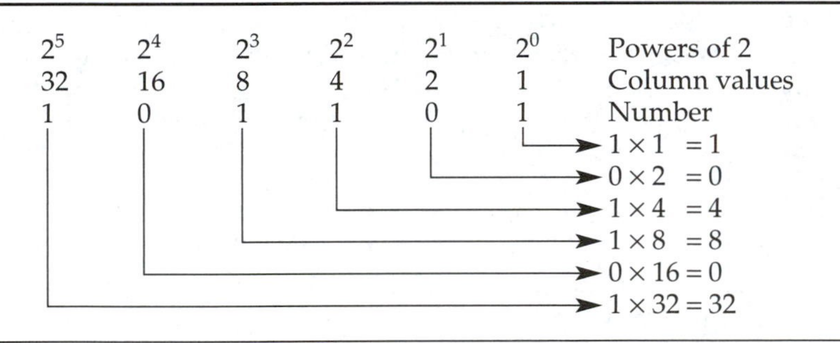

Figure 1–2
Binary number system

Binary column values					Decimal
8	4	2	1		
0	0	0	0	=	0
0	0	0	1	=	1
0	0	1	0	=	2
0	0	1	1	=	3
0	1	0	0	=	4
0	1	0	1	=	5
0	1	1	0	=	6
0	1	1	1	=	7
1	0	0	0	=	8
1	0	0	1	=	9
1	0	1	0	=	10
1	0	1	1	=	11
1	1	0	0	=	12
1	1	0	1	=	13
1	1	1	0	=	14
1	1	1	1	=	15

Figure 1–3
Binary progression from 0 to 15

BASIC BINARY FUNCTIONS

Binary to Decimal Conversion

To convert a binary number to its decimal equivalent, we merely add the column values of the columns with 1s in them, as shown in Figure 1–4.

Decimal to Binary Conversion

This conversion routine is one of several that can be used for this function. To convert decimal to binary, we start with the column values in binary up to at least one value larger than the decimal number we are converting. We start by subtracting the highest binary column value we can from the decimal number. For each column value we can subtract, place a 1 in that column position. For each column value we cannot subtract, place a 0 in that column. An example is shown in Figure 1–5.

Note that in binary as in decimal we omit preceding zeros from the final number. One quick way to check your answer would be to see if it is an odd or even number. If the decimal number you converted was an odd number, then the least significant bit of the binary number should be a 1. If the decimal number that was converted was an even number, then the least significant bit of the binary number should be a 0.

Binary Addition

In decimal addition a carry is needed any time the column value sum is greater than 9. In binary addition a carry is needed any time the column value sum is greater than 1. Several examples are shown in Figure 1–6.

Note in example D the second column is equal to 3, and that translates to a 1 carried and a 1 down in the sum.

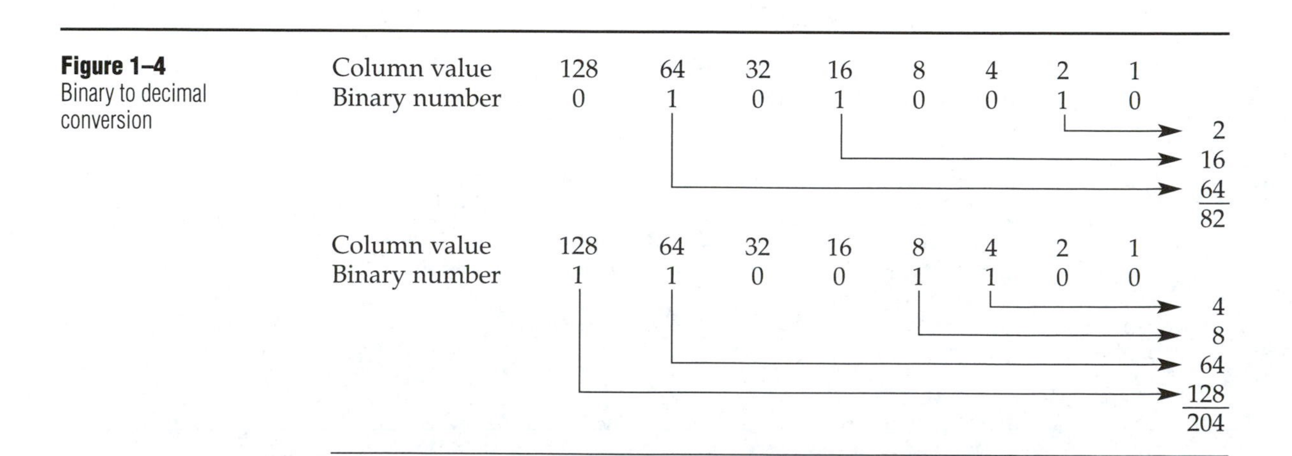

Figure 1–4
Binary to decimal conversion

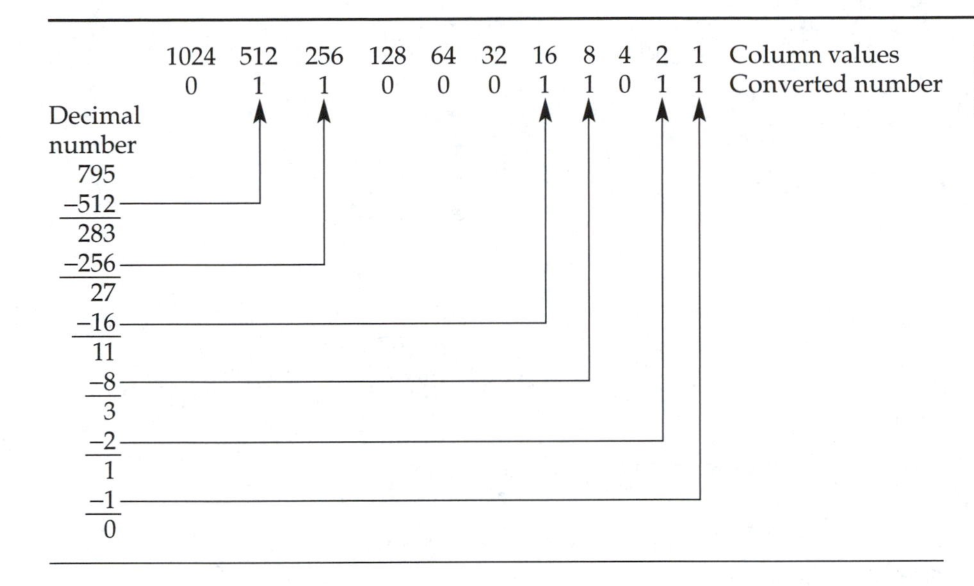

Figure 1–5
Decimal to binary conversion

```
            A                      C C C      B
                                   1 1 1
   1  0  0  1  =  9               1  0  1  1  = 11
 + 0  1  1  0  =  6             + 0  1  1  0  =  6
   1  1  1  1  = 15             1  0  0  0  1  = 17

  C         C                        C C      D
  1                                  1 1
   0  1  0  1  =  5                0  0  1  1  =  3
   0  1  1  0  =  6                1  0  1  1  = 11
 + 1  0  1  1  = 11              + 1  1  1  0  = 14
```

Figure 1–6
Binary addition

Binary Subtraction

In decimal subtraction when the column value to be subtracted is larger than the number above it, we borrow 10 from the column to the left. In binary subtraction we also borrow from the column to the left, but the value that is borrowed is 2. Several examples are shown in Figure 1–7.

In example A no borrow is necessary. In example B a borrow is needed in the third column. Note that two 1s are brought over to the third column from the fourth column. The 1 is subtracted, resulting in a 1 dropping down to the bottom of the column. In example C we have to borrow in the first column, but the only available place to borrow from is the fourth column. Note how each borrow brings two 1s to the column to the right. One of these borrowed 1s is then borrowed again to the next column until we end up with two 1s in the first column. In the first column, subtracting 1 from the two borrowed 1s results in a 1 moving down into the bottom of the column. In the second column, subtracting 1 from the one remaining borrowed 1 results in a 0 moving down to the bottom of the column.

Figure 1–7
Binary subtraction

A

	1	0	1	0	= 10
−	0	0	1	0	= 2
	1	0	0	0	= 8

B

	B→	11			
	~~1~~	0	1	1	= 11
−	0	1	0	1	= 5
	0	1	1	0	= 6

C

	B→	B→ 1~~1~~	B→ 1~~1~~	11	
	~~1~~	0	0	0	= 8
−	0	0	1	1	= 3
	0	1	0	1	= 5

D

	B→	B→ 1~~1~~	11		
	~~1~~	~~1~~	0	1	= 13
−	0	1	1	0	= 6
	0	1	1	1	= 7

BINARY LOGIC FUNCTIONS

We cover four binary logical functions in this text: INVERTER, AND, OR, and XOR.

Inverter Function (Complement)

The symbol shown in Figure 1–8 is the inverter or complement logic function. To the right of the logic symbol is a truth table, which shows the output state for any given input state.

Complement functions are used in binary math. The examples in Figure 1–9 illustrate the complement of a register's contents.

To find the complement of any binary number you must invert each bit in the number.

Figure 1–8
Inverter (complement) logic symbol and its truth table

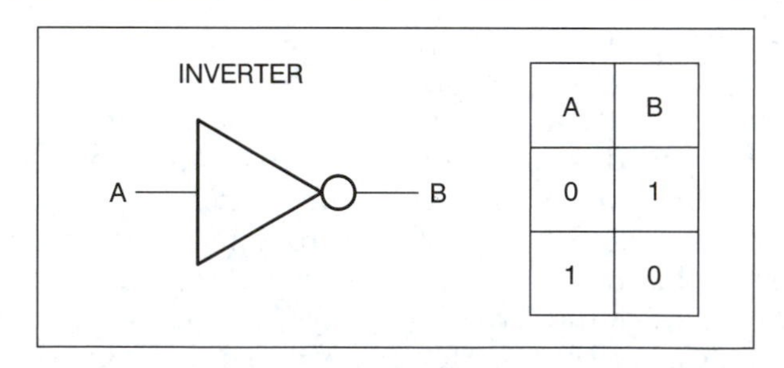

Figure 1–9
Complements

Register X	01100101	10111100
Complement	10011010	01000011

AND Function

Figure 1–10 shows the AND logic symbol and its truth table. This logic symbol has two inputs, A and B, and one output, C. From the truth table we can see that output C only switches to 1 if both A and B are 1. As before, the truth table shows all the possible states of the inputs and output.

AND functions can have more than two inputs in logic design, but the output will only be 1 if all inputs are 1s. This AND function is also used in binary mathematics. This function is most often used when the programmer wants to clear (all 0s) a portion of a register without changing the rest of the register's contents. That is to say, we can clear a single bit or group of bits in a register without affecting the rest of the register. Any bit ANDed with 1 will remain unchanged; any bit ANDed with 0 will become 0. Consider the examples shown in Figure 1–11.

From these samples we can see that in the leftmost example we cleared the low 4 bits of register X and left the high 4 bits unchanged. In the right example, we cleared only a single bit (5) and left the other 7 bits unchanged.

Another use of the AND function is to test a bit position in a register to see if it is set to a 1. To use the function in this way, we AND the register with a word that has all 0s in it except for the position we wish to test.

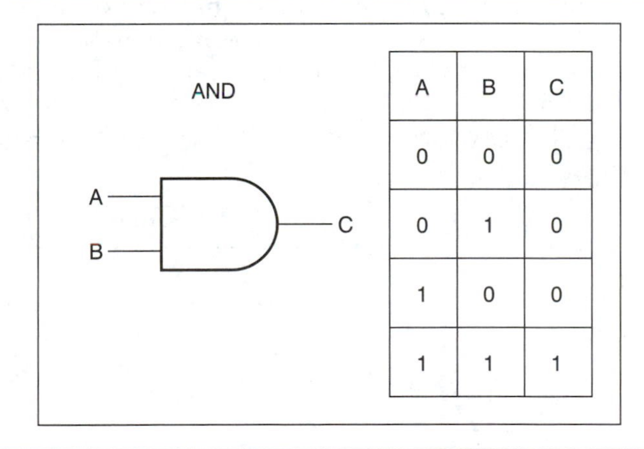

A	B	C
0	0	0
0	1	0
1	0	0
1	1	1

Figure 1–10 AND logic symbol and its truth table

	Bits to clear	Bit to clear
Register X	10110110	01101100
AND function	11110000	11011111
Result in register X	10110000	01001100

Figure 1–11 Binary AND function

Figure 1–12
AND function used as a bit test

	Bit to test ↓	Bit to test ↓
Register X	10110110	01101100
AND function	00010000	00010000
Result	00010000 Not zero	00000000 Zero

In that position we place a 1. If the result of the function is all 0s, the bit was not set. If the result is anything above 0, the bit was set. Examples of this use are shown in Figure 1–12.

In the left example, bit 4 was set and the result was not zero. In the right example, bit 4 was not set and the result was zero. By checking if the result is zero after an AND instruction, you can determine whether or not the bit in question was set.

OR Function

The symbol shown in Figure 1–13 is the OR logic function. It is shown with two inputs, A and B, and one output, C. Looking at the truth table we can see that the output C changes to 1 when either A or B or both are 1. The truth table shows all the possible input and output conditions for this logic function.

In logic design the OR function can have more than two inputs. Regardless of the number of inputs, the output will change to 1 if any input is 1. The OR function is used in binary mathematics to set a bit, or group of bits, in a register without changing the rest of the register contents. Several examples of this logic function are shown in Figure 1–14. Any bit ORed with a 1 will be changed to 1; any bit ORed with a 0 will remain unchanged.

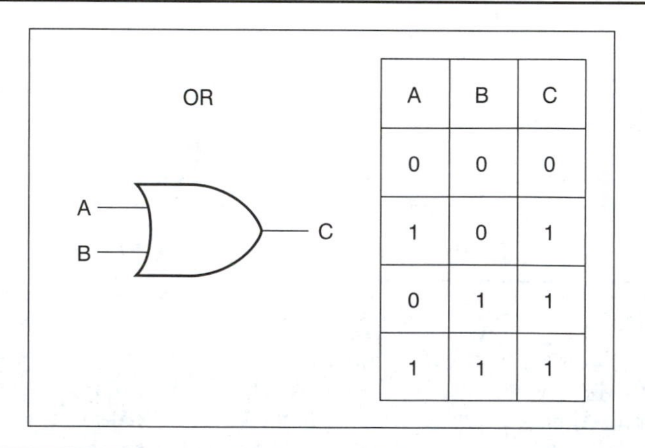

A	B	C
0	0	0
1	0	1
0	1	1
1	1	1

Figure 1–13
OR logic symbol and its truth table

	Bits to set (left)	Bit to set (right)
Register X	10110010	01001100
OR function	00001100	00100000
Result in register X	10111110	01101100

Figure 1–14
Examples of the OR logic function

In these examples 2 bits were set in the left example and 1 bit was set in the right example.

Exclusive OR (XOR) Function

Figure 1–15 shows the XOR logic symbol and its truth table. Notice that it is the same as the OR function except in the last state. The XOR output is 1 if A or B is 1, but not both.

The XOR logic function is used in binary mathematics to compare the values of two registers to determine if they are equal. For example, if we have a counter register and we need to know when the count reaches a preset value, we would use the XOR function. If we XOR the counter value with the preset value and the result is zero, the registers are equal; any other value means the values are not equal. Two examples are shown in Figure 1–16.

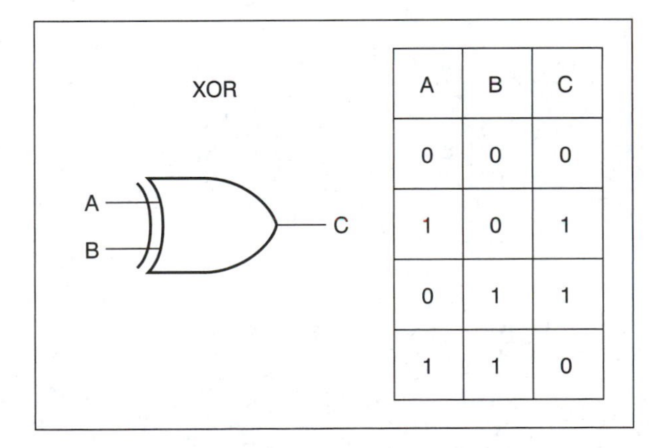

A	B	C
0	0	0
1	0	1
0	1	1
1	1	0

Figure 1–15
XOR logic symbol and its truth table

Register X	10110010	01001100
XOR function	10001100	01001100
Result	00111110	00000000
		Equal

Figure 1–16
Examples of the XOR logic function

NUMBER SYSTEMS

When we are working with binary numbers most people have a difficult time dealing with long strings of 1s and 0s. Most addresses in PLCs are 16 bits long. Pretend you need to tell a coworker to look at a certain PLC address, and try to say the following binary number out loud:

1001 0111 0011 1010

If you're like most people, you found it hard to express long binary numbers by speaking 1s and 0s. For this reason and to make displaying binary numbers easier, engineers have developed an alternative number system called *hex*. It is very important to understand that this alternative number system does not change the binary number value, it only makes it easier to display and express numbers.

Hex Number System

In the hex number system, all binary numbers are arranged into 4-bit groups. Each group is then given a single digit value. Look at the hex number progression shown in Figure 1–17. Notice that after 9 we change to alpha characters.

To convert a binary number to hex, first group the binary number in groups of 4 bits, then assign each group its hex value. Examples are shown in Figure 1–18.

Figure 1–17
Binary to hex tables

Binary Hex	Binary Hex
0 0 0 0 = 0	1 0 0 0 = 8
0 0 0 1 = 1	1 0 0 1 = 9
0 0 1 0 = 2	1 0 1 0 = A
0 0 1 1 = 3	1 0 1 1 = B
0 1 0 0 = 4	1 1 0 0 = C
0 1 0 1 = 5	1 1 0 1 = D
0 1 1 0 = 6	1 1 1 0 = E
0 1 1 1 = 7	1 1 1 1 = F

Figure 1–18
Binary to hex conversion

1 0 1 1	0 0 1 0	0 1 0 0	1 1 0 0	Binary
B	2	4	C	Hex
A	D	F	7	Hex
1 0 1 0	1 1 0 1	1 1 1 1	0 1 1 1	Binary

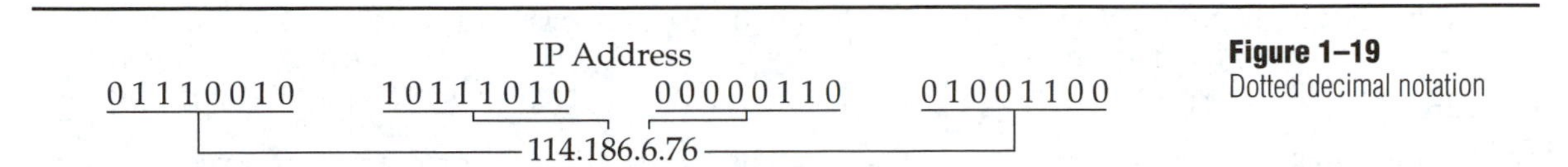

Figure 1–19
Dotted decimal notation

Again, I wish to emphasize the fact that the binary value has not been changed, it has merely been expressed in a much cleaner form.

To convert from hex back to binary, reverse the process and convert each hex digit into its binary bit pattern.

Dotted Decimal Notation

As we gain in our understanding of PLC systems we will come to a point where we will need to interface several PLC systems together. One method used to allow several PLC systems to talk to each other is to network them together using Ethernet. Each Ethernet device requires a unique IP address. IP addresses are 32-bit binary numbers and are normally expressed in dotted decimal notation. You express a 32-bit binary number in dotted decimal notation by first breaking the binary number up into four 8-bit binary groups. Each 8-bit group is then converted to a decimal number between 0 and 255. The four decimal numbers are then separated by placing a period between them (see Figure 1–19). Network IP addresses often have to be converted between binary and dotted decimal and back again.

CONCLUSIONS

In concluding this chapter, I should state that this material is not meant to be an all-inclusive study of binary mathematics. If the student desires a comprehensive understanding of this subject, there are many good books devoted entirely to this topic. This chapter should serve as a review for those who have used binary math in the past and a basic starting point for students new to the material.

REVIEW QUESTIONS

1. Convert the following binary numbers to decimal:
   1011 0110 0111 0101 1010 0011
2. Convert the following decimal numbers to binary:
   1592 279 763
3. Convert the following binary numbers to hex:
   1010 0111 0101 1011 1000 0011 1111 1100
4. Convert the following hex numbers to binary:
   FA67 EDA7

5. Add the following binary numbers:

1000 0110	0111 1101
+ 0110 0101	+ 1001 1011

6. Subtract the following binary numbers:

1000 1101	1110 0010
– 0011 0110	– 0111 1110

7. Show the results of the following AND logic functions:

1011 0111	1000 0011	1010 0011
AND 0000 1111	AND 1111 1101	AND 0010 0000

8. Show the results of the following OR logic functions:

1100 0011	1110 0000	1010 1010
OR 0001 0000	OR 0000 1111	OR 0000 0101

9. Show the results of the following XOR logic functions:

1010 0011	1110 0110	1101 0011
XOR 1010 0011	XOR 1111 0101	XOR 1101 0011

10. Convert the following binary number into dotted decimal notation:
    11001010 10101110 10000011 01110101

11. Convert the following dotted decimal number into its binary form:
    245.134.85.18

CHAPTER

2

Input/Output Wiring

LEARNING OBJECTIVES

After completing this chapter the reader should understand:

1. Standard symbols used to identify I/O devices.
2. Interfacing to input modules.
3. Interfacing to output modules.
4. The interface concept of current sinking and sourcing.
5. Benefits of using opto-isolation on all I/O ports.

One of the main advantages of designing control systems using programmable controllers is the ease of connecting input/output (I/O) devices to the system. The first step in control system design is to define the type and number of I/O devices the system will have to control. All I/O devices must be categorized as to their voltage and current requirements. Programmable controllers have many different I/O modules available to the designer and his or her job is to match up the right module to the I/O types being controlled. It is very important for the reader to understand that I/O modules do not provide any power of their own. Instead, they provide a connection between the I/O device and one side of the power supply rail.

INPUT/OUTPUT SYMBOLS

Input Symbols

Before we start our introduction to input and output modules we need to explain some of the terms and symbols we will be using. The first set of symbols we investigate are switch symbols. The list of switches shown in Figure 2–1 is only a partial list and shows only the normally open symbol for each switch type. In addition to those shown, many other special sensor devices are available to the control systems engineer. Examples include optical sensors, capacitive sensors, and laser level sensors. The symbols for these are usually shown in their product installation books.

One might ask why we have so many symbols for a simple switch contact. Although we could use just one symbol to cover many switch

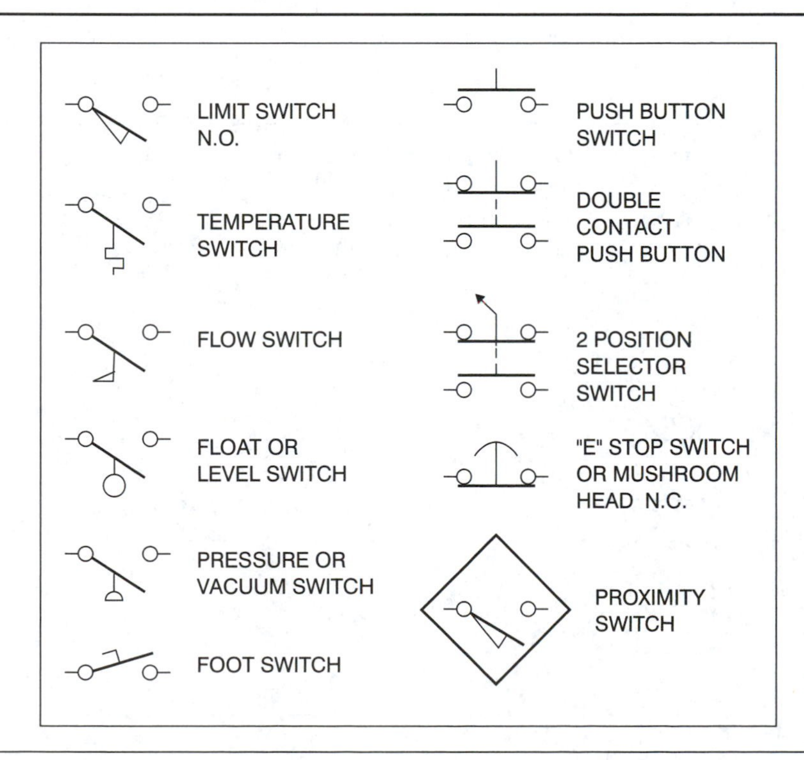

Figure 2–1
Typical switch symbols used in control systems

types, it would make identification of a particular switch element very difficult. On large automation equipment, it is often difficult to find an exact electrical component when the machinery fills a large room or is several floors high. By using the correct symbol for each type of contact, it makes the job of finding that exact switch element considerably easier. When designing control systems, the designer should always use the correct symbols and a description that exactly defines the element or its function. If a maintenance person is trying to troubleshoot an equipment problem and the wiring drawing lists 58 switch inputs as SW-1 to SW-58 it is going to be a long difficult process. If, on the other hand, the contact in question is listed as "Cooling tank #1 high-level float switch," the maintenance person will have a much better idea of where to look for the problem switch and what it will physically look like. If you, as the control system designer, use descriptions like SW-22, rest assured that you will be receiving lots of 3 A.M. phone calls from maintenance department personnel when they have trouble reading your drawings or identifying certain elements from the drawings.

Output Symbols

The list of output symbols, shown in Figure 2–2, is smaller but no less important. Again, it is important always to use an accurate description to go along with each symbol.

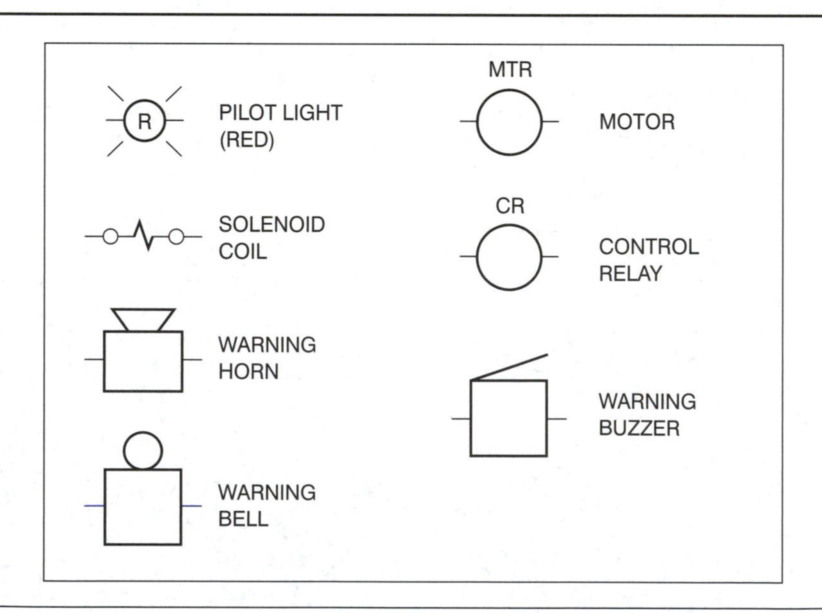

Figure 2–2
Typical output symbols used in control systems

INPUT MODULES

AC Input Modules

Figure 2–3 shows a typical 16-position input module wiring schematic for 120V AC operation. Notice that one side of the AC line is connected to the module's common input terminal. The other side is common to all of the input devices. This simplifies the input wiring because each device has one side connected to AC hot and one wire to the interface module. Except for the common wire, no wire numbers are needed because the I/O address can serve as a wire marker.

Notice also the typical input circuit drawing shown at the right in Figure 2–3. It indicates that each input is optically isolated from the internal computer circuits. Because each input is optically isolated from the internal circuits of the controller there should be very few problems with voltage spikes on the input lines. This also ensures that any catastrophic event, like shorting high voltage to an input line, will damage only the input module it is connected to. The opto-isolation stops the damage at the input module instead of allowing the voltage spike to damage the whole PLC system.

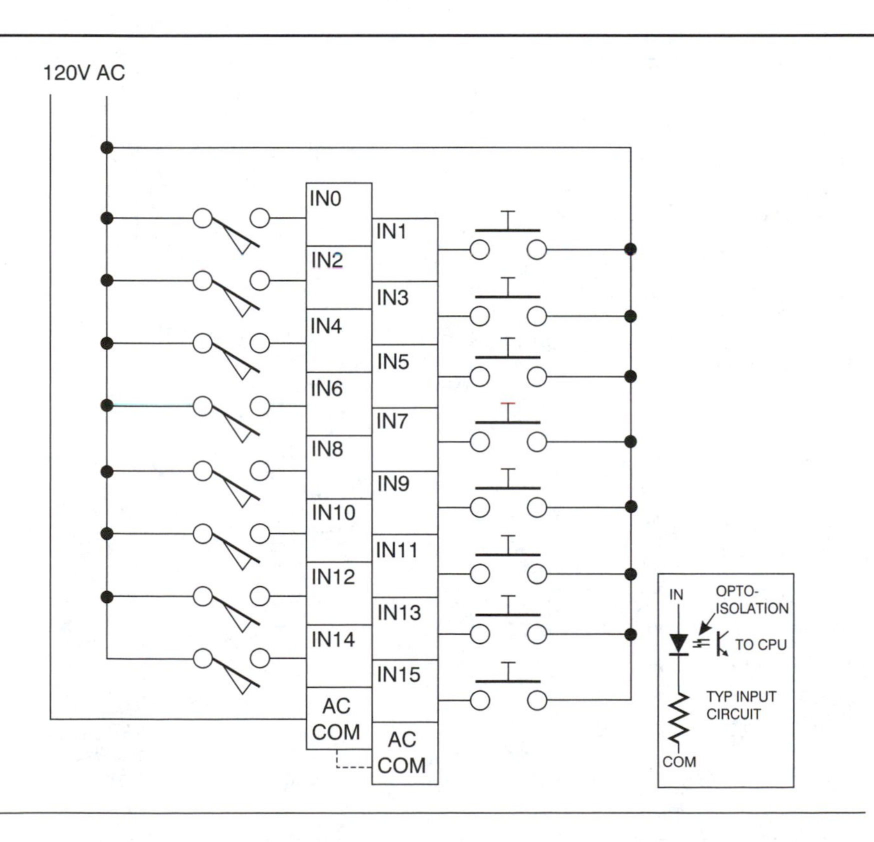

Figure 2–3
Typical AC input module wiring schematic

Which side of the AC line you connect to the input module's common terminal depends on the designer's preference. Some designers prefer that the hot side be connected to the common terminal because this allows them to run the common side to all the switch devices. Connecting power in this manner prevents most of the exposed wiring on the machine from carrying a high current or voltage.

DC Input Modules

When interfacing DC input modules, an additional parameter must be considered. Most DC input and output modules come in two types, *sinking* and *sourcing*. This designation refers to the direction of current flow into or out of a DC module. Figure 2–4 shows the schematic wiring diagram for a 16-input, 24V DC sinking module.

Figure 2–4
Schematic wiring diagram for a 16-input, 24V DC sinking module

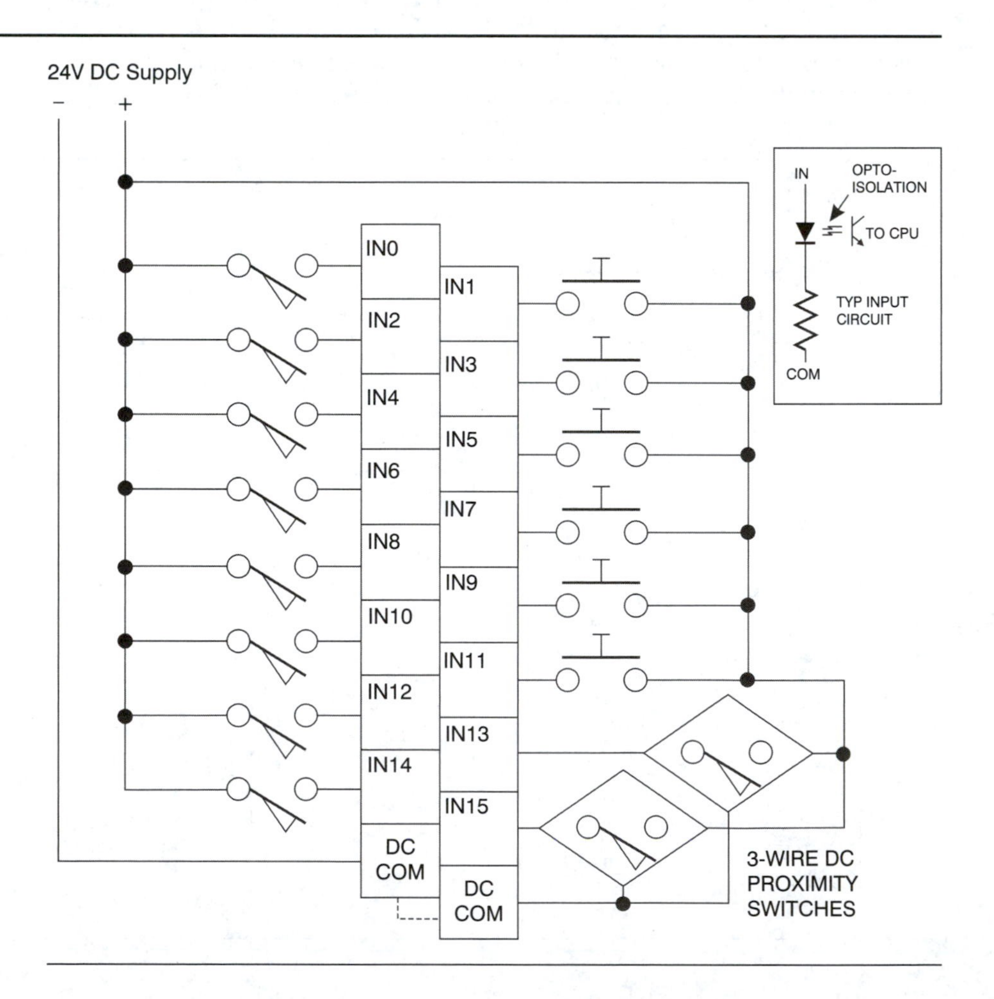

From Figure 2–4 you can see that current flows from the positive supply rail through the input device and is sinked to ground through the LED inside the module. Connecting the module backwards will cause the diode to block current flow, and the module will not sense any input current.

In Figure 2–5 we show the wiring schematic for a sourcing-type input module. Notice how the power supply rails have been reversed. The designer must be sure the power supply is connected correctly so that the LED is forward biased. All I/O modules are optically isolated between the external connections and the PLC's internal computer inputs. This was implemented for two reasons. First it prevents noise and voltage spikes picked up on the machine wiring from being coupled into the PLC's computer circuits. Even if a large spike destroys the input module, the isolation will keep it from killing the main PLC computer and other I/O modules. Second, it allows multiple power supply circuits to be interfaced to the PLC and still be isolated from each other. In a PLC system with many input modules, the designer may have both AC and DC inputs and yet have no physical connections between them.

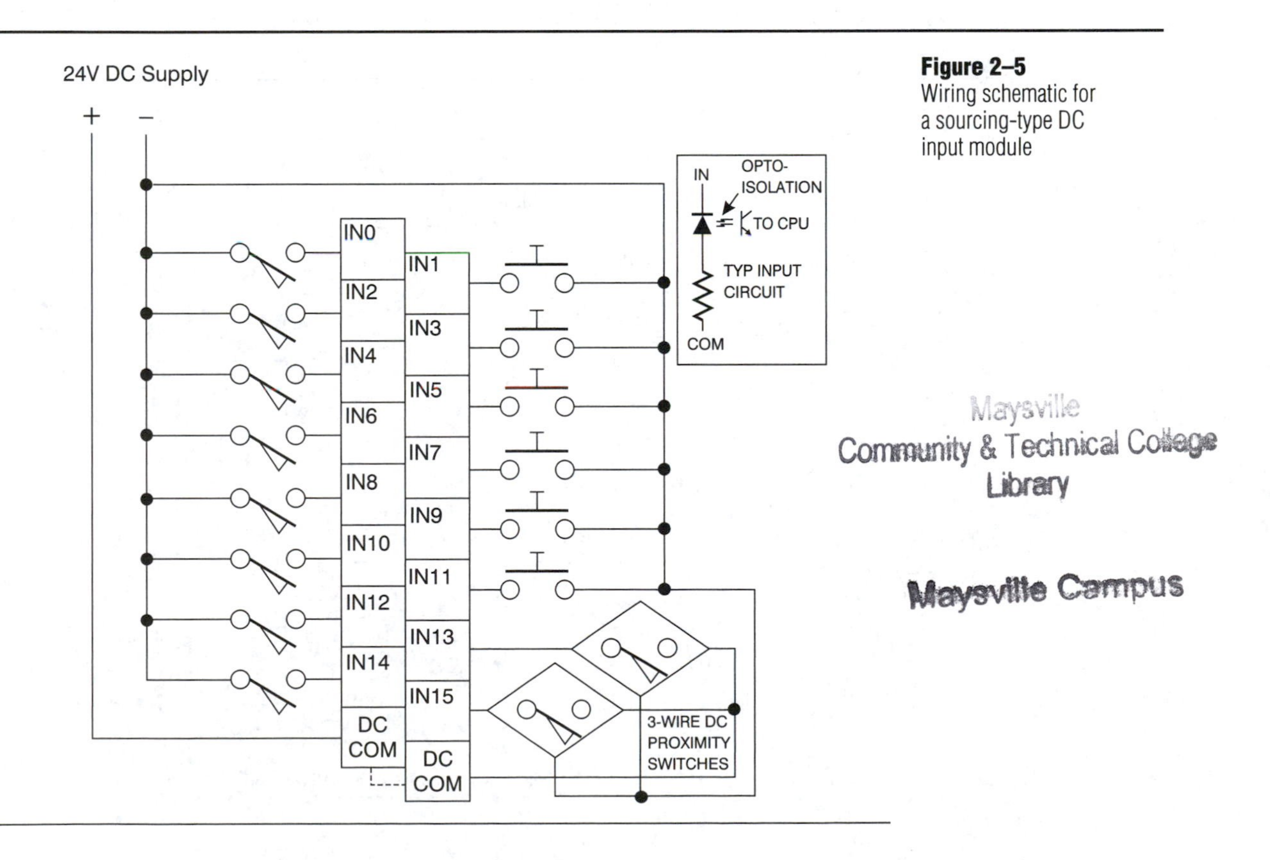

Figure 2–5
Wiring schematic for a sourcing-type DC input module

Proximity Switches

In both of the DC input wiring diagrams we have shown *proximity switches* that act as sensor devices. These devices come in several types. The most common use magnetic or capacitive sensing. I will refer to these devices as *prox switches* throughout the rest of this book. They work by projecting a magnetic field and sensing when a metal object disturbs the field. When interfacing these solid-state devices to the PLC, we have to match the prox switch output type to the input module we are using.

Figure 2–6 shows both types of prox switch outputs. They are also called sourcing (PNP) or sinking (NPN) type switches. This designation refers to the type of output transistor used. The prox switch output type should be the opposite of the input module type. If the designer is using sinking-type input modules, sourcing-type prox switches must be used. Likewise, if sourcing-type input modules are used, the designer must specify sinking-type prox switches. Some optical switches have both output types available in the same package. In this case the designer must specify what output wire to choose when installing these devices.

Figure 2–7 shows the proper wiring diagrams for interfacing both sinking- and sourcing-type prox switches to PLC input modules. Notice that the current flows into the common terminal on the sourcing input

Figure 2–6
Proximity switch outputs

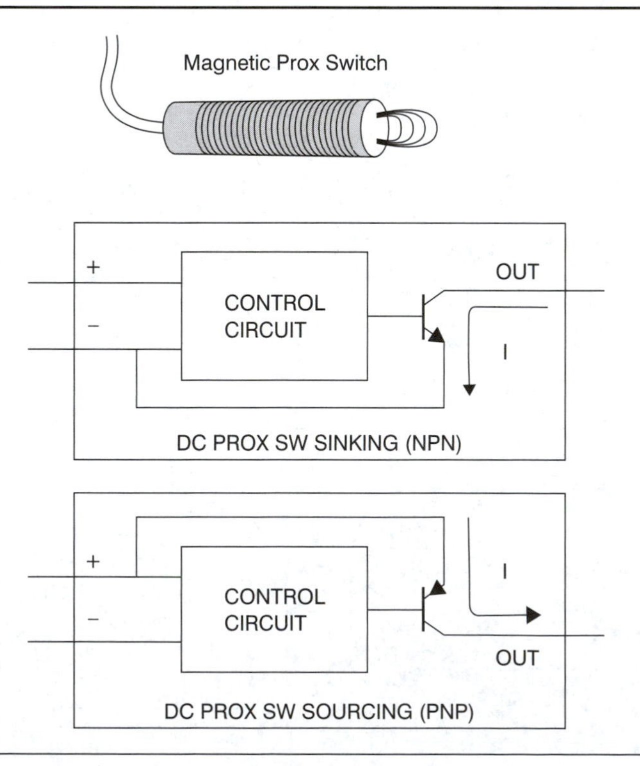

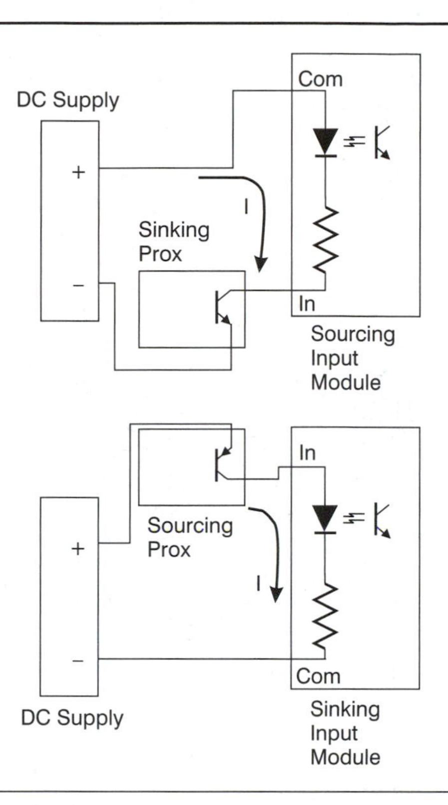

Figure 2–7
Prox switch wiring for sinking and sourcing input modules

module and exits from the actual input port. This way the port sources current to the sensing device, and the sensor must sink it to ground. In the sinking module the current enters the input port and exits the common terminal. This module sinks the input current to ground.

OUTPUT MODULES

Output modules are very similar to input modules in that they also do not provide any power of their own. Output modules provide a selective means of connecting output devices to one side of a power rail.

AC Output Modules

Figure 2–8 shows a 16-output triac-driven 120V AC module. Notice that, as with input modules, all output modules are optically isolated from the internal PLC's circuits. Also not shown in this example is the fact that each output is fused at 1.5 amps in most cases. If more than 1.5 amps are needed, two modules can be operated in parallel or an external high-current relay can be added, as is done when controlling large AC motors.

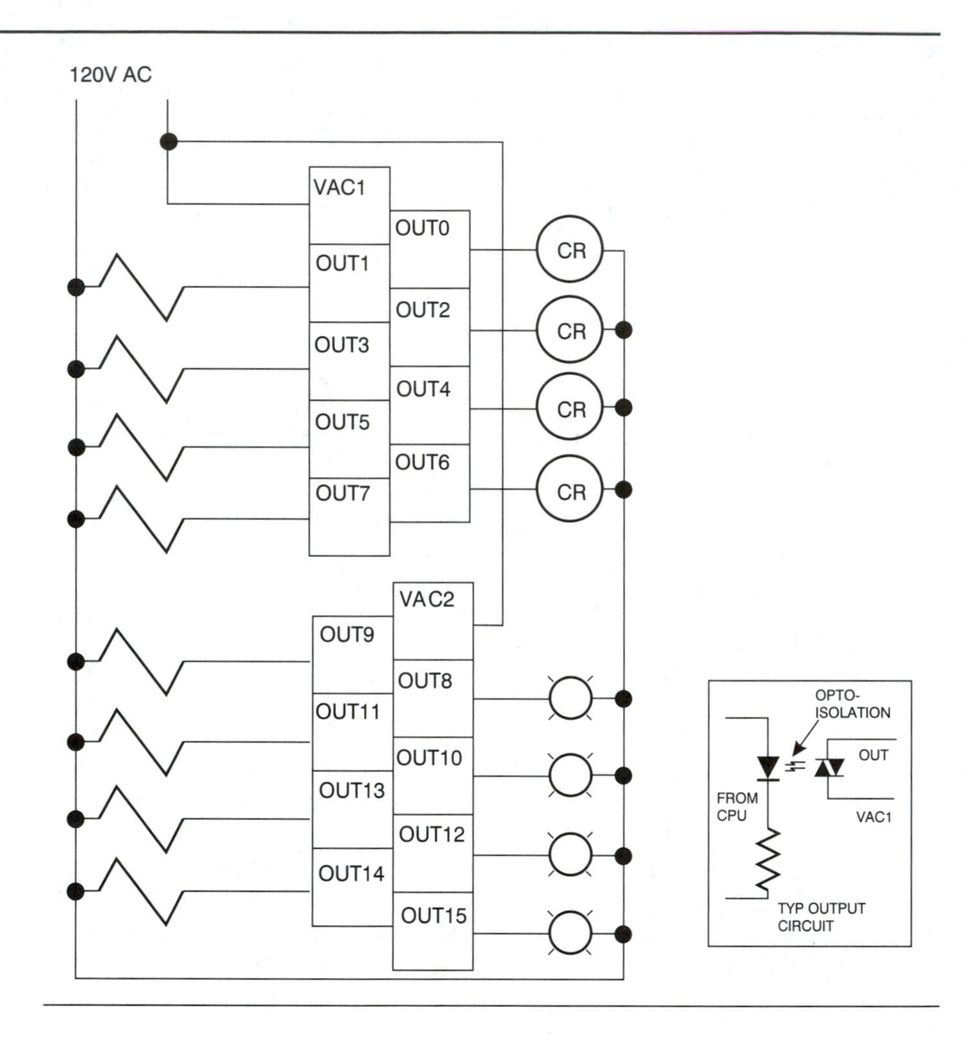

Figure 2–8
16-output, triac-driven 120V AC module

Although the triac output is a great advance over relay-type contacts when considering mechanical wear and cycle life, it does have one small drawback. When a relay contact is open, no current can flow out of the terminal. With triac outputs, even when the output is off, a small (microamps) leakage current can still flow. Although this will not cause any problems when connected to a high current load, such as a solenoid valve coil, it will give a person a mild tingle if you make contact with the output terminal. It may also cause some problems when driving LED-type lights or low-current sound devices like Sonalerts. In such cases, the device may be very dimly lit or slightly audible when the output is off. Placing a resistor load across the device will usually shunt out this leakage current.

When using AC output modules, deciding which side of the AC line to connect to the VAC common terminals is a matter of personal choice. Because AC circuits have no polarity, the module can be connected to either side of the

AC line. Most designers prefer to connect the hot side of the AC supply to the VAC common terminals. This allows the ground side of the AC supply to be connected to the common side of all the driven devices on the machine. This means that most of the wiring run through the wiring channels on the machine will be the ground side of the line and therefore less likely to short or cause safety problems to people working around and on the equipment.

DC Output Modules

DC output modules are much like input modules in that they come in two types, sourcing and sinking. All DC modules use optically isolated transistors as output drivers. As you can see in Figure 2–9, the power supply connections are very important. The power transistor must be forward biased to work.

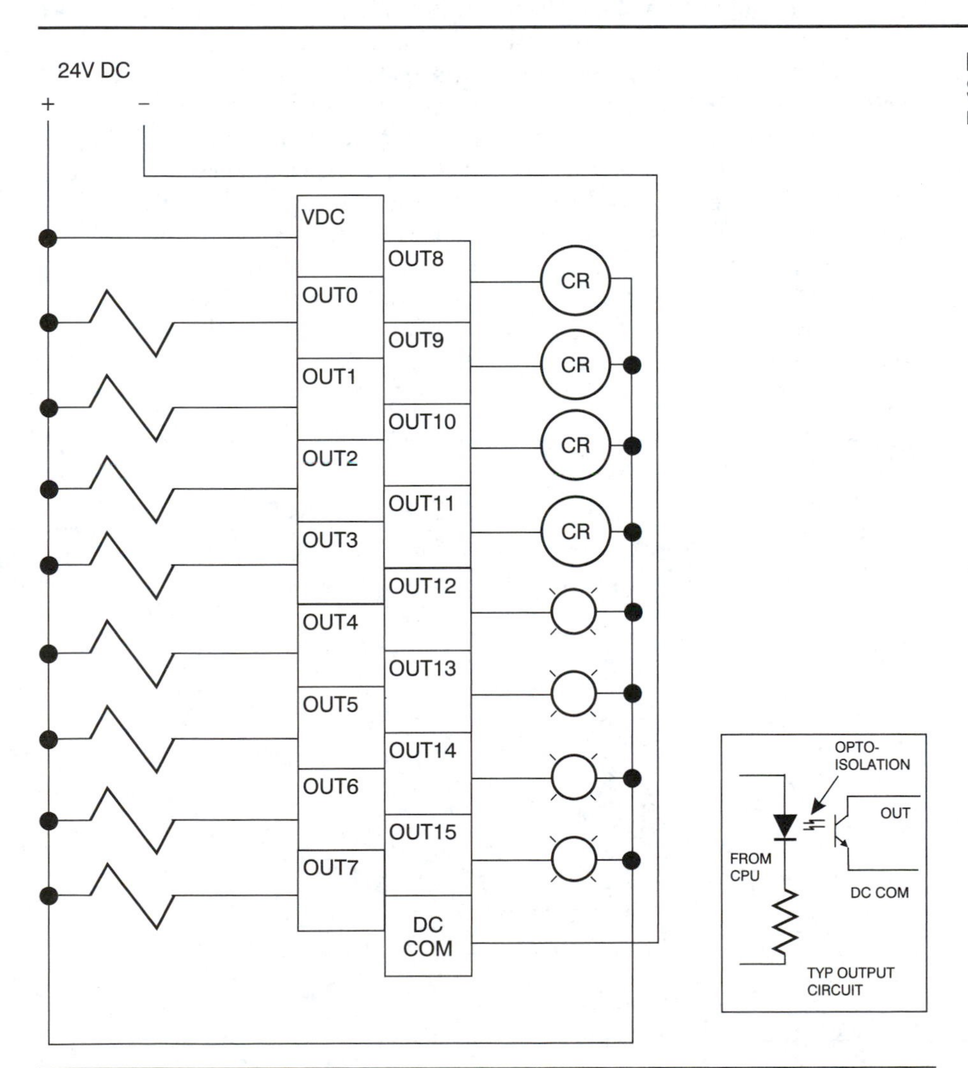

Figure 2–9
Sinking-type 12V DC output module

If you're using a sourcing-type DC output module, the PNP transistor's emitter is connected to the VDC terminal and its collector is connected to the output terminal. All the driven devices would then have to have one terminal connected to the negative (–) side of the DC power supply.

Most DC modules are also fused at 1.5 amps. More than one output can be combined to get higher drive currents. When using DC output modules to control solenoid valves, the designer must take into account the current drop on long runs. If you are using 120V AC to power outputs and you lose 10 volts to wire loss, you still have 90% left at the output. If you use 24V DC and have the same 10V loss, your power to the load is only about 60%. For this reason we recommend using 24V DC only for input devices, because that practice allows for the use of much smaller wire bundles and connectors. All outputs that require any power should be run from 120V AC to minimize the voltage drop problems.

A second problem created when you use DC to drive solenoids is the electrical noise caused by the fly-back effect of the coil when it is shut off. When DC current is used to drive an air solenoid, a large magnetic field is generated when the coil is turned on. When the coil is turned off the magnetic field collapses, and as the lines of magnetic force cut the coil windings a large voltage is generated in the opposite direction of the driving voltage. This voltage spike can be three to four times the magnitude of the driving voltage. If this voltage spike is not suppressed it can cause damage to the driving circuit and also to the DC power supply. Because the spike is of a very fast rise time it can also propagate to other circuits and cause noise

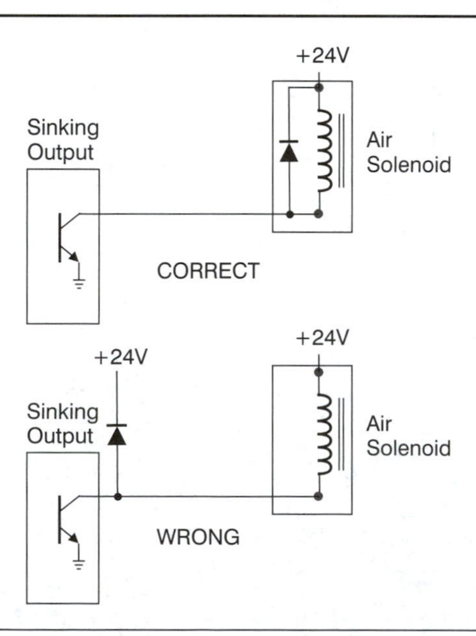

Figure 2–10
Connection diagrams for diode suppression on DC solenoid coils

problems all around. The solution is to install a fly-back diode across the coil to short out the voltage spike (see Figure 2–10). It is very important that this diode be mounted as close to the coil as possible. Many DC solenoid valves have a diode already installed by the vendor. If you must install your own diode be sure not to install it on the output module terminals. If the diode is installed improperly the wiring between the coil and diode will act as an antenna and cause even worse noise problems. Install the diode as close as possible to the actual coil. This will keep noise propagation to a minimum and enhance the stability of your control design.

SYSTEM DESIGN

As we said earlier in this chapter, defining the I/O mix and amount is the first step in PLC system design. The number and type of I/Os will often be the major factor in deciding which PLC to use in a particular design project. Allen-Bradley has a whole line of PLC controllers to pick from, as do most other suppliers.

The design engineer's job is to pick the least expensive system that will do the job at hand. In doing so, the engineer must keep in mind a few rules of thumb in this business. If this is the original design for a machine, the designer should put together a system that has at least 40% room for expansion. It is rare that first-time designs are accurate in their I/O count. In most cases, the I/O list is generated months before the machine is even under construction. In the time between conception and system debugging, many changes and additions are usually made.

If the machine is small in nature (under 40 I/O points) and all of the I/Os are of the switch contact type, the designer may pick one of the fixed I/O type controllers. Allen-Bradley (A-B) MicroLogix 1000 controllers are designed for the low end of the control market, up to 30 I/O points. The next step up is their SLC 500 product line, which handles up to 40 I/O points in fixed form and up to 4096 I/O points in multiple-rack systems in the top-of-the-line SLC 505 controllers. For large systems, that is, those over 1024 I/O points, A-B offers the PLC-5 and Control Logix 5000 families, which can handle several thousand I/O points.

There is a great cost advantage in staying with the fixed I/O size systems, but the designer must understand the limitations of these small controllers:

- Limited or no expansion capability
- Limited or no special function I/O modules (millivolt, thermocouple, analog)
- Small memory size
- Limited or no interface capability for diagnostics or data collection.

Unless the design engineer is very confident that the I/O quantity and types will remain fixed, we do not recommend using fixed I/O systems.

The advantage of using an open rack system is that the designer can add additional I/O modules at a later date and, if necessary, an entire additional I/O rack if space was left for it in the original panel layout. It is often good practice to leave empty space in a panel layout for additional racks. The cost of empty panel space is cheap compared to the cost of rebuilding the whole panel when additional I/O points are requested and they overrun the available I/O rack space.

Optimum System Design

For a number of reasons, we believe the best I/O design should use 24V DC to power all inputs and 120V AC to power all outputs. Most control system designs have 70% of their I/O assigned to inputs and 30% assigned to outputs. Because 70% of our wiring will be input wiring, anything we can do to reduce its size will have a great effect on the system cost.

A sample system power wiring schematic is shown in Figure 2–11. Note that the input DC power supply is always powered up with the PLC power. Only the output power source is controlled by the MCR relay contacts. Note also the E stop buttons are always hard-wired to the power system. Listed next are the main advantages of using 24V DC to power inputs in a PLC system:

1. Lead wire gauge can be reduced from 16 gauge to 20 gauge.
2. Lead wire insulation can be lowered from 300 volts used on 120V systems.
3. Smaller wire duct or conduit can be used to contain the same number of lead wires.
4. Smaller panel switches can be used and therefore smaller panel enclosures.
5. When button panels are open they are safer because only low voltage is present.
6. If connectors are used they can be smaller and can contain more contact points.

Outputs as we said before should be run off the 120V power system. On large automation projects, the distance between the control panel and output devices can be long. This can lead to a considerable voltage drop in the wiring. Keeping the supply voltage high and the current low will help to minimize this problem.

As an example, let us look at the following design problem. Our system design has an output that requires 20 watts of power to operate. The wiring between the control panel and the device is of such a length that 3 ohms of resistance are present in each lead. This means we will have a total of 6 ohms of resistance. Here are the calculations for a 24V power system:

$$20 \text{ watts}/24 \text{ volts} = 0.833 \text{ amp}$$

$$6 \text{ ohms} \times 0.833 \text{ amp} = 4.99 \text{ volts line drop}$$

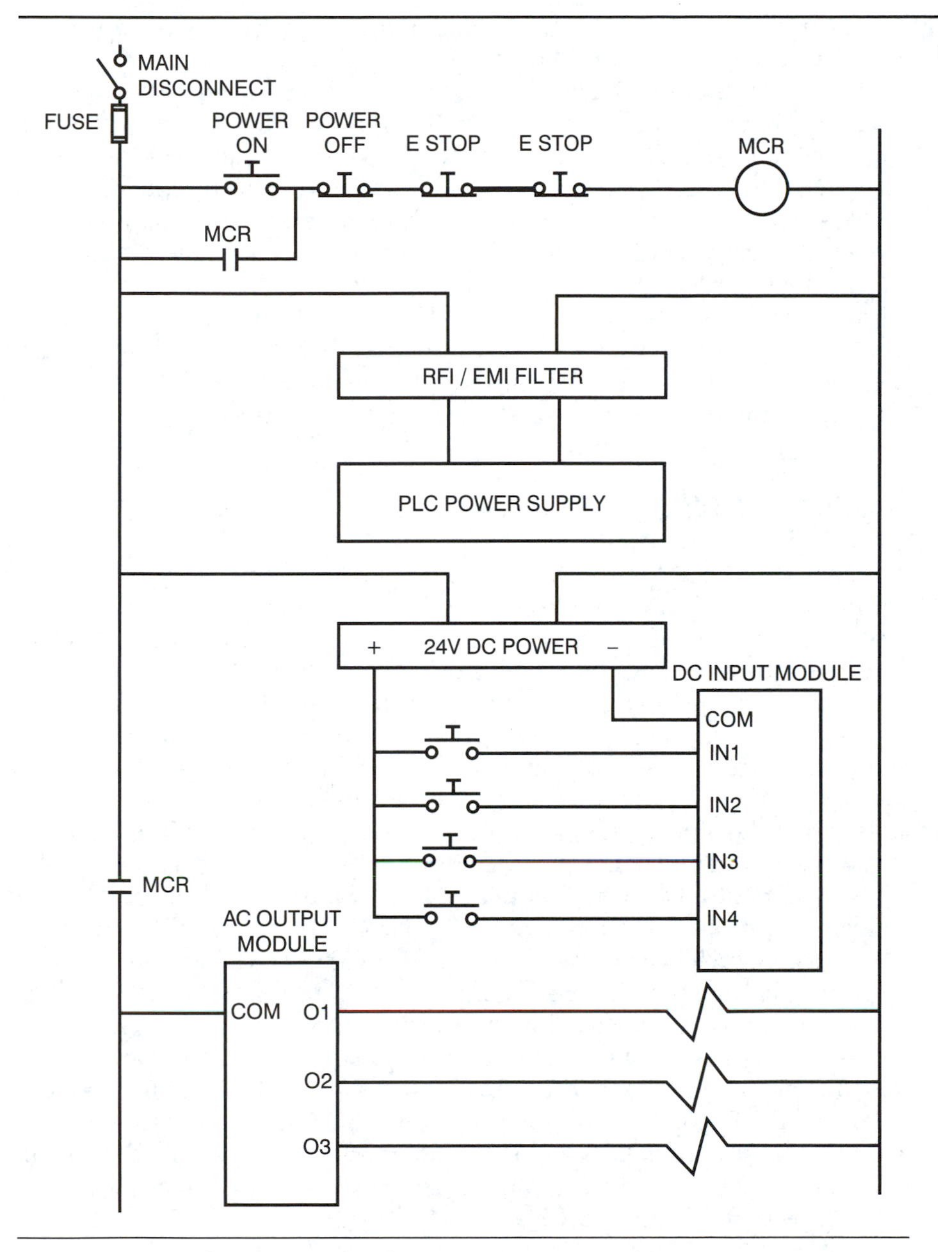

Figure 2–11
PLC power wiring diagram

This means that more than 20% of our energy is wasted in the wiring and our output device will see only 19 volts of the 24 volts supplied.

The calculations for a 120V system are as follows:

$$20 \text{ watts}/120 \text{ volts} = 0.166 \text{ amp}$$
$$6 \text{ ohms} \times 0.166 \text{ amp} = 0.99 \text{ volt}$$

In a 120V system only 1 volt is dropped in the lead wiring. This represents less than 1% loss of power in the wiring.

As we have demonstrated, powering the outputs from 120V AC is the preferred design method if long wire runs are involved. This will avoid marginal operation of the high-power outputs. A sample of this electrical design can be seen in Figure 2–12. If DC solenoids are used in the design, be sure to diode suppress the coils at the solenoid or purchase devices with diode suppression already designed in. If this is not done, noise generated by the collapsing DC magnetic field will induce voltage spikes into the adjacent wiring. It is also a good idea to keep the 120V and 24V wiring separate in the control panel, conduit runs, and square duct to prevent noise from the output wiring from inducing voltage spikes into the input wiring. Under some extreme conditions the input wiring may have to be run in shielded cable.

HARDWARE WIRING

We have discussed the I/O wiring of our PLC system but not its power wiring. Figure 2–11 is a sample power wiring drawing. Note that the PLC is connected directly to the main power wires, and not through the MCR relay. The PLC should be powered any time the main disconnect is on. Only the AC output modules are disconnected by the MCR relay. If the E stop push button is pressed, all outputs are dropped, but the PLC and the inputs stay powered on. This is necessary if you intend to use a diagnostic display device in the design. If the input power were dropped, the diagnostics would not be able to read the input status and report the fault. This example highlights the reasons for designing a control system using 24V DC for input power and 120V AC for output power. Not only is 24V DC safer to work on, but the 24V DC power supply will shut down if an overload current problem arises.

It is good practice always to use a filter to remove RFI interference and line spikes between the AC line and the PLC power supply. A Corcom RFI filter is shown between the AC line and the PLC power supply in Figure 2–11.

For safety reasons all E stop buttons must be hard wired to the electrical system and cannot be wired to PLC input modules. If the PLC processor fails, and the E stop buttons are input contacts to the PLC, they will be of little use to shut down the system. For diagnostic purposes, the E stop buttons can be ordered with an additional set of N.O. contacts, which can then be used as inputs to the PLC system.

Panel Wiring

To ensure noise-free operation, the designer should locate the PLC rack away from high-current cables and switching relays. The designer should also lay out the wire runs so as to keep input wiring away from high-current output wiring. If, as recommended, a split power system is used, care should be taken to keep the 24V input wiring away from the 120V output wiring.

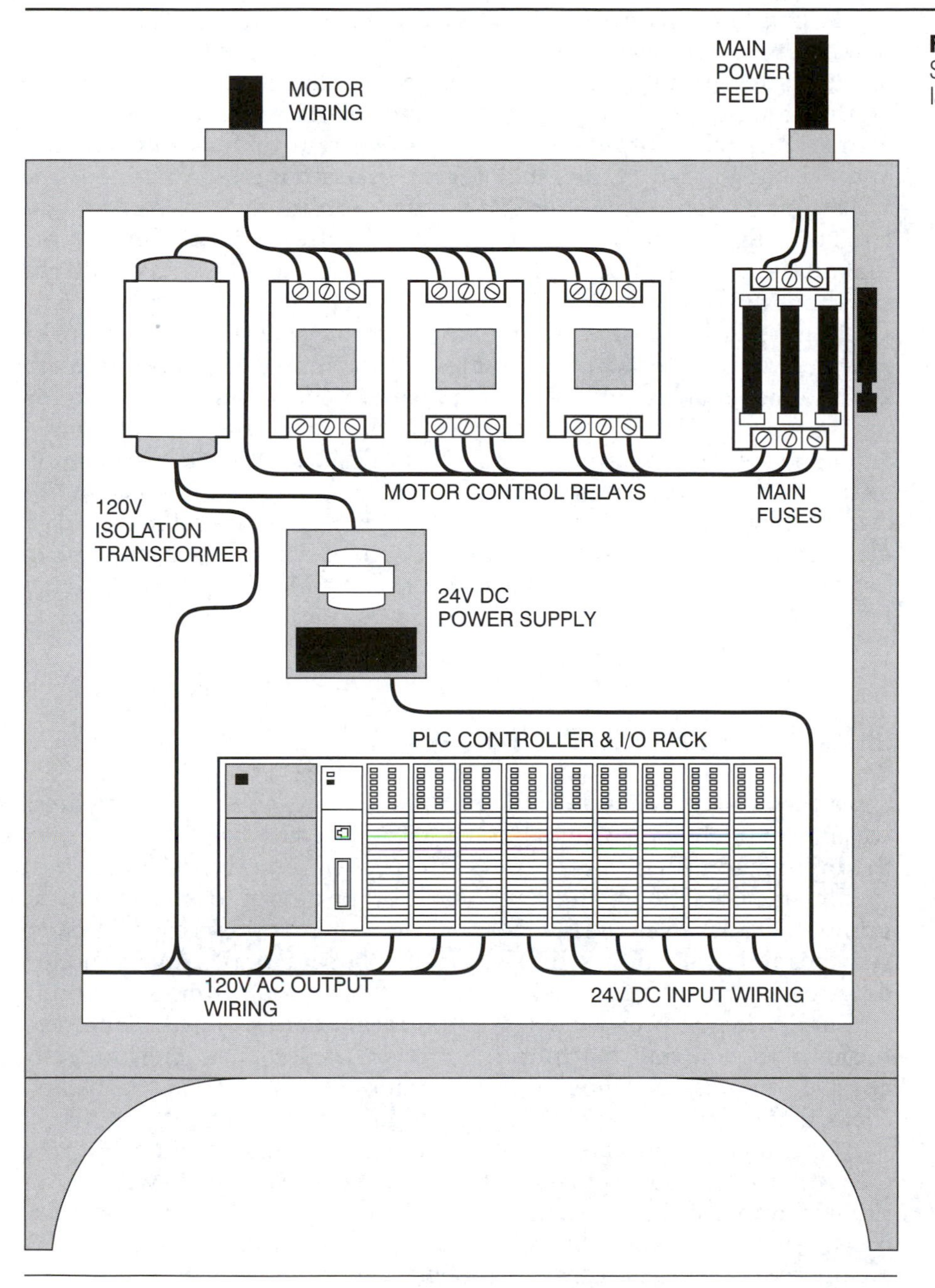

Figure 2–12
Simplified control cabinet layout drawing

A simplified control cabinet drawing is shown in Figure 2–12. Note that main power runs into the top right corner of the cabinet and directly to the disconnect mechanism and fuse elements. From there it is run to all the high-power motor start relays and the step-down control transformer. The wire outputs from the high-current relays run to the left and exit the top left of the cabinet. The isolation transformer feeds the 120V control circuits and the 24V power supply. On large control systems this transformer may be located outside the control cabinet. Note that all the 120V AC output modules are located together, as are all the DC input modules.

With this layout the wiring can be kept separate and, as in the sample, even exit from opposite sides of the cabinet. The closer you can keep to the ideal layout, the fewer noise problems you will encounter. If you must cross high-power and low-power wiring, do so at right angles. Never run high-power and low-power wiring in close proximity for long parallel runs. Another source of electrical noise is servo motor cables. If possible, place servo controllers in a separate cabinet. If they must be in the same cabinet, keep the wires as short as possible and have them exit the cabinet as close to the controllers as possible. Never run servo motor cables in the same square duct or conduit runs with PLC signal wiring.

INITIAL SETUP AND TROUBLESHOOTING

Once the design is complete, the panel built, and all I/O points wired, we enter the debug phase of the project. The first step is to power up only the input system with the outputs left unpowered. The designer must systematically activate and check each input device to verify that the proper I/O address in the PLC is activated. Whereas this is easy for panel switches and push buttons, the same cannot be said for position or motion detector devices. The designer must be sure to schedule time for aligning all limit switches, prox switches, and optical devices to make sure they perform their actual machine functions. After each is mechanically aligned, the designer must activate each device in both positions and check that the proper I/O address is activated in the PLC.

Once all the inputs are verified to be correct, we can begin debugging the outputs. In an air-operated system, it is often best to begin with the air pressure turned down to minimal pressure. This will help prevent physical damage to mechanisms if the outputs are wired incorrectly or the air hoses are not connected to the correct air solenoid ports. The designer must now activate each output address and verify that the correct solenoid, light, or relay comes on. In the case of air valves, he or she must also verify that the air cylinder moves in the correct direction because the hose connections may have been attached to the ports incorrectly.

In a complex machine or process control system this verification process can take several weeks. If errors are found—and they almost always will be on a new design—it is important to keep a set of marked-up prints nearby to note all the changes. These prints must be meticulously maintained, because they will be invaluable at the project end when completing the final documentation. If the changes are not recorded as they happen, it will be impossible at the end of the project to remember all of the changes and problems that were discovered. This will doom you to debugging the same set of problems on the next system.

CONCLUSIONS

This concludes our initial instruction in I/O modules. Many specialized I/O modules are available to the designer that we have not covered. As we grow in our understanding of the PLC design process, we will introduce more complicated I/O modules. Understanding the interface of discrete I/O devices to the PLC system allows us to cover 80% of our machine control needs. In Chapter 4 we will begin to show you how to control the I/O modules and internal memory locations to accomplish a specified sequence of events.

REVIEW QUESTIONS

1. When wiring DC I/O modules, what precautions must be taken if you are interfacing with DC prox switches?
2. What are the two main types of DC prox switches?
3. Why can we have several different power voltages present in an I/O rack without concern for grounding problems?
4. What are the main advantages of using 24V DC for powering input modules?
5. Why should 120V be used for powering outputs if long wire runs are necessary?
6. On new mechanical design projects, why should the engineer allow for 40% I/O expansion?
7. Why are suppression diodes required on DC solenoid coils?
8. What do the terms “sourcing” and “sinking” mean when used in reference to DC I/O modules?
9. What programming precautions must be taken when output terminals are tied in parallel to achieve higher drive currents?
10. If your design calls for using sinking-type DC input modules, what type of prox switches must be used?
11. Why should 24V DC and 120V AC wiring be kept separate in conduit runs or square duct enclosures?
12. Why must E stop buttons always be part of the hard-wired electrical system?
13. Why should the PLC remain powered up even if the E stop circuit has been dropped?
14. When designing a control panel, why should high-power wiring be kept separate from the PLC signal wiring?
15. What is the purpose of the EMI / RFI filter placed between the main power line and the PLC’s power supply?

CHAPTER

3

Switches and Position Sensor Devices

LEARNING OBJECTIVES

After completing this chapter the reader should understand:

1. How to determine the total power supply current needed for a control project.
2. How to pick the proper switch devices to fit a particular design constraint.
3. Interfacing and applications for different types of magnetic proximity switches.
4. Interfacing and applications of capacitive proximity switches.
5. Interfacing and applications of optical position sensors.
6. Interfacing and applications of ultrasonic level sensors.

Before we start designing a PLC control system we must have a good understanding of the basic sensing elements used in machine controls. As we begin the design of even a small machine control system we need to understand, in great detail, the basic sensing and switch elements that must work together to make up the input portion of our control system. We will start this discovery process by studying some of the simpler input devices used to sense mechanical position and that allow human interaction. These devices usually have only two states, ON and OFF. Inputs are the "I" in the computer term "I/O devices," with the "O" standing for outputs. Most machine control systems have a ratio of three to four inputs for every output. Inputs are defined as active elements that provide information to the control system. These elements can be manually activated, such as a push-button switch, or they can be automatically activated, as is the case in a position sensing switch or pressure sensing switch. In most cases the input device provides a connection between one side of the power rail and an output device, or to a PLC input module.

If the device is OFF then no current flows to the output or PLC control module. If the device is ON then electrical current is allowed to flow from the power rail to the device connected on the other side of the input device. Several example circuits are shown below.

Figure 3–1 shows a 24V DC power supply wired to three parallel circuits. Note that none of the push buttons are closed, so no current flows in the circuit. Switches PB1, PB2, PB3, and LS1 are the input devices and lamps L1, L2, and L3 are the outputs in this circuit. The first two circuits have a single input device controlling the output. The last circuit has two input devices in series that control the output device. In this case both switches must be closed before current will flow in that branch of the circuit. This type of electrical circuit is also called a ladder circuit or ladder diagram. The vertical power rails supply power to all the circuits

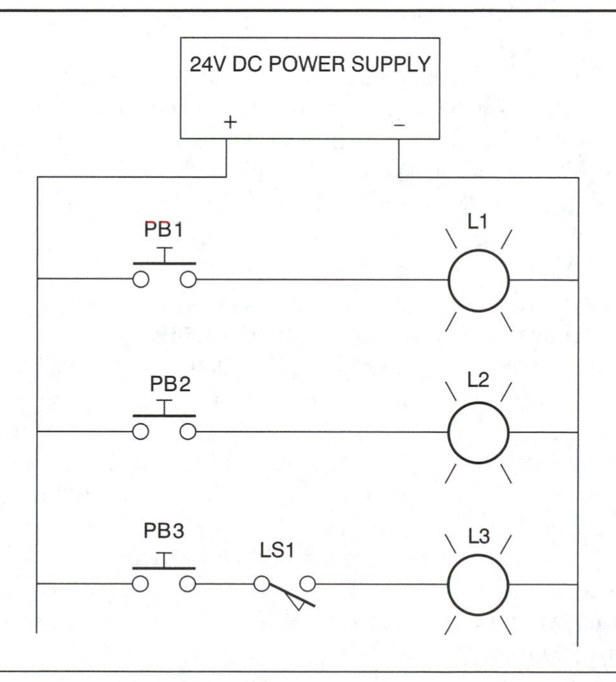

Figure 3–1
Sample ladder circuit with all contacts open

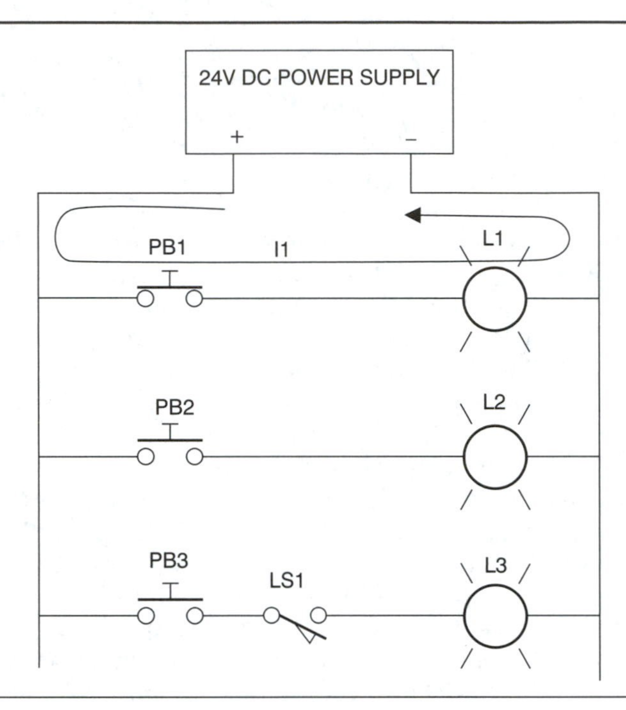

Figure 3–2
Sample ladder circuit with current through one rung

between them. The circuits between the power rails are called the rungs of the ladder diagram.

In Figure 3–2 we have changed the first leg of the circuit by closing the PB1 push button. This allows current I1 to flow from the + terminal of the power supply through PB1 and into the lamp L1. The current then flows back to the – side of the power supply through the negative power rail. The resistance of the lamp's filament wire limits current I1. For our example let's say the lamp current is specified at 100 mA at 24V DC.

It is important to note here that the switch does not provide any power of its own; it only allows current to flow through it to the load. The power for this circuit comes from the DC power supply.

In Figure 3–3 we have changed the circuit again by closing both PB3 and LS1. This now allows current to flow through the third leg of our ladder circuit. Let us assume for this example that lamp L3 also is rated at 100 mA. From this we can see that if I2 is 100 mA and I3 is 100 mA then I1 must supply both of these currents and therefore is 200 mA. In like manner I4 must also be 200 mA as it is the return path for both currents to the power supply. Take note of this fact since the power feed wiring must always be suitable to handle the total current drawn by the entire circuit. If we were to close PB2 and lamp L2 was also rated at 100 mA then currents I1 and I4 would rise to 300 mA.

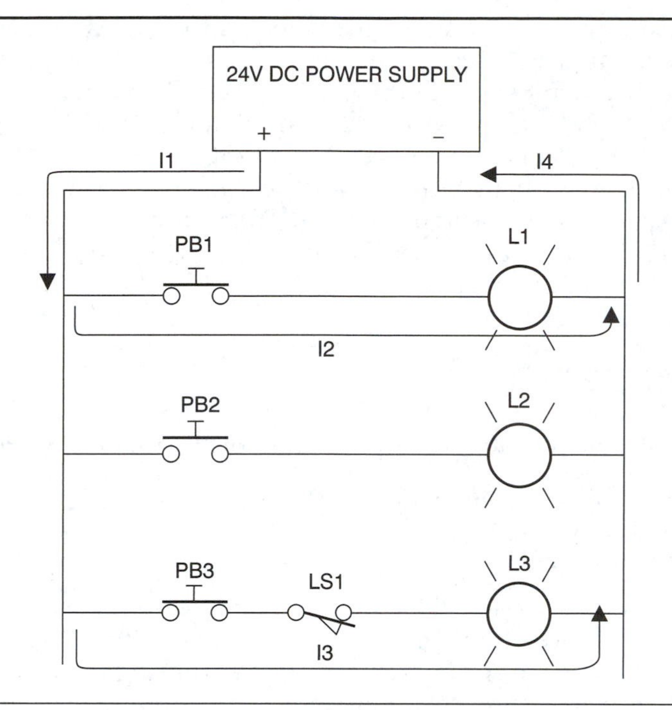

Figure 3–3
Sample ladder circuit with current in two out of three rungs

When designing ladder circuits you must be sure the power supply can support all the rung currents in the design. In addition to the total circuit current it must have some reserve power in case we later decide to add a few more rungs to the ladder. Note that 120V AC could also have powered this circuit. All we would have had to do is change the lamps to work on this higher voltage. The standard control voltages that most systems work on are:

- 12V DC
- 24V DC
- 120V AC
- 240V AC

As previously explained in Chapter 2, when designing programmable controllers, it is good design practice to have all the inputs running on 24V DC and all the outputs powered by 120V AC. Many input devices, such as proximity sensors and optical sensors, use solid-state outputs instead of switch contacts. With these solid-state output devices we must match up the device's output type with the input type of the PLC module it will be connected to. DC sensors generally have two output driver styles that are determined by the type of output transistor that is used. They are PNP, also called *sourcing* type, and NPN, which is called *sinking* type. As we stated in Chapter 2 the PLC's DC input modules are also rated as sourcing- or sinking-type modules. It is important to know that you must have opposite types connected to make the system work. Connecting the wrong style input device to a PLC input module will not result in any damage; it just will not function at all. Some 4-wire sensors have both output types and you can choose which output device to use in your design. The reader should keep these interface restrictions in mind as we work our way through the many sensor devices explained in this chapter.

SWITCH DEVICES

Most actual switch contact devices fall into three major categories.

- Push-button switches
- Selector switches
- Limit switches

All of these devices will have at least one set of real switch contacts. Most will have more than one set and you must specify the type when you place your order. As an example, when ordering push buttons you must specify if the contacts are maintained or momentary, if they are normally closed, normally open, or some combination of both types, and what the button configuration looks like. There are three mounting hole standards for most panel mount switches. They are 30.5 mm, 22.5 mm, and 16 mm. The 30.5 mm size is the old standard for control systems (Figure 3–4). The 22.5 mm switches are a more up-to-date style and can be mounted closer together on a panel (Figure 3–5). The newest addition to the switch standard is the 16 mm mini switches (Figure 3–6). These new smaller devices are mostly used in low-current applications such as inputs to a PLC controller or some other type of controller system. The standard contact ratings for the three different types are

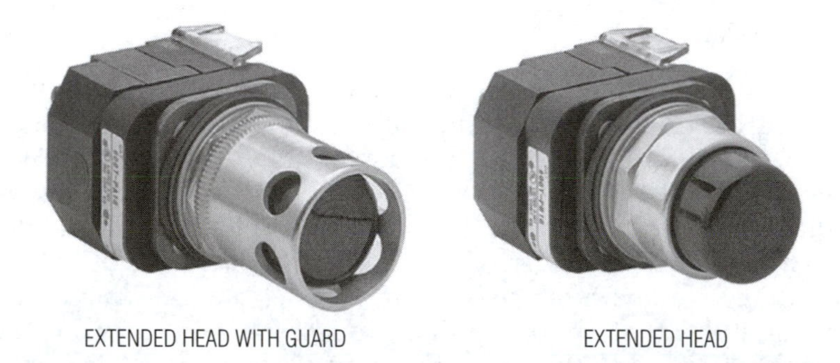

Figure 3–4
Two 30.5 mm push buttons.
Courtesy: Rockwell Automation, Milwaukee, Wisconsin.

Figure 3–5
Selection of 22.5 mm devices: Two push buttons, two selector switches.
Courtesy: Rockwell Automation, Milwaukee, Wisconsin.

Figure 3–6
Selection of 16 mm push buttons.
Courtesy: Rockwell Automation, Milwaukee, Wisconsin.

listed below in both AC and DC voltage ratings. Special high-current contacts can be ordered for the larger mounting sizes.

30.5 mm = 10 A AC, 2.5 A DC Continuous

22.5 mm = 10 A AC, 2.5 A DC Continuous

16 mm = 1 A AC , 1 A DC

As the system designer you must first decide what functions the equipment's operators will need to perform from the various panels and subpanels in your system. Functions that require activation once for each contact closure will probably require a push-button input device. Functions that require the contacts to stay in one position for a long time will require some kind of selector switch. Functions that require you to sense the position of some object will most likely require some type of limit switch.

Push Buttons

The following is a list of the questions that must be answered for each device when ordering push buttons.

Mounting style

Which of the three mounting sizes fit the application? Often this decision is made because of space constraints. If you can use low-voltage DC for your inputs and they are driving a low-current PLC input module then 16 mm buttons will allow you to design the most compact size panels. One must be careful not to place the small buttons too close together as some operators may have problems pressing only one button if they are wearing gloves or some type of hand protection. The larger buttons are a better choice if the operator is making contact to higher current devices like relay coils or if the equipment must operate in an extremely dirty environment.

Contact type

This is determined by the electrical function the switch performs. For making connection to PLC inputs one normally open contact is all that is required. For some relay operations it is often necessary that one of the buttons in the circuit have a normally closed contact if it acts as the release device for a holding circuit. In some special cases, like an E stop push button, you may need one of each type. Contact blocks must be ordered separately in the 30.5 mm and, in some cases, in the 22.5 mm buttons.

Button style

This is often determined by the type of control function the push button will be required to perform. The button's color often indicates its function type. Red is most often used for stop or reset functions. Green and black are most often used for starting some functions. Other colors are also available for special functions. Lighted push buttons are often used to display the system's status to the operator who may have to see the status from a short distance away from the panel. Lighted buttons also give the operator positive feedback that the function he pressed is really working. The shape of the button is also important. For most applications the standard extended head button is a good choice. For dangerous functions, a protected head button will prevent an operator from accidentally pressing the button. For E stop functions a large mushroom head button will make it easy to press the button when the operator is in a hurry and has to hit it the first time.

In some cases you can get two push buttons mounted in the same device that work in a latch-unlatch operation. In Figure 3–5, the second device from the left is an example of a double operator device.

Legend plates

These must also be ordered for the 30.5 mm and 22.5 mm button styles. In addition to ordering the right size and shape of plate you must indicate what is to be printed on the plate. The 16 mm buttons are designed to have the legend placed inside the clear button cover.

Selector Switches

Selector switches are used when the input condition must stay on for a longer period of time (Figures 3–7 and 3–8). Many of the same parameters that we saw in push buttons are also present on selector switches.

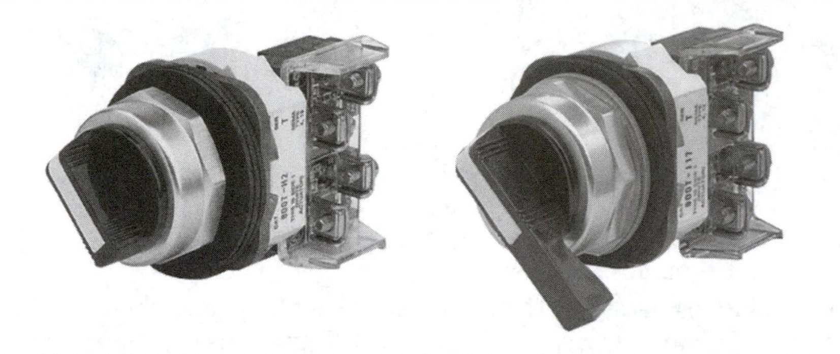

Figure 3–7
One standard selector switch and one extended operator selector switch.
Courtesy: Rockwell Automation, Milwaukee, Wisconsin.

Figure 3–8
Selection of 16 mm selector switches.
Courtesy: Rockwell Automation, Milwaukee, Wisconsin.

Selector switch mounting hole sizes and contact block arrangements are virtually the same as those for push buttons. Knob-style selector switches have a few different elements that must be specified.

Figure 3–7 shows the standard 30.5 mm knob and the extended lever knob. Looking back at Figure 3–5 you can see examples of the 22.5 mm standard knob and a special keyed switch. Keyed selector switches are often used as a safety switch bypass function when a machine is in manual mode or to enable run mode on equipment where a trained operator must be present to start up the equipment.

Selector switches can also be lit if desired. This gives the operator feedback that the selected function is active. The size of the selector knob is often determined by the operator's needs. While the standard selector knob is fine for most applications the lever knob may work better if the operator wears gloves or some other kind of hand protection.

Limit Switches

Limit switches are used to detect the presence or absence of an object or condition. This is normally an object that moves during the machine's cycle. When you're programming to control a machine's cycle the control program needs to know when mechanisms are in a certain position. In most new control system designs limit switches have been replaced with proximity switches or optical devices due to their better long-term performance.

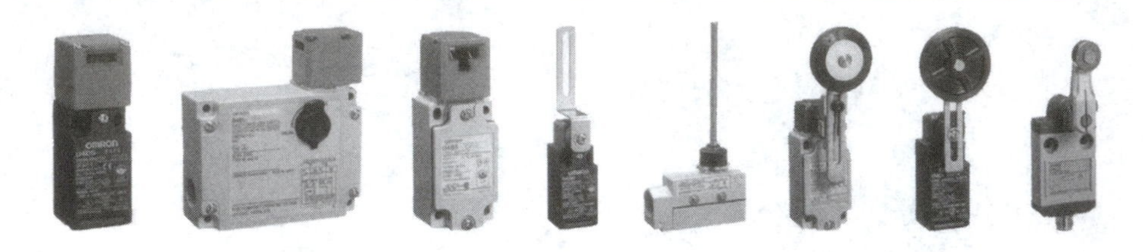

Figure 3–9
Safety and limit switches.
Courtesy: Omron Corporation.

Limit switches have actual moving contact points and therefore have a limited life span. A variety of different configurations of limit switches are available on the market depending on the type of object that is to be detected (see Figure 3–9).

In Figure 3–9 the three devices on the left are safety switches. They are mounted in such a way that a tab or tang is inserted into the slot on the device when the door or cover is in place. If the door or cover is opened or removed, the tab moves out of the slot and the contacts in the switch are opened. The switches on the right side of Figure 3–9 are more typical of the types of limit switches used to sense a moving element on a machine control application.

Figure 3–10 shows the three major actuator types of limit switches in both their open and closed positions. The three switches shown on the left of the figure are in their non-activated states. On the right the same three switch types are shown activated. Of the three types, the push rod type is the least forgiving of overtravel. Great care must be taken not to allow the moving device to use the switch rod as an end stop. Placing excessive pressure on the switch rod will damage the switch and shorten its life. Some type of physical stop must be used to limit the travel of the moving member. The other two limit switch types, rod lever and wheel lever, are much more tolerant of overtravel. When mounting limit switches the mechanical engineer needs to use elongated mounting slots or some other method to allow the switch to be adjusted. As equipment ages it is often necessary to realign switch positions in order to compensate for the mechanical wear on moving elements. A good

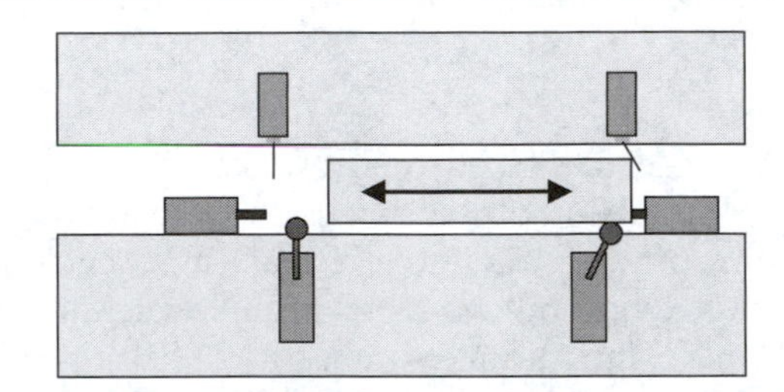

Figure 3–10
Three major limit switch actuator types shown in both open and closed positions

mechanical design will make switch alignment an easy procedure for the maintenance person to perform.

In addition to mechanical position sensing limit switches we have another category of switches that sense conditions in the process or machine. These conditions include pressure, flow, temperature, levels, and vacuums. Most of these switches have a contact that closes when the setting or limit is reached. In the case of the flow switch the contact closes and opens for each increment of material that passes the switch (in ounces, cubic feet, or pounds).

INDUCTIVE PROXIMITY SENSORS

Proximity sensors have for the most part replaced limit switches in most industrial applications where position sensing is required. Proximity switches, hereafter called *prox* switches, work by sensing change in a magnetic field when a metal object moves close to it. Because prox switches have no moving contacts to wear and are in a sealed container they work well in dirty or oily environments and have a long life expectancy. The system designer must take into account a number of technical considerations when using these devices. We will now examine each element and the limitations it imposes on the design engineer.

Operation

Prox switches come in many different sizes and shapes. The basic proximity device looks like a threaded rod with leads extending from one end. As shown in Figure 3–11, there is a coil at one end of the prox switch that is driven by the oscillator circuit. When a metal object disturbs the magnetic field of the coil, the sensor circuit detects this change and turns on the output driver circuit.

Detection Distance

The maximum distance at which an object can be detected is an important rating of most prox switches. This is determined by three factors: the diameter of the prox switch, the use of shielding, and the type of metal that is being

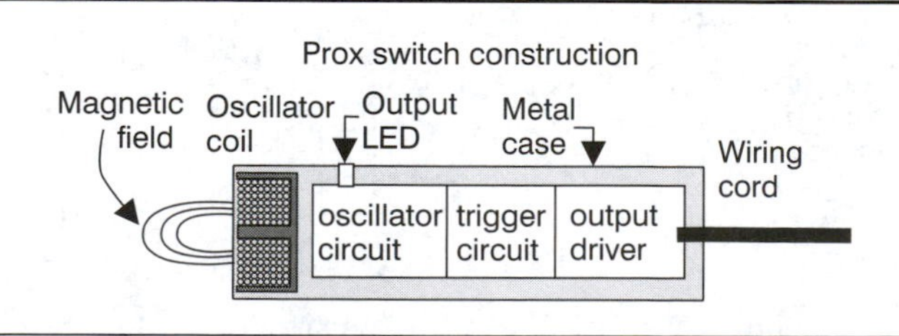

Figure 3–11
Inductive proximity switch showing internal layout

Figure 3–12
Chart showing the relationship between prox switch size and sensing distance

Prox Switch Sensing Distance		
Diameter	Shielded	Not Shielded
8 mm	1 mm	2 mm
12 mm	3 mm	4 mm
18 mm	5 mm	8 mm
30 mm	10 mm	15 mm

detected. In most cases the larger the diameter of the switch, the longer the sensing distance is. Most manufacturers will provide a chart showing the diameter and sensing distance for both shielded and non-shielded devices (Figure 3–12). The term "shielded" means that the prox switch has a metal shield that extends all the way up to the edge of the coil. Non-shielded prox switches have sensing ends that are completely made of plastic.

Prox Switch Mounting Constraints

Using prox switches that have shielding has several effects on determining the proper operating procedures. The first of these is mounting positions and clearances. Non-shielded prox switches must be mounted with their coils above any metal surface. If these switches are not mounted correctly the surrounding metal will false trigger the switch (see Figure 3–13). Note that unshielded prox switches have magnetic flux lines emanating from all around their coil area. Shielded prox switches can be flush mounted because the shield keeps the flux lines concentrated directly in front of their coils. The shield also has an effect on the minimum distance between prox switches. Since the shield prevents the flux path from extending around the sides of the coils, shielded switches can be mounted much closer together than the non-shielded devices. The rule of thumb in mounting prox switches is that shielded prox switches should be mounted with

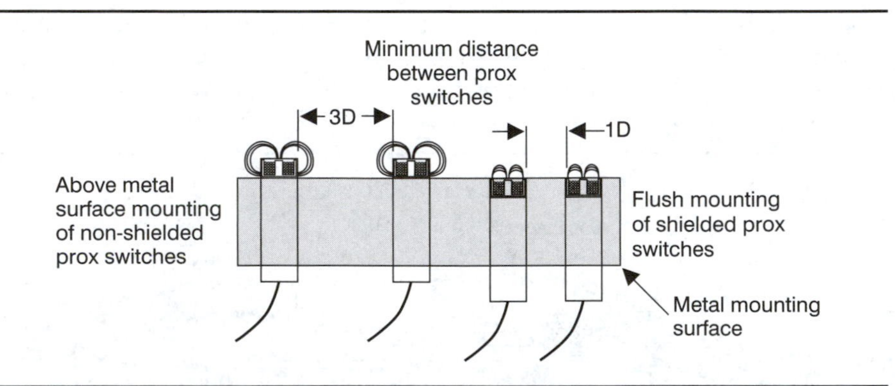

Figure 3–13
Mounting requirements for shielded and non-shielded prox switches

at least the length of their diameter between the sides of their metal enclosures, whereas non-shielded prox switches should have three times their diameter between the sides of their enclosures. You need to add an additional diameter if the switches are mounted facing each other. Which type of prox switch to use will be determined by how close the switches must be mounted to each other, what the minimum sensing distance must be, and how they will be mounted if the bracket is made of metal.

Sensing Problems

We now cover a few additional application considerations that should be kept in mind when using prox switches for position sensing. The first is that prox switches have a certain amount of hysteresis built into their trigger circuits. This is to prevent a flutter effect when the sensed object gets close to the trigger point. Because of this phenomenon it is not recommended to have the prox switch sense an object moving toward its tip. A better arrangement would be to have the sensed object move across its tip. In either case, the designer must be sure the moving object will traverse a distance greater than the hysteresis distance. This hysteresis distance changes with the diameter of the prox switch and also between shielded and non-shielded units (see Figure 3–14).

The designer must also be careful not to let the prox switch act as a physical stop for the moving element. This will damage the prox switch and in time cause it to stop working. A prox switch that is equipped with an LED indicator should be mounted in a position where it can be seen easily for debugging purposes.

Another consideration when deciding if your sensing application can be performed using a prox switch is the type of material that is being sensed. The sensing distance values given in Figure 3–12 were developed using a ferrous material. The sensing distance will change dramatically if the material being sensed is not of a ferrous type. Some typical materials are listed below along with the multipliers needed to get the correct distance.

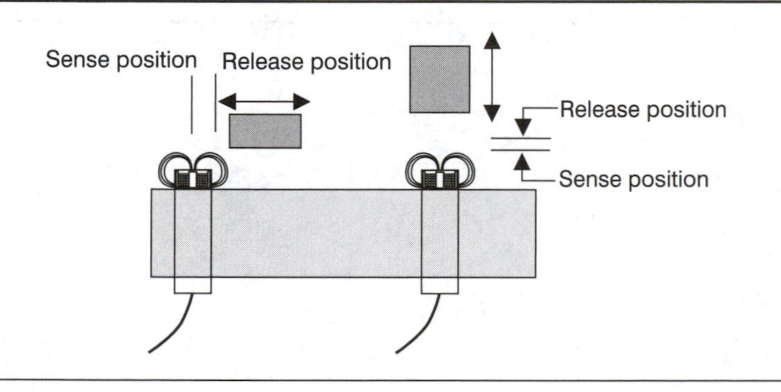

Figure 3–14
Hysteresis considerations when mounting prox switches

Type	Multiplier
Stainless steel	.85
Brass	.50
Aluminum	.45
Copper	.40

Sensing Frequency

The last element we need to consider is the prox switch's frequency response, or the maximum speed an object can be moved past the sensor and still be reliably detected. Figure 3–15 shows two examples of where frequency response maybe a problem. In the case of the index cam the speed of the shaft must be considered. Coil winding machines often have shaft speeds from 1,000 to 5,000 rpm. Using a prox switch with a frequency response of 10 Hz on this application would be a problem, as the frequency of this application is 16 Hz at the low end, and at the top end would be 83 Hz. The designer would need a device with a frequency response of at least 100 Hz to work reliably in this application. In the examples shown in Figure 3–15 the prox switch is being used to count positions as the device moves along a sense strip. Here again we need to be sure that the maximum speed at which the strip moves is below the frequency response limits of the device that was picked for this application. Failure to stay within the frequency response of the sensing device will lead to unreliable operation. The problem will be difficult to detect because it will only happen at high speed, and even then only a few counts will be missing.

One last sensing problem that we should mention is the use of prox switches in welding operations. Because prox switches detect the presence of a metallic object with a magnetic field one would think they would have problems around TIG, MEG, or plasma arc welding operations due to strong stray magnetic fields. There are special "weld immune" prox switches for this purpose, but even they have to be kept at a distance from high-current wiring and magnetic fields. The rule of thumb in the business is to keep .5 inch of space between the sensor face and wiring for every

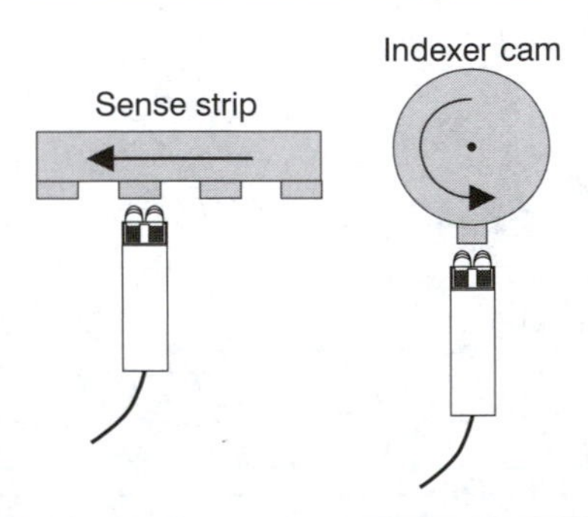

Figure 3–15
Frequency considerations when using prox switches in high-speed applications

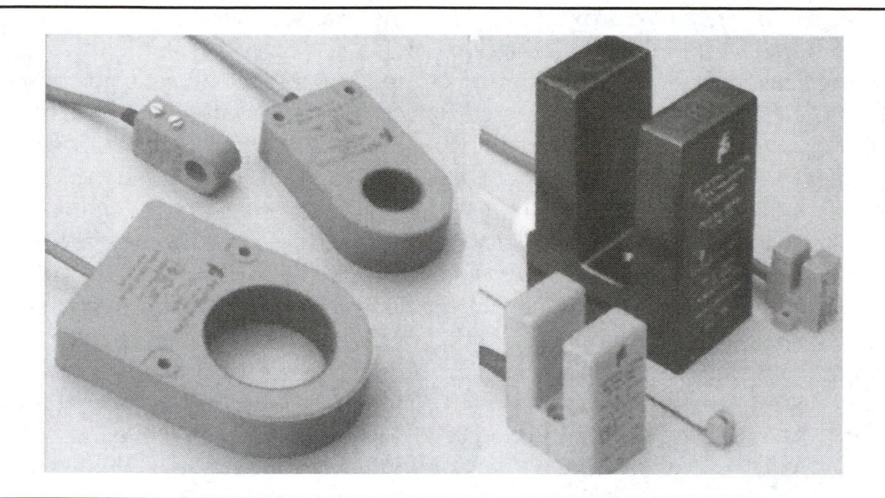

Figure 3–16
Samples of slotted and ring-type proximity sensors.
Courtesy: Pepperl-Fuchs Inc., Twinsburg, OH.

10 kA of current. It is also important to be sure none of the welding cables are run close in parallel with the sensor wires. If the welding and sensor wires must cross they should do so at a 90-degree angle. This is also good advice for high-current motor wiring and servo motor wiring as well.

Several special proximity sensor configurations are used for special sensing problems. Slotted and ring-type sensors are shown in Figure 3–16. Slotted sensors are good for sensing metallic sheets being fed from rolled material. A punch press operation is a good example of this type of application. Ring-type proximity sensors are used to detect small rod-shaped material like solder, wire, or metal rod being fed into some kind of assembly or coating operation.

CAPACITIVE PROXIMITY SENSORS

Many capacitive proximity sensors on the outside look much the same as the inductive proximity sensors we previously described. The main difference is that capacitive proximity sensors detect an object with an electrostatic field instead of a magnetic field. Figure 3–17 shows the internal

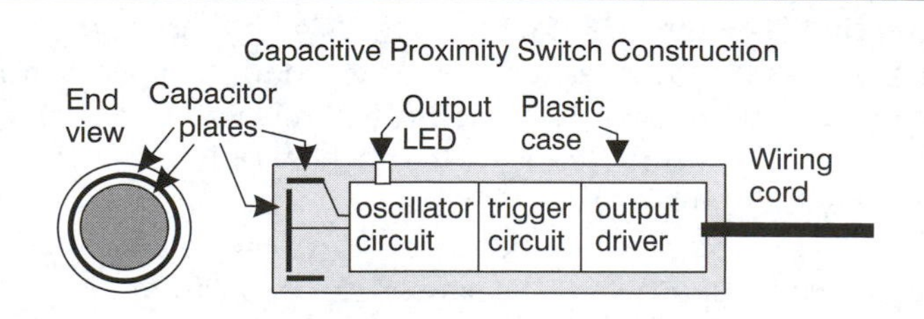

Figure 3–17
Construction of a capacitive proximity sensor

design of a capacitive prox switch. Note the similarity to the inductive prox switch in design. This device has two plates of a capacitor in the front face of the sensor. The oscillator circuit is tuned to sense an object by a change in the dielectric material presented to the capacitor plates. The sensitivity of the device is determined by the size of the object, its distance from the front surface, and the dielectric constant of the material being sensed. Every material has a dielectric constant value, and those with larger values are easier to detect. Complete information on dielectric constants can be found in the *CRC Handbook of Chemistry and Physics* (CRC Press).

The real beauty of the capacitive sensor is that it can detect a substance right through a container wall. For this reason capacitive sensors are often used to detect the level of a liquid or powder inside a tank or container. An example would be sensing the level of a water-based solution through a sight glass in a tank. Because glass has a dielectric constant of about 4 and the water-based solution will be close to 50 or 60 the sensor can detect a level right through the sight glass even though the solution is clear. Another example would be sensing the level of liquid in plastic containers moving down an assembly line. Since the plastic container would have a very low dielectric value of 2 or 3 and the material inside the container (milk, water, or soap) would have a high dielectric value of 60 to 80, the sensor can detect the fact that the container is full right through the container material. This technique works only if the difference in dielectric values of the container and the material is significant. If we tried to sense sugar inside a plastic container the process would not work as well because sugar only has a dielectric value of 3. Capacitive sensors can also detect metal targets, but the considerably higher cost of capacitive sensors usually makes inductive sensors a better choice for this function.

Just like their inductive counterparts, capacitive sensors come in shielded and non-shielded formats. The shielded devices are more immune to electrical noise and can be flush mounted in a metal or plastic support device just the same as the inductive types. Unlike inductive sensors, many capacitive sensors have a sensitivity adjustment in the rear of the device. This allows the designer to adjust the level of sensitivity needed for the material type and size of target he is working with. The designer must also take care to provide the proper mounting distance between each sensor just the same as when you mounted inductive sensors. Each manufacturer gives detailed spacing values for parallel mounting or face-to-face mounting of their devices. Failure to follow these rules will lead to unreliable operation of the sensors in your system. When mounting these sensors in a tank to sense liquid levels it is often a good idea to use mounting wells for the sensors (see Figure 3–18). A mounting well is a plastic threaded device that screws into the tank wall and has a blind threaded hole in it for mounting the sensor. With this plastic plug in place, if the sensor becomes damaged or needs to be replaced for some reason, the tank will not have to be emptied before the sensor can be removed.

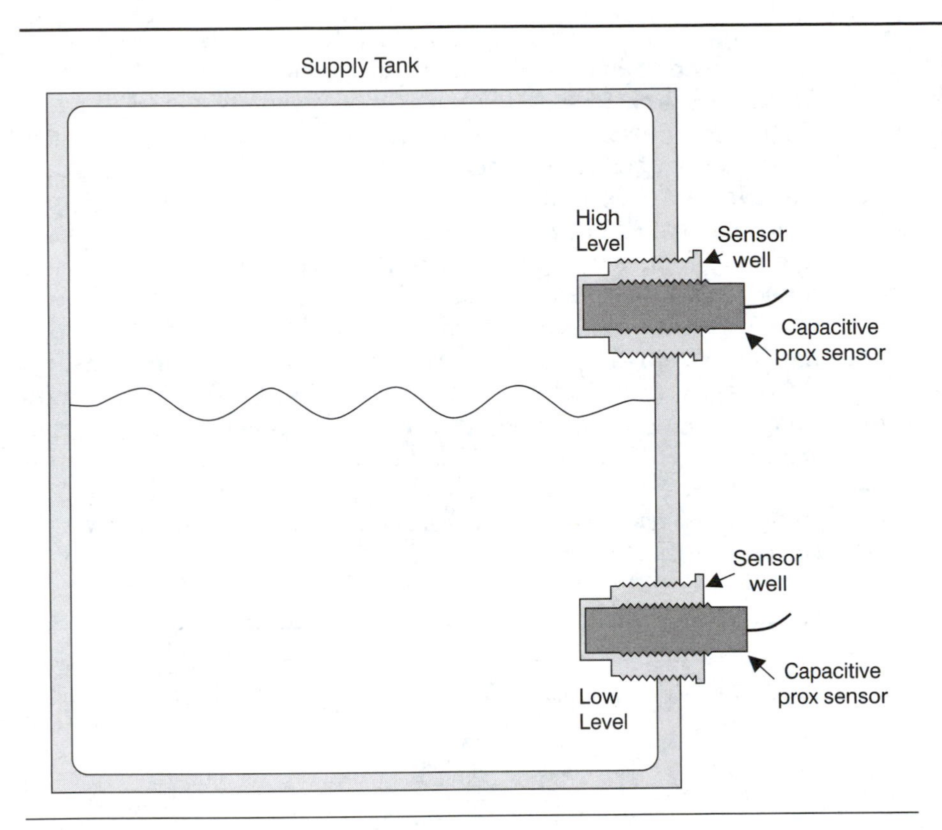

Figure 3–18
Application using two capacitive proximity sensors to sense the high and low levels of a tank

Another limitation we must consider when using capacitive sensors is dust or dirt buildup on the sensor face. Unlike inductive sensors, any material that builds up on the face of a capacitive sensor will have a dielectric value and therefore cause the sensor to change sensitivity in response to the buildup. Inductive sensors would only be affected by metallic dust buildup. Most unshielded capacitive sensor have compensation probes built into the face of the sensor to overcome some of the contamination problems. But even these devices cannot overcome large quantities of material buildup on the face. For this reason the designer needs to be sure the sensor is covered or protected in some way from material building up on the face of the device. Failure to provide protection for the sensor can lead to unreliable operation later on in the life of the equipment.

OPTICAL SENSORS

Optical sensors use some form of light beam transmitter and receiver to sense the presence of an object that either breaks the beam or causes it to be reflected back to the receiver. Most of these devices use an LED, or *Light Emitting Diode,* as their light source. In the more advanced devices that are used to actually measure distances this may be a laser diode. We will begin the discussion with the less expensive LED devices used to detect the presence or absence of an object in their light path. These devices are generally divided into two main groups: thru-beam types and reflective types. Figure 3–19 shows the basic types of photoelectric modules that can be used in control systems. The slotted sensor shown in Figure 3–20 is a special case of the thru-beam type that has a fixed opening size between the sensors. It is often used as a home position sensor because it has a small beam size and a very repeatable trip point. Another use for this slotted-type sensor is as a counting sensor if the sensed object can be made to pass through the slot. This is common for products that have a flat edge like a vacuum packed item on a cardboard backing sheet. This type can also be used to check for the presence of feed material in a production process. Examples of this kind of sensing would be sensing the feedstock in a press operation or wrapping material on a packaging machine.

Figure 3–19
Photoelectric sensor types

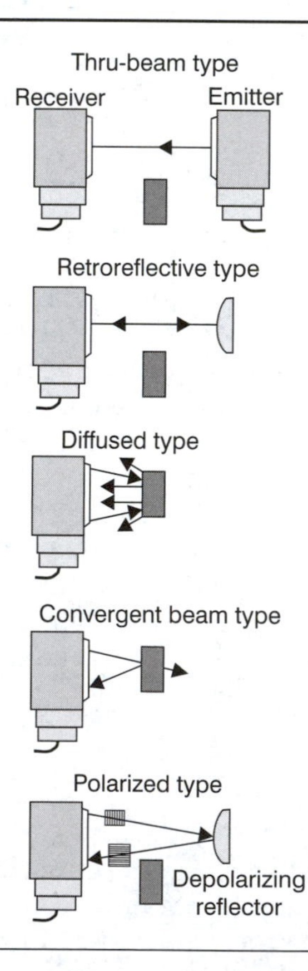

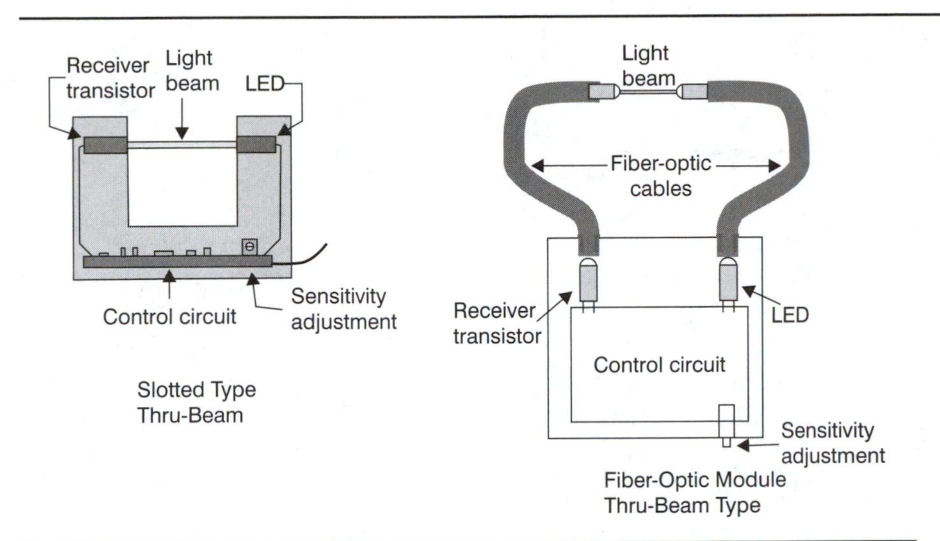

Figure 3–20
Slotted and fiber-optic thru-beam types of optical sensors

The fiber-optic module type can be used in a number of sensing operations. Its one big advantage is that the control circuit can be quite a distance away from the sensor point. This allows the fiber-optic cables to be mounted in very tight places or in areas of high electrical noise because only the cable ends must be present at the detection point. These fiber-optic modules come in both thru-beam and reflective types. The emitter and receiver ends can be mounted a considerable distance apart (usually 6″ to over 1.5 feet) and still function well. The mechanical design engineer must make some provisions for mounting and aligning the ends of the fibers.

Reflective optical modules work by transmitting a beam of light from the end of the sensor and looking for the beam to be reflected back to the receiving transistor. Many different versions of these devices are available on the market (see Figure 3–19). In the diffused type, the beam is aimed straight out the end of the module and any reflection will produce an output. In the convergent type, the beam is at a 45-degree angle and only an object at the correct distance will bounce the beam back at the correct angle. The designer must pick the module that will give the best performance under the conditions he has to work with. Figure 3–21 shows two examples of reflective type fiber-optic modules. Note that the fiber-optic module has both fiber-optic cables joined into one cable with a single end. This again makes it a good candidate for use in tight areas or areas of high electrical noise since the control module can be mounted a good distance away.

Many frequency ranges are available to the sensor designer. They range from the infrared range through visible red, green, and blue and finish up in the ultraviolet range. The ultraviolet units have the best penetrating power and can see through dust and oil deposits on the sensor. This has one bad side effect in that the material of the target needs to be of a type that

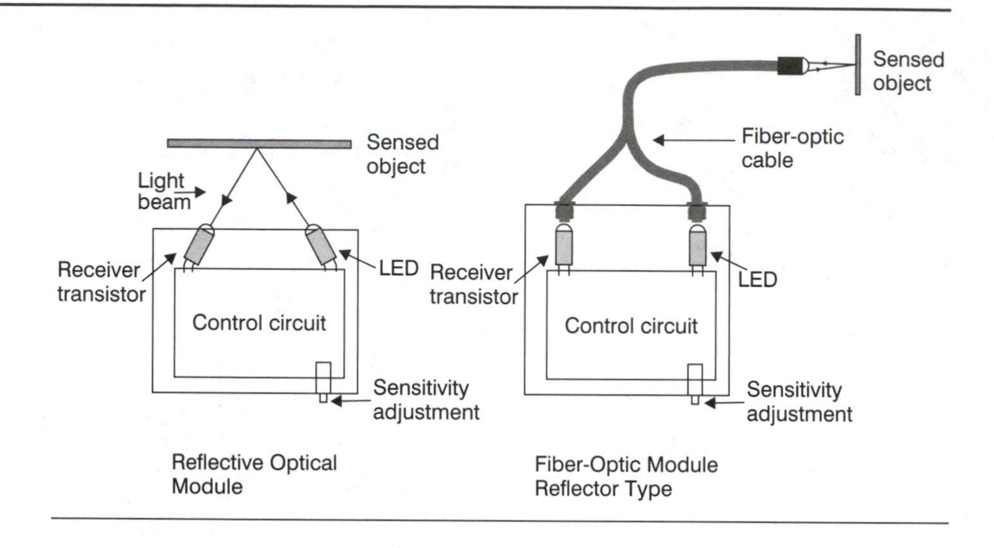

Figure 3–21
Examples of two different types of reflective optical sensors

ultraviolet light cannot pass through. Many plastic materials are completely transparent to ultraviolet light even though they look opaque to our eyes.

In some cases it is desirable to use visible light frequencies for the light source as sensors are much easier to align and set up if we can see the light beam. Red, green, and yellow light sources are used for these units. Infrared units have the best light output per watt of input power used of any of the ranges and are used when a high level of light is required. Some special sensors can discriminate different colors at the target. These can be used to test wire harnesses or plugs to verify that the correct color wire is placed in the correct hole of the plug. Most photoelectric sensors also have an option to change the on condition from dark to light operation. By this we mean that the sensor's output can be activated when the sensor sees light or when it does not see any light. Which setting you use will depend on the control function you are trying to perform.

Ambient light can be a problem when using optical sensors if it floods out the sensor's own source of light. The sensor manufacturers have solved this problem by modulating the light source. The light source is flashed on only for short pulses and the receiver devices are timed to accept light only during this pulse time. A modulated light source tends to override any outside light source that is on constantly or flashes, such as welding arcs, and makes the sensors much more reliable.

Target material must also be considered when using photoelectric sensor devices. We have already noted this problem when working with ultraviolet light sources, but target material can be a problem with visible light sources as well. With thru-beam type units we have to be sure the target material is opaque to the light frequency that is being used. When using reflective units we have to consider the type of surface the beam will bounce off. If the surface is flat and shiny, the beam will return to the sensor with almost the same intensity as it had when it left the LED. If the target surface is dull, rounded, or rough, a lot of the light energy will be scattered and not returned to the sensor. This may cause problems with sensitivity in high ambient light conditions or in dusty or heavy oil mist environments. For the same reasons, reflective devices don't always work well on liquid level sensing.

Another problem with reflective units is encountered when the beam is reflected from some object other than the desired target. This problem is called second surface reflection. This can often happen when a diffused beam device is used and the beam travels a much greater distance than desired for the application to be performed. An example might be a device to count cartons as they move down an assembly line. If the beam is allowed to travel past the side of the conveyor, then anyone passing by the assembly line, wearing a bright shirt, could cause the sensor to miss count. The solution to this kind of problem is to place a deflector on the other side of the belt to absorb or scatter the beam or deflect it up out of the normal path. This will ensure that no false triggers are counted.

Another way to solve this problem is to use polarized light sensors. They work by polarizing the light as it leaves and enters the sensor. Because the filters are 90 degrees out of phase with each other any reflected light will not be sensed. We now place a depolarizing reflector across from the sensor and as the sensor's light beam hits the reflector it is depolarized and returned to the sensor. In this case the sensor will respond only to its own light beam if it is bounced off the depolarizing reflector, but not to light from any other source.

The type of photoelectric sensor you choose to use in your application will depend on many factors. The major conditions to consider are the size of the intended target, the material the target is composed of, distance to the target, repeatability of the target distance, operating environment, and space available to mount the sensors. Careful consideration of all these factors will lead to a stable operating system. Failure to consider these options may cause unstable operation or erratic problems with counts or control functions.

ULTRASONIC SENSORS

Ultrasonic sensors operate by emitting a pulse of sound and measuring the time it takes the sound to return to the sensor. Because sound has a very constant speed of travel we can determine the distance to the target (see Figure 3–22). Ultrasonic sensors are less affected by moisture or light dust buildup than would be the case for photoelectric sensors. This makes them good candidates for level sensing in tanks of liquid. Ultrasonic sensors do have a problem if used on material that absorbs sound, such as sand or flour. But if used in the proper application they can be very reliable and trouble-free devices.

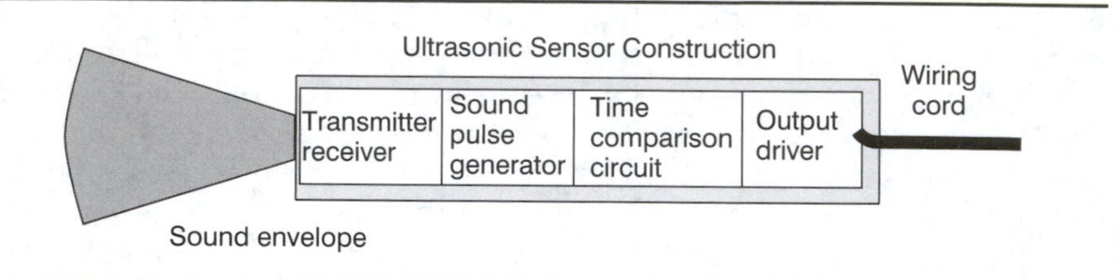

Figure 3–22
Ultrasonic level sensor construction

Ultrasonic sensors have both digital and analog output formats. In the digital type of device, adjustments can be made on the back of the sensor to create a window in the sensing range. The output is activated if the device receives a return that is inside that distance window. The analog version of the device will provide a voltage or current level that corresponds to a set amount of distance measurement. These sensor outputs can be read by a PLC controller using either a digital input module or an analog input module.

Ultrasonic sensors emit sound at frequencies between 25,000 Hz or 25 kHz and 500,000 Hz or 500 kHz. In general, the lower the frequency, the longer the sensing range. Low-frequency units may be able to measure accurately up to 40 ft, while high-frequency units may be limited to less than a foot. All the units also have a blind area directly in front of the emitter that extends a short distance, less than a inch in most cases. This blind area is specified by the manufacturer and must be adhered to for proper operation.

The type of material that is being detected also limits the maximum sensing distance. Hard materials such as steel or plastic will give good readings at the far end of the range; soft materials such as cardboard or cloth will greatly limit the maximum sensing range because they absorb much of the sound pulse.

Mounting and environment problems

Mounting more than one sensor in the same enclosure presents a problem unless the sensors are mounted far enough apart that their sound cones do not overlap. If the sound cones cross, then you will have crosstalk, with each sensor picking up sound from the other devices, and the readings will become unstable. The sensor's manufacturer will have detailed mounting instructions as to how far apart sensors must be for stable operation. Ultrasonic sensors can also experience problems reading hot surfaces or surfaces that are not parallel to their face because these surfaces tend to break up sound waves and distort them. Also, if an object is between the sensor and the target, such as a support rod in a tank, the object can be covered in a sound-absorbing material and become invisible to the sensor. Most units should be mounted so that liquid or dust will not cling to their face. This will help the sensor stay stable and maintain your original settings. Loud sounds like air hiss or banging may distort the sensor's own sound and give you bad or wildly fluctuating readings.

CONCLUSIONS

In this chapter we have presented the reader with descriptions and examples of most of the common types of input-sensing devices available in the controls marketplace. Each device has its own strengths and weaknesses when applied to a particular control's application and work environment. It is up to the design engineer to pick the device that is the most cost-effective for the application as well as one that will give reliable service and have a long working life. We can, however, give some guidelines as to which devices best fit particular applications. Figure 3–23 shows a table of common control applications and the suggested sensor types best suited for each application. You will note that some applications have more that one choice. In these cases the least expensive sensor is normally chosen. But be careful to take account of the life of the sensor and its working environment in your considerations. An unfortunate consequence of price being the only deciding factor in vendor selection may be numerous maintenance problems and high down times on your equipment.

A good example might be selecting a switch for an application that senses a cam lobe on a machine's main driveshaft. If the shaft turns at just 20 rpm the sensing device will be changing state 1,200 times an hour. Should you choose a limit switch for this application or an inductive proximity switch? Let's also suppose that the limit switch's cost is half the price of the proximity switch. Even the best limit switch has only about 1 to 2 million cycles between failures. At 1,200 cycles per hour, running 2 (8 hr) shifts a day 5 days a week, we will have well over 1/4 million cycles per month. At this rate the limit switch would reach its failure point in only 3 to 6 months. This is not a good application for a mechanical switch unless you're willing to replace the switch up to 4 times a year.

The extra cost of a prox switch is justified because it will provide several years of problem-free running between failures, while the mechanical switch could cost you many times its cost in repair down time and lost production over the same period.

Another consideration when picking a sensor for an application is how easy it will be to align and adjust. The mechanical designer must give a lot of thought to how he will mount the sensor in a stable way so that it does not drift or move due to the normal vibrations of the equipment. Failure to do so will make the equipment unreliable and cause it to have excessive down time. From the maintenance department's point of view, sensors that are stable but also easy to replace and realign in the event of a failure are best. For this reason most sensors are equipped with cables that can be

Figure 3–23
Application chart for standard sensing devices

Sensor/Application Matching Chart

Application	Target Material	Work Environment	Sensor Type
Safety doors and covers	Any	Any	Safety switches or limit switches
E stops		Any	Mushroom head E stop switches or cord pull E stop switches
Manually activated control system inputs		Clean	Normal-use push buttons and selector switches
		Dusty or oil mist	Oil-tight push buttons and selector switches
		Explosive atmosphere	Explosionproof push buttons and selector switches
Position-sensing moving mechanical parts at close tolerances	Metal	Any	Inductive proximity switches
	Nonmetallic	Any	Capacitive or photoelectric, thru-beam and reflective
Sensing fill levels in containers or levels in tanks containing liquids	Nonmetallic and liquids	High moisture	Capacitive proximity or ultrasonic
Sensing objects randomly placed on a moving belt or objects moving down a line on pallets	Nonmetallic	Clean or dusty	Photoelectric thru-beam or diffused reflective
Sensing presence of metal rod being fed to an operation from roll stock (solder, wire, rod)	Metallic	Any	Inductive proximity circular sensors
Checking the presence of nonmetallic items in a container or that caps or covers are present	Nonmetallic	Clean	Ultrasonic or photoelectric fixed focus / sharp cutoff
Sensing the presence of flat roll stock material	Metallic	Dusty/oil mist	Inductive proximity
	Nonmetallic	Dusty	Slotted photoelectric thru-beam or capacitive proximity

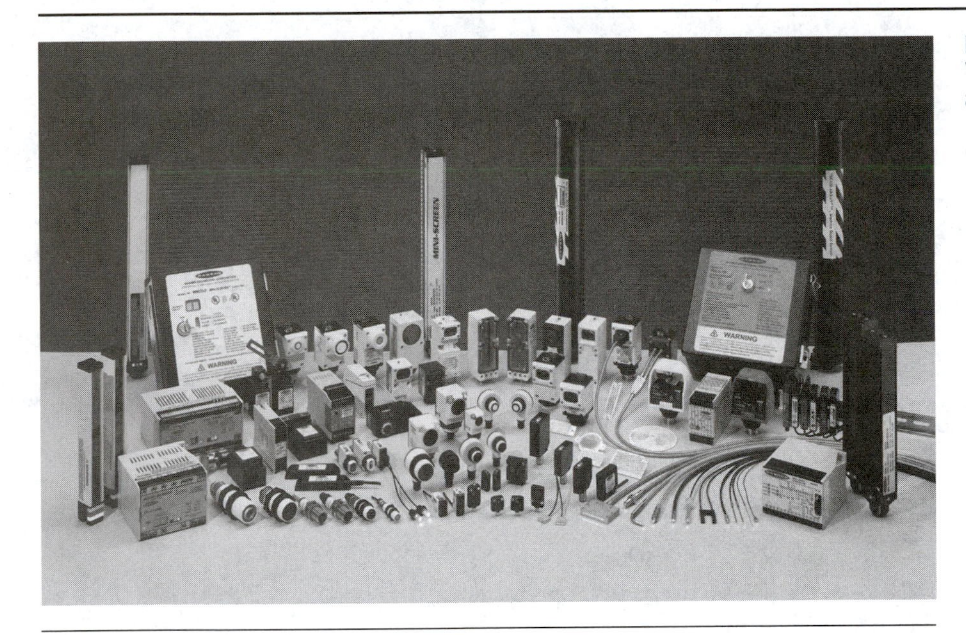

Figure 3–24
Various optical, proximity, and safety sensors.
Courtesy: Banner Engineering Corp., Minneapolis, MN.

plugged and unplugged instead of hard wiring (Figure 3–24). Hard wiring often requires the maintenance person to unlace the wire harness all the way back to the main cabinet. In addition, the mechanical designer should use mounting brackets that have slots for easy adjusting and alignment.

REVIEW QUESTIONS

1. If a design uses sourcing-type input devices, what kind of PLC input module would have to be used?
2. Give some examples of industrial considerations that might influence the type of push buttons used on a control system's panel layout.
3. Give some examples of functions that might require a keyed selector switch to be used on a panel layout.
4. Why should push button limit switches never be used as a mechanical stop for a mechanism?
5. What is the single biggest factor that makes prox switches superior to contact-type limit switches?
6. What is the functional difference between shielded and non-shielded prox switches?
7. What does the term "hysteresis" mean when applied to prox switches?
8. What types of prox switches should be used to sense nonmetallic objects?
9. What are the five main types of optical sensors?
10. What type of optical sensor should be used in dusty or oily environments?
11. Which type of reflective optical sensor has the least problem with background reflections?
12. When designing with ultrasonic sensors, what effect does changing the frequency of the emitter have on the device's performance?

CHAPTER

4

Ladder Rung Logic

LEARNING OBJECTIVES

After completing this chapter the reader should understand:

1. The symbols used in ladder rung programming.
2. The relationship between Boolean equations and rung logic.
3. How to analyze complex ladder rungs.
4. The difference between latched and normal output functions.
5. How to create a state diagram to track the I/O for a mechanism sequence.

In PLC controllers the control function comes from the ladder rung logic. In the introduction, in Figure I–1, we showed a simulated relay control drawing and explained why it was referred to as a *ladder diagram.* We also stated that each horizontal section was referred to as a *rung.* In PLC control systems we still use the rung and ladder concept except that now it is a software function instead of hard wired as was standard in relay controls. We use some of the same symbols as relay logic but they have different meanings.

BASIC LADDER LOGIC FUNCTION SYMBOLS

Our two main input symbols are shown in Figure 4–1. In relay logic these symbols were denoted as *normally open* and *normally closed.* If you have previous experience with these terms, it is best to wash these terms out of your head for present.

The reader should get used to referring to these symbols by their new descriptions, as shown in Figure 4–1, at least for now because it will be much less confusing. Allen-Bradley uses a different description for each, EXAMINE ON and EXAMINE OFF. Which term you find comfortable to use is up to you. It is important to understand that the symbols can refer to real input terminal addresses or an internal memory location. For this reason HIGH and LOW or ON and OFF are better descriptions to use than "normally open" and "normally closed" because these old terms do not really relate to software functions or memory bits.

What is extremely important to understand is that the symbols refer to the status of a memory location or input bit address and, in the case of an input address, the symbols do *not* refer to what device is physically connected to the input terminal.

Figure 4–2 shows the three main output function symbols we will be using throughout most of this book. For the first part of this chapter, we use only the full output instruction (the first symbol shown in Figure 4–2). Latched and unlatched instructions are explained later in the chapter.

The purpose of a ladder rung is to detect a specific set of input conditions and turn on the rung output only if these specific conditions are true. What we are really constructing is digital logic statements using our now familiar AND and OR logic functions. One of the simplest types of ladder rungs is shown in Figure 4–3.

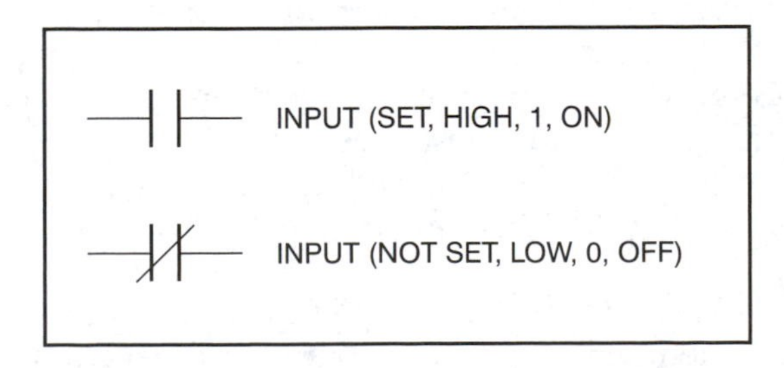

Figure 4–1
Two main input function symbols

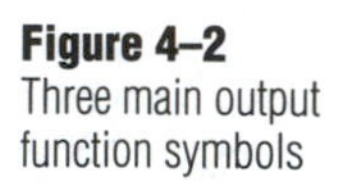

Figure 4–2
Three main output function symbols

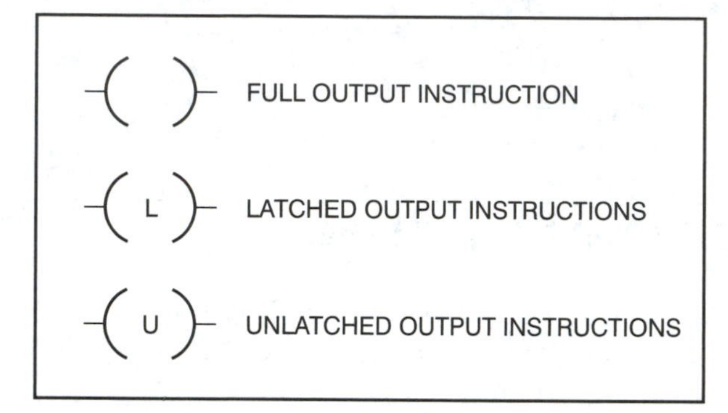

Figure 4–3
A simple ladder rung

To decrease the level of confusion, we will not use real I/O addresses for the time being. Instead we will use letters to define each element of the rung logic. In Figure 4–3, if the A input is ON, then the B output will also be ON. If A is turned OFF, then B will also be turned OFF. We could say A = B.

A man named George Boole developed a branch of mathematics devoted to these types of logic functions in England around the year 1864. This type of math is called *Boolean algebra* and is heavily used in computer circuit design. Boolean algebra has two main symbols: · and +. The dot · is used to denote the AND logic function, and the + sign is used for the OR logic function. The dot can sometimes be dropped, thus (A · B) reads the same as (AB).

In the following examples we show each ladder rung function, its digital logic diagram, and the Boolean equation that describes it.

Figure 4–4 shows the basic AND rung diagram and equation. Note that C only goes ON if both A and B are ON. Figure 4–5 illustrates the basic OR logic rung. In this function output C will be turned ON if A or B or BOTH are ON.

Figure 4–6 shows a combination of both logic functions in one rung. Note the Boolean equation that describes this function. When dealing with very complex rung logic, it often helps to work out the rung's Boolean equation to prove that the function you created will give you the results you wanted.

In Figure 4–7 we introduce the NOT function. In Boolean algebra the NOT function is denoted by placing a bar over the term that is inverted.

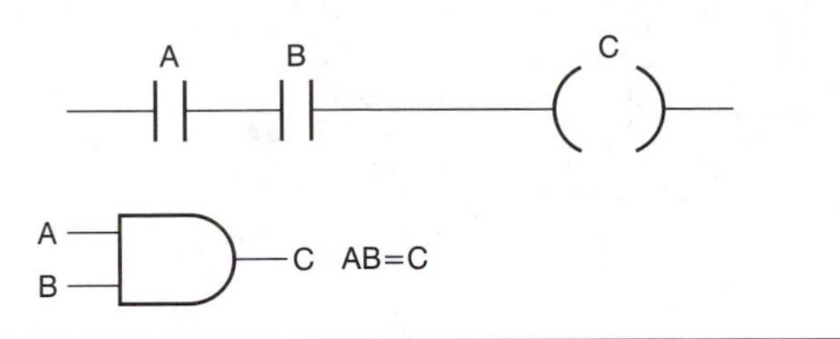

Figure 4–4
Basic AND logic rung diagram

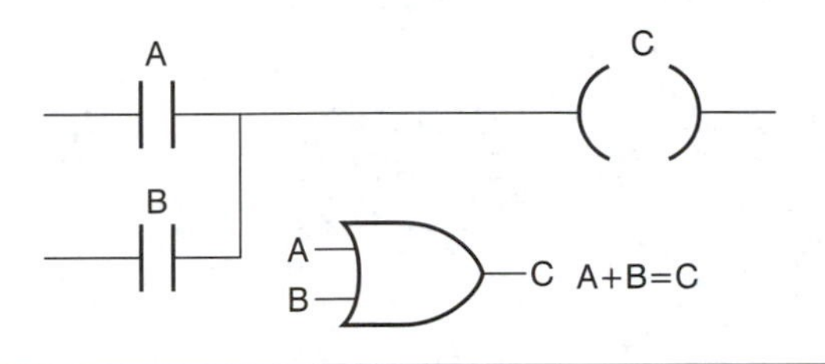

Figure 4–5
Basic OR logic rung diagram

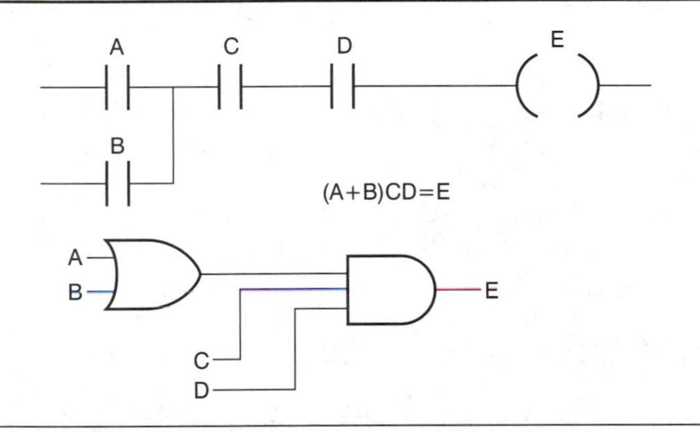

Figure 4–6
Both logic functions in one rung

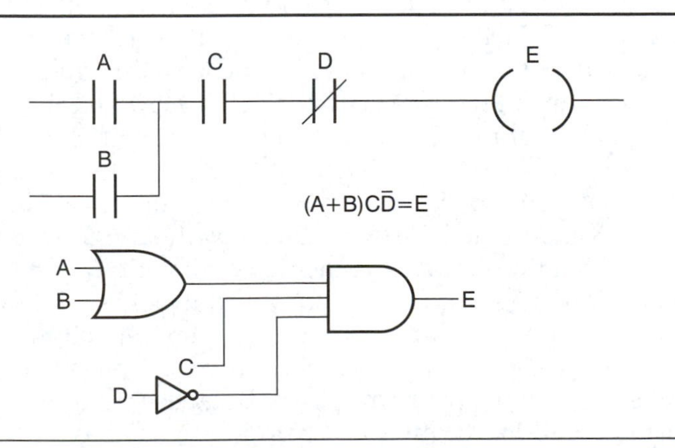

Figure 4–7
The NOT logic function

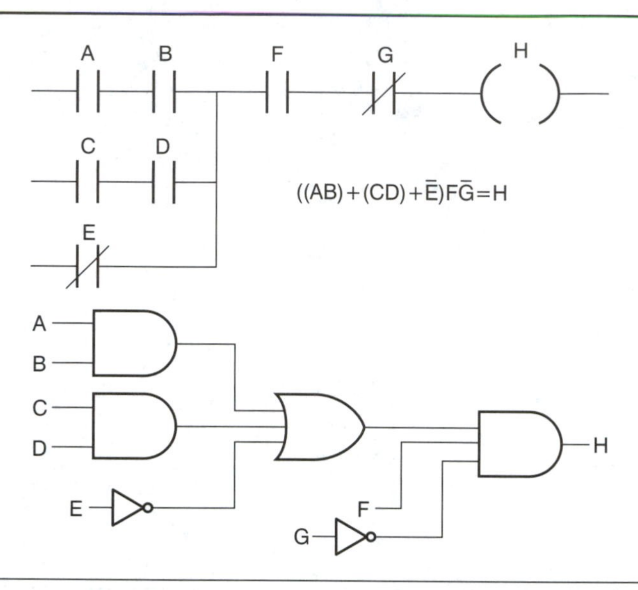

Figure 4–8
A complex logic rung and its equation

The equation in Figure 4–7 is interpreted in this way: The quantity A or B and C and NOT D equals E. Note that an inverter symbol has been added to the digital logic diagram. The NOT function is true if the input or memory bit is not set HIGH.

Figure 4–8 demonstrates one final example of a complex logic rung and its equation. It incorporates multiple branches as well as the NOT functions.

LATCH AND UNLATCH OUTPUT TYPES

Figure 4–2 shows two additional output symbols not discussed until now. They are called *latch* and *unlatch* functions. They are subsets of the full output instruction. When you use the full output instruction, the rung logic has full control over the output. That is, it can turn the rung output ON or OFF as the logic function changes from TRUE to FALSE. The latch output function has only half the ability of the full function: It can only turn the output ON but NOT OFF. The unlatch output instruction has the other half of the full output instruction: It can only turn the output OFF but NOT ON.

These latch and unlatch instructions allow the programmer to design separate logic equations for each transition of the output bit. In Chapter 8 we will learn how the separation of output control has advantages when we are working with zones and subroutines. The unlatch function is also commonly used to clear outputs when error conditions are detected that cause us to shut part of a program down. Because this can happen in the middle of an operational sequence, the program may have left some of its

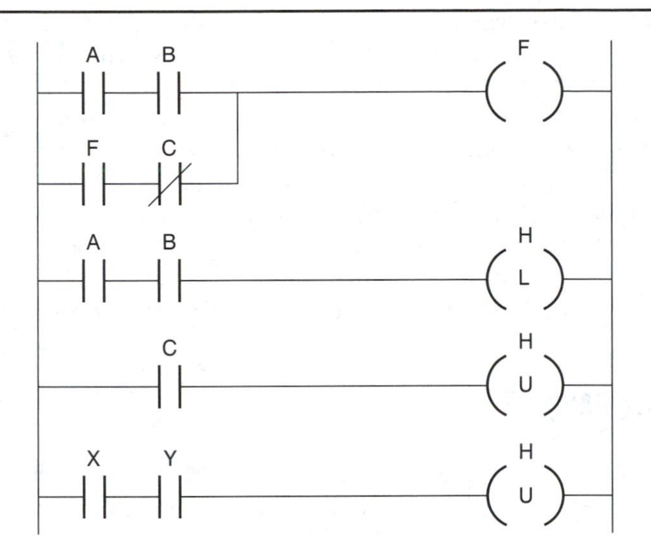

Figure 4–9
Latch function designs

outputs ON. It can also be used to reset outputs when we change operational modes in machine control programs.

In the ladder rungs shown in Figure 4–9 we demonstrate two ways to design latch functions. In the first rung we have created a latch function by using the F output as an input on the holding branch of the rung. If A and B are both ON, then output F is turned ON. This latches the function on through the lower branch. The output will stay on if the top branch goes false and can only be reset if input C goes on to release the latch.

The second and third rungs accomplish the same function using the latch and unlatch instructions. One additional advantage of using the latch and unlatch instructions is that they give us the ability to create a second and totally different rung condition to unlatch the same output. The fourth rung is an example of having a second unlatch function. This would be very difficult to integrate into the first rung's holding branch rung-type latch.

Now that we have an understanding of how logic functions are constructed, let's look at how we can use them to control mechanical functions. Every mechanical system has a number of distinct states in which it can exist. The programmer's first task is to define the I/O status for each mechanical position in which the mechanism can exist. Once this I/O list is compiled, the programmer can decide what I/O state changes will trigger each output change. In some cases several I/Os may have to change state before the next output changes are made.

We are creating what is called a *state diagram.* A ladder program is nothing more than a series of logic statements that define each state through which the mechanism moves. By looking at the I/O chart, the programmer can easily tell what I/O states he needs to incorporate into his logic statements. One problem with this state system occurs when the same I/O pattern exists for several states in the mechanical system. In this case a latch bit can be used to differentiate between the two positions. This is a common use for latch bits. In many mechanical systems, the start and end positions are the same and therefore the I/O pattern is the same.

We will write our first real program in Chapter 8 and it will use the same logical concepts as we are about to investigate here. At this time we are not able to write a real program due to our lack of knowledge about the programming software and I/O addressing. But we can at least go through the logical steps that will be necessary to control a piece of equipment.

EQUIPMENT SAMPLE

Let us create a very simple fictional piece of equipment and look at the I/O states we will need to work with and control. Let's assume we have a drill press that has a clamp mechanism for holding the part to be drilled. Our I/O list would be as follows.

Inputs:	**Outputs:**
Start PB 1	Clamp close solenoid
Start PB 2	Clamp open solenoid
Clamp open LSW	Drill down solenoid
Clamp closed LSW	Drill up solenoid
Drill down LSW	
Drill up LSW	

Our machine sequence is as follows. We assume the machine is in home position to start for this example.

1. When both start push buttons are pressed, activate the clamp closed solenoid.
2. When the clamp closed limit switch is detected, activate the drill down solenoid.
3. When the drill down limit switch is detected, turn off the drill down solenoid and activate the drill up solenoid.
4. When the drill up limit switch is detected, turn off the clamp closed solenoid and activate the clamp open solenoid.
5. When the clamp open limit switch is detected and both start push buttons are released, the cycle is complete and we will unlatch the down latch bit.

Notice that in step 5 we require both push buttons to be off before we complete the cycle. This is necessary to prevent the cycle from repeating if both push buttons are still held down at the end of the cycle (see Figure 4–11).

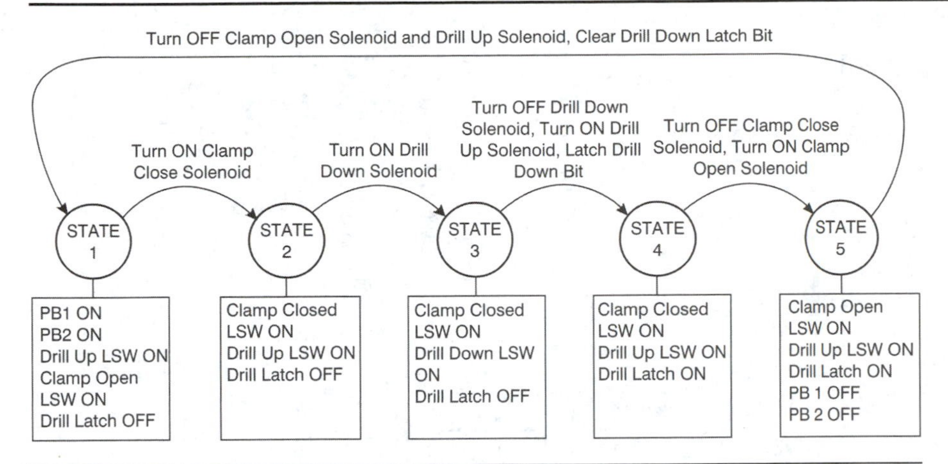

Figure 4–10 State diagram of our machine's cycle

Figure 4–10 shows a state diagram of the proposed machine function. The circles define each state the machine can exist in. The box of conditions below each circle describes the conditions that must be present to move to the next state. Shown above the transition arrows are the output conditions that we need to change to move to the next state.

State diagrams are very useful for organizing all the conditions you need to monitor during the machine cycle. You may wonder how I knew I needed to latch a bit when the drill was in the down position. If you look at the conditions listed for states 2 and 4 you will see that they would have been exactly the same had it not been for the drill down latch. One of the first tasks a designer must perform when working on a new state diagram is to check each state to be sure its conditions are different from those of all the other states in the cycle. Anytime you have exactly the same set of conditions for more than one state you must provide some means for the program to differentiate which is the correct state. One way to accomplish this is to use a latch bit to keep track of where you are in the cycle.

Another way to work out the I/O sequence is to make a chart of all the I/O and then step the mechanism through its cycle by setting and clearing the output bits and defining the input bit conditions for each step (see Figure 4–11).

Now that we have defined the I/O status for each step of the machine sequence, we can begin to design rungs to detect each state and turn on the proper outputs.

To move from the home position to the first step, we must design a logic function that looks for the two push buttons to be activated and the down latch bit to be OFF. (The function of the latch bit will be explained shortly.) As we can see from the output section of the step 1

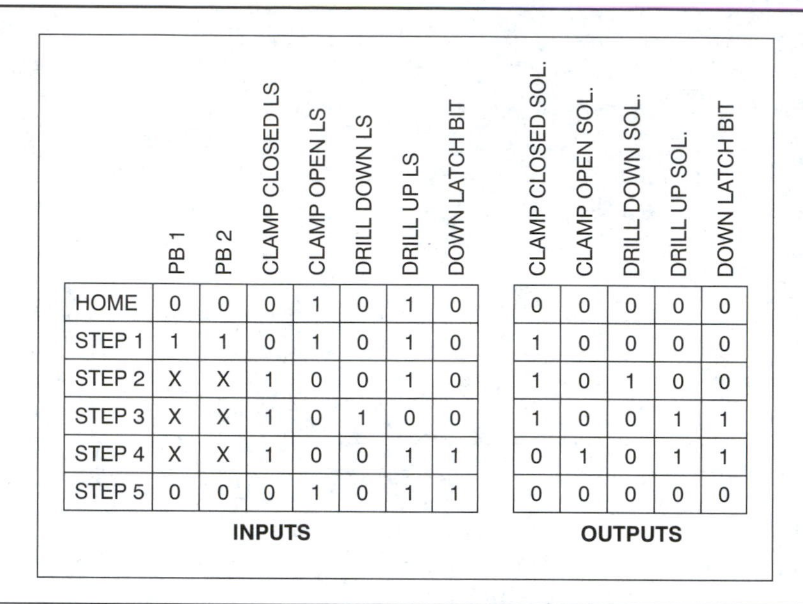

	PB 1	PB 2	CLAMP CLOSED LS	CLAMP OPEN LS	DRILL DOWN LS	DRILL UP LS	DOWN LATCH BIT	CLAMP CLOSED SOL.	CLAMP OPEN SOL.	DRILL DOWN SOL.	DRILL UP SOL.	DOWN LATCH BIT
HOME	0	0	0	1	0	1	0	0	0	0	0	0
STEP 1	1	1	0	1	0	1	0	1	0	0	0	0
STEP 2	X	X	1	0	0	1	0	1	0	1	0	0
STEP 3	X	X	1	0	1	0	0	1	0	0	1	1
STEP 4	X	X	1	0	0	1	1	0	1	0	1	1
STEP 5	0	0	0	1	0	1	1	0	0	0	0	0
	INPUTS							OUTPUTS				

Figure 4–11
State diagram for sample machine

position in Figure 4–11, the result for this logic condition will be to activate the close clamp solenoid. The logic rung would look like that shown in Figure 4–12.

The next step in our sequence is activated when the clamp closed limit switch is detected. In Figure 4–11 note that the two push-button inputs have an "X" showing, which means we do not care what their state is at this point in the cycle. The problem with this state is that if the latch bit is eliminated, steps 2 and 4 have the same I/O conditions. This presents a problem in that we cannot differentiate between the step that requires the drill to go down and the step that opens the clamp once the drill is back up again because the I/O status is the same.

We can get around this problem by introducing an additional logic element, the drill down latch bit. We will use this bit to keep track of the fact that the drill has been down and also to prevent the cycle from repeating if both push buttons are still held down at the end of the cycle. The logic statement for step 2 is shown in Figure 4–13.

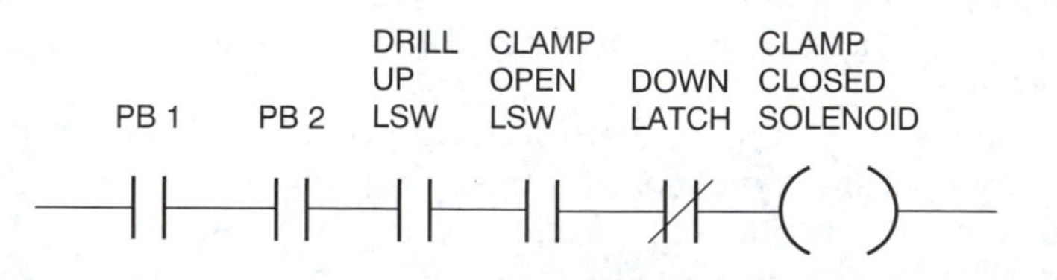

Figure 4–12
Logic run for closing the clamp

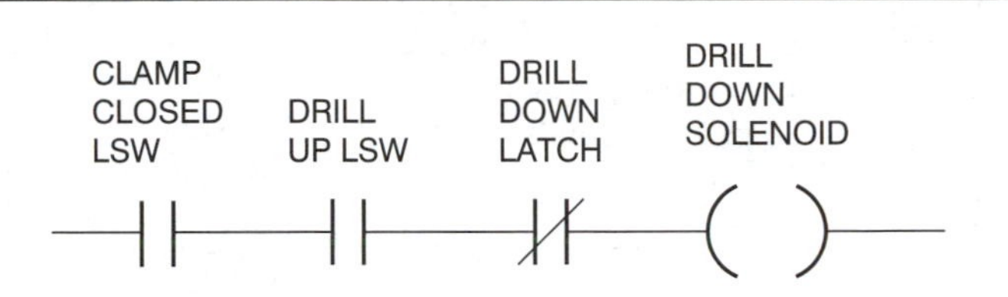

Figure 4–13 Logic statement for step 2 of our machine sequence

Step 3 looks for the drill down limit switch to be detected as its only element of change. The output for this step will be to activate the drill up solenoid and to latch the drill down latch bit. Figure 4–14 shows the rung logic for this step.

Note that latching this bit deactivates the previous rung, which turns off the drill down solenoid and also deactivates the first rung to prevent the push buttons from holding the clamp closed solenoid on. In the next step, we will be looking for the drill up limit switch to be activated and the drill down latch bit to be set to activate the open clamp solenoid. Figure 4–15 shows step 4's logic.

Step 5, our last step, is used to detect the end of the cycle and will reset the latch bit only if both push buttons are released and the clamp

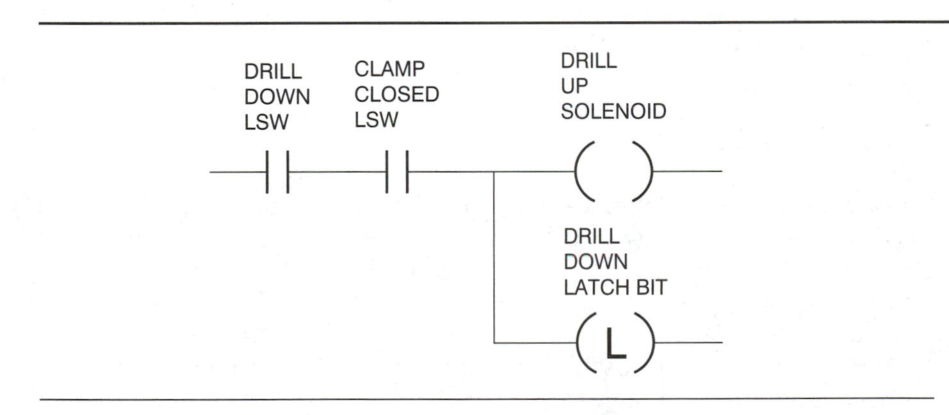

Figure 4–14 Logic statement for step 3 of our machine sequence

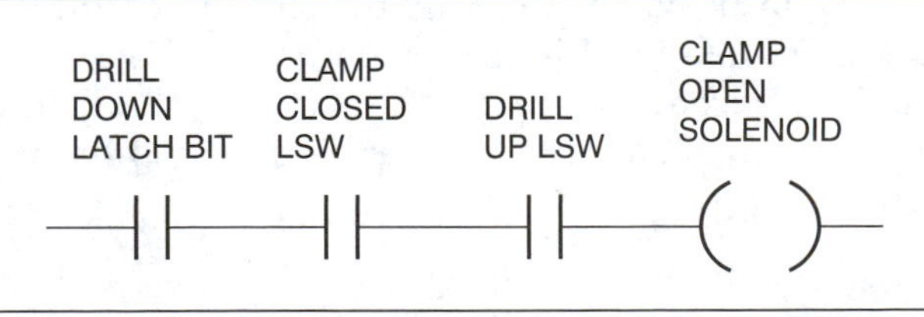

Figure 4–15 Logic statement for step 4 of our machine sequence

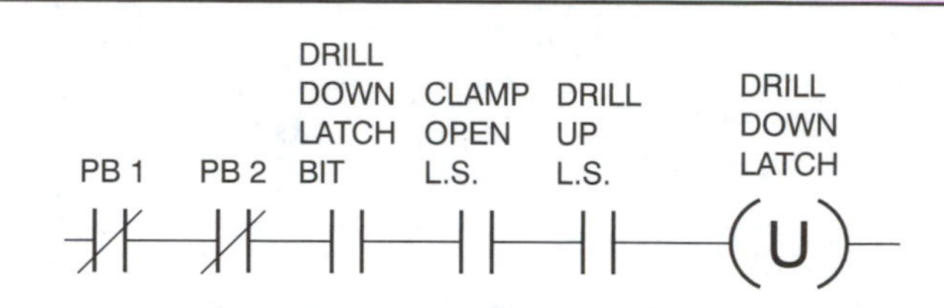

Figure 4–16 Logic statement for step 5 of our machine sequence

open and drill up limit switches are made. In addition we need to add in the drill down latch bit; otherwise the rung would come on if the push buttons were released before the clamp closed limit switch closes in step 2 (see Figure 4–16).

Even very complicated mechanical devices can be programmed easily if each step is worked out in detail and the sequence charted. The programmer must carefully check for any steps in the sequence that have identical I/O states. If any rungs having identical states are found, an additional element must be added to differentiate the states. Note that latch bits are to be used any time we need to remember an event has occurred, after the input event is no longer present. In the program discussed here, we needed to know that the drill was in the down position after the down limit switch was no longer active. This is one of the typical uses for latch bits. Because real I/O addresses have not been discussed yet, this is as far as we can go with this example. A complete program for this simulation is given in Chapter 8.

CONCLUSIONS

To summarize this chapter, we started by defining the basic logic elements used in most control system programs. We then combined AND and OR functions into complex logic functions and used Boolean expressions to better define their exact logic function.

Latch and unlatch output functions were introduced and we explained how they were really only half of the normal output function. We also showed how it was possible to use several unlatch rungs with completely different conditions to unlatch the same latched address.

We then created a fictional machine and charted all the mechanical states and their I/O status. We introduced the concept of building a state diagram to help organize the I/O requirements for each state in the sequence. From this chart and state diagram we were able to create the logic functions and rungs that allowed the PLC to step the mechanism through its cycle.

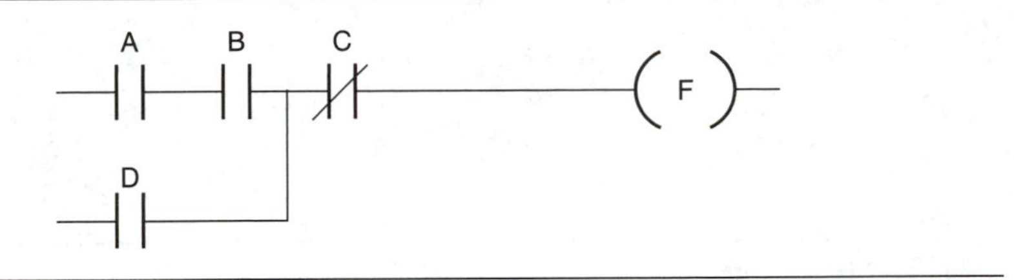

Figure 4–17
Ladder rung for Review Question 2

REVIEW QUESTIONS

1. Create a ladder rung to match the following Boolean equation: (A + B + C)DE = F.
2. Show the Boolean equation for the ladder rung shown in Figure 4–17.
3. What is the difference between a normal output instruction and the latched output instruction?
4. Why is it advantageous to be able to unlatch addresses from several places in the program?
5. Add a second horizontal drill axis to the sample program and program it to cycle after the vertical drill is complete and before the clamp opens. Be sure to rework the I/O sequence chart (Figure 4–11). Do you have to add another latch bit?

CHAPTER

5

PLC Models, Modes, I/O Scans, and Memory Layout

LEARNING OBJECTIVES

After completing this chapter the reader should understand:

1. The Allen-Bradley family of PLC controllers.
2. How to pick the proper PLC controller for a particular application.
3. How to use the program and run modes of the PLC system.
4. How data is transferred between the real I/O ports and the I/O image tables.
5. Addressing formats used in PLC programming.
6. How the PLC's memory and file structures are arranged.

When programming industrial controllers, the designer must have a good understanding of the PLC's hardware and software parameters and functions. We will be investigating the SLC-500 controller in depth in this chapter. Many different controllers are available on the market today and they all have certain advantages and disadvantages depending on the requirements of any particular project.

ALLEN-BRADLEY PLC SYSTEMS

In the Allen-Bradley family there are no less than seven different PLC systems. The SLC-150, PLC-2, and PLC-3 families are now obsolete but there are still thousands of them running equipment every day in the workplace. The current Allen-Bradley product line is comprised of the MicroLogix 1000, 1200, and 1500, SLC-500, PLC-5, and Control Logix 5000 families. Which PLC system we use on a particular project will depend on how much I/O and memory are needed and what special modules are required. By special modules I am referring to modules other than the traditional digital input and output modules. These may include thermocouple, millivolt, servomotor controllers, and high-speed counter modules to name just a few.

MicroLogix 1000

The MicroLogix 1000 family (Figure 5–1) is designed for very small control projects when 32 I/O points or less are required. The MicroLogix 1000 controllers are all fixed I/O types. By that, we mean the type and number of I/O points are fixed because no rack-style units are available. These fixed

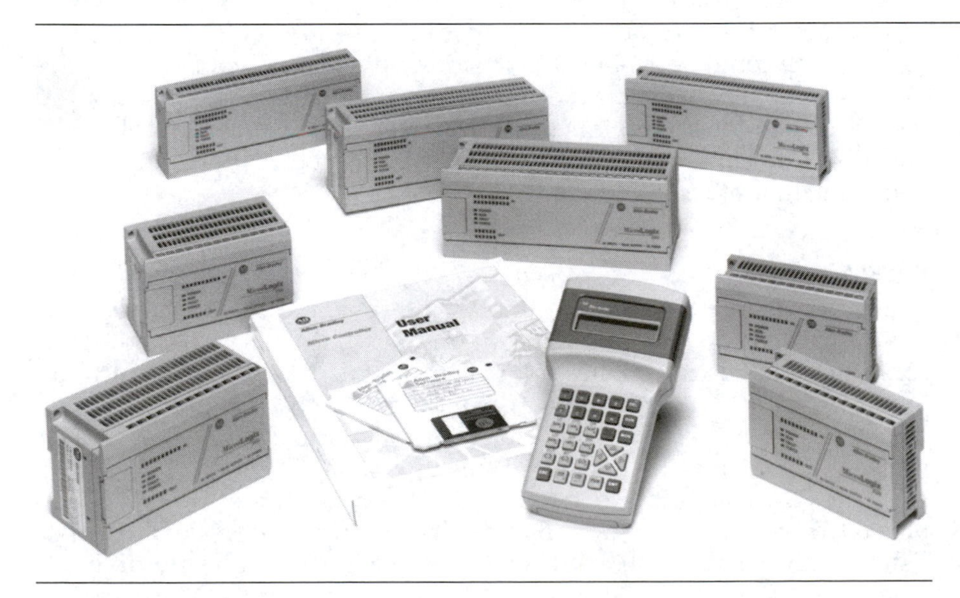

Figure 5–1
The MicroLogix 1000 family of controllers.
Courtesy: Allen-Bradley, Rockwell Automation, Milwaukee, Wisconsin.

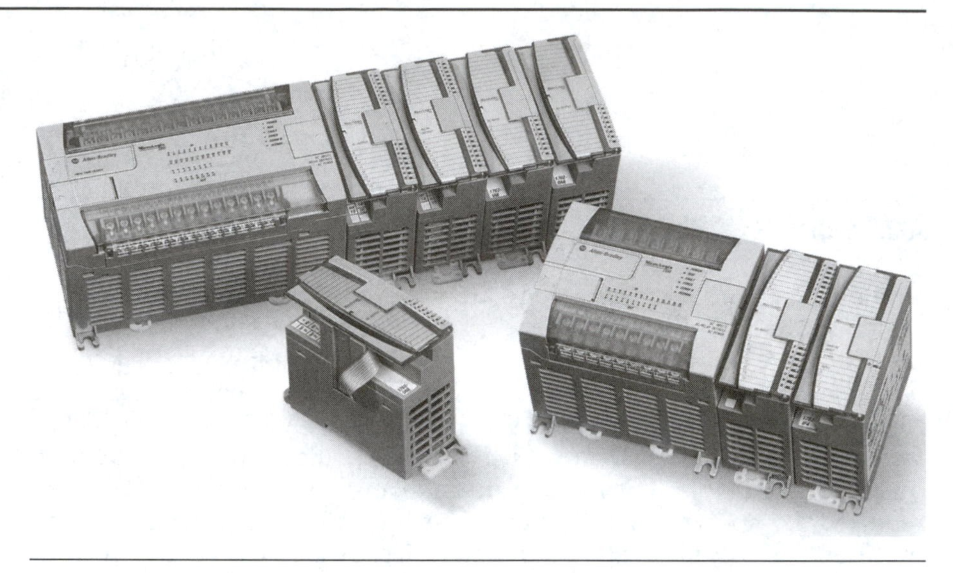

Figure 5–2
MicroLogix 1200 controllers and add-on modules.
Courtesy: Rockwell Automation, Milwaukee, Wisconsin.

I/O controllers are programmed through an RS232-C cable connected to a PC running RSLogix-500 software.

This line was expanded in the last few years to include the MicroLogix 1200 (Figure 5–2) and 1500 families. These are still small controllers, but they have numerous add-on I/O modules. Although these are not traditional rack-based systems, the main controller can be connected to a series of add-on modules by means of a flat cable and plug assembly. Additional modules can be connected in a daisy-chain arrangement. With the additional capability of add-on I/O modules this line of small controllers now has considerably more power than the original 1000 line. The 1200 has the ability to handle up to 88 I/O points and the 1500 up to 156 I/O points using add-on I/O modules. Both the 1200 and 1500 also have analog capability, real-time clocks, PID loop control, and some motion capabilities.

SLC-500

The SLC-500 family (Figure 5–3) is an intermediate controller that can handle up to 960 I/O points in a single rack and has a very powerful controller set. It uses a smaller I/O rack format than the PLC-5 family and therefore requires considerably less control panel space than the full-size racks used in the PLC-5 family. This is the family of choice for most new projects. It is also the main focus of this text.

This PLC comes in a number of fixed I/O models of up to 40 I/Os plus a two-slot add-on rack and in several multiple-slot rack-style units. The rack-style SLC-500/03, /04, and /05 models can handle multiple I/O

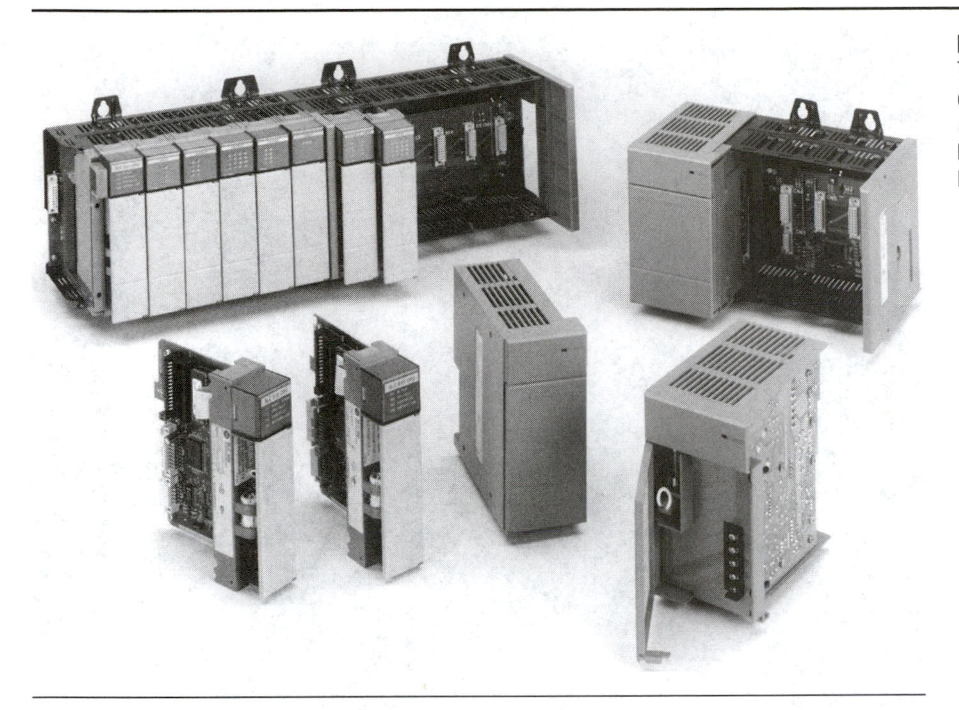

Figure 5–3
The SLC-500 family of controllers.
Courtesy: Allen-Bradley, Rockwell Automation, Milwaukee, Wisconsin.

racks, which increases the number of I/O point counts they can handle to 4096. They also come in memory sizes of 16K, 32K, and 64K. Most units have a DH-485 or RS232 programming port for interfacing to a PC used as a programming terminal. Two software packages are available for programming this family, the PLC-500 AI series for DOS systems and the RSLogix-500 software for Windows systems. All of the programming examples in this text will use the RSLogix-500 software.

PLC-5

Before the introduction of the Control Logix family, Allen-Bradley's most powerful controller was the PLC-5 family (Figure 5–4). It is able to handle several thousand I/O points and has the widest assortment of control modules in the line. Most of the programming functions we are going to learn about in the SLC-500 are usable in both the SLC-500 and PLC-5 families. There are some differences in file formats and special instructions between the families, but for the most part the ladder programming is the same. The PLC-5 family uses the full-size 1771 I/O racks (Figure 5–5) used in previous versions of Allen-Bradley controllers. This makes it an ideal replacement for the older PLC-2 family of processors as well as new designs requiring large I/O counts and multiple-rack systems.

Figure 5–4
The PLC-5 family of controllers.
Courtesy: Allen-Bradley, Rockwell Automation, Milwaukee, Wisconsin.

Control Logix 5000

This is Allen-Bradley's newest and most powerful PLC to date. It is built in the small frame size of the SLC-500, but it is much more powerful. One of the biggest advantages it has is its ability to have multiple processors in the same rack. You can have a PLC, a motion control processor, and a process control processor all in the same rack and all sharing the same I/O points. This is a true high-end PLC control system for projects that require a mix of control requirements. Control Logix uses a new programming software package called RSLogix-5000 that has symbolic programming with structures and arrays in addition to the standard ladder logic programming, as well as special instructions for motion control and process control. This control system family will eventually replace the PLC-5 product line, at which point all of Allen-Bradley's controllers will be in the small frame size package (see Figure 5–6).

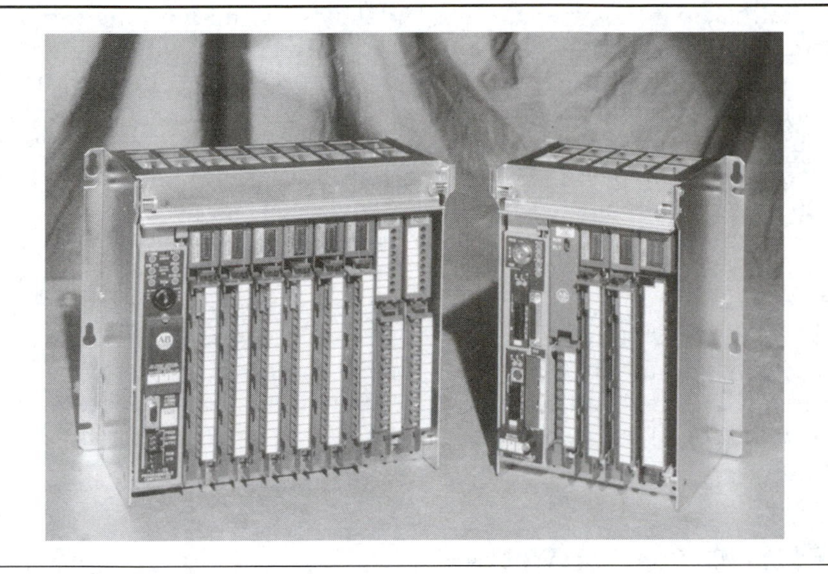

Figure 5–5
The 1771 I/O racks with processors in slot 0.
Courtesy: Allen-Bradley, Rockwell Automation, Milwaukee, Wisconsin.

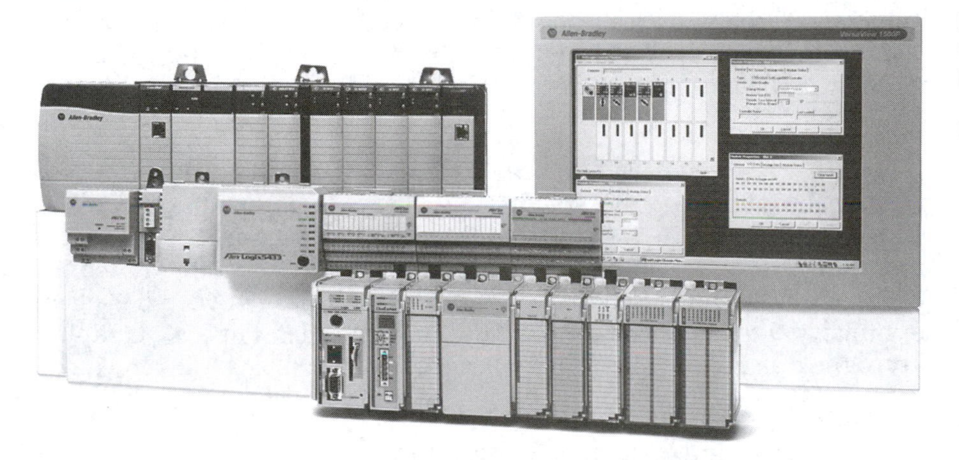

Figure 5–6
Control Logix 5000 processors and racks.
Courtesy: Rockwell Automation, Milwaukee, Wisconsin.

SELECTING A PROCESSOR SYSTEM

When starting a new project, the programmer must take into consideration the number and type of digital I/Os that will be used, what special modules will be needed, and how much data, if any, will be transferred between the controller and peripheral data displays or data collection systems. One of the costliest mistakes first-time control system designers make is to build a system with too little room for expansion. This is especially true if the project is a completely new mechanical design. On a completely new design, the experienced programmer will design in 40% or more empty I/O space. This will allow room for the numerous I/O additions most first-time mechanical designs require during their evolution and the subsequent debugging process. In a well thought out system design, we should be able to retrofit these I/O additions to the control design without running out of physical space.

It is often a good idea to leave room in our panel layout for the installation of an additional I/O rack. Once the control panel is completely built and wired, it becomes very expensive to move elements and rewire them or, in the worst case scenario, to have to disassemble and build a completely new and larger control panel. Leaving an empty space for an additional I/O rack can be a very cost-effective insurance policy.

Let us now begin our study of the SLC-500 PLC controller and its I/O system.

THE SLC-500 CONTROLLER SYSTEM

Software

The SLC-500 controller really has only two modes of operation: the *program* and *run* modes. We will start with the program mode because it is in this mode that all ladder programs must be created.

Program Mode

When the PLC is switched into the program mode, all outputs from the PLC are forced off regardless of their rung logic status, and the ladder I/O scan sequence is halted. (Note that the I/O scan sequence is explained in detail in the section on the run mode.) Program mode allows the programmer to upload and download program files between her PC or terminal and the PLC and to edit those program files. Here is a complete list of the functions available in program mode:

- Monitor and edit ladder logic.
- Add or edit I/O address descriptions.
- Add or edit rung descriptions.
- Set up I/O force bits without enables being executed.
- Search for specific ladder rungs or addresses to display for editing.
- Monitor and edit data files.
- Transfer program files to and from backup memory modules.

The system programmer switches to the program mode any time she wants to use the preceding functions. We should note here that only the ladder logic is downloaded to the PLC controller during a download. The program's address descriptions and rung descriptions only exist in the PC that was used to monitor and edit the program. If a file is uploaded from

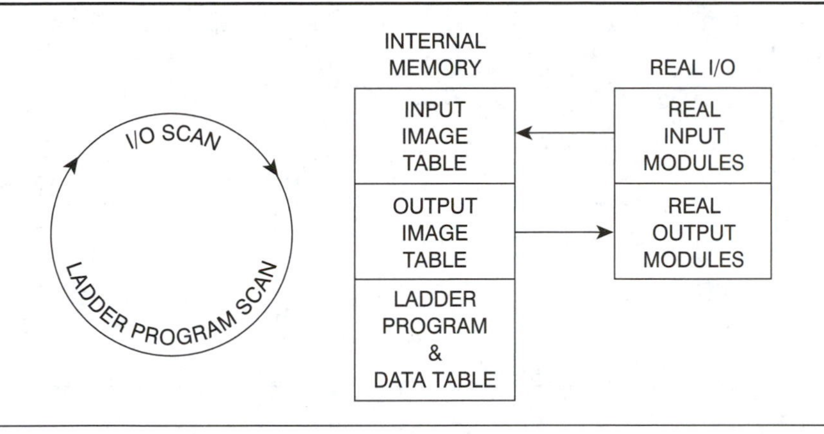

Figure 5–7
The PLC scan cycle

a controller to a different PC system and the original program files are not present, only the ladder and addresses will appear. All the descriptions will be lost. For this reason, it is usually a good idea to make all program corrections on a common PC or keep copies of the programming files on all the PCs being used to monitor and edit controller programs in your shop.

In conclusion, the program mode is used any time the programmer wishes to edit, upload files, download files, document (print out) programs, or change any software configuration file in the program. We will talk a great deal more about this mode in the section dealing with Allen-Bradley's programming software.

Run Mode

When the PLC is switched to the run mode, the PLC begins a continuous cycle of scanning the ladder rungs and then updating the real I/O addresses. Real I/O addresses are the actual connections to our system wiring made through the various I/O modules installed in the PLC rack. It is very important to understand this scan process. Figure 5–7 shows a diagram of the PLC scan cycle and its memory-to-I/O and I/O-to-memory data transfer function.

When run mode is activated, the processor transfers the status of all of the real inputs into the input image table in memory. The PLC then begins to scan each rung in the ladder program beginning at rung 0. It compares each input element of each rung and sets that rung's output ON or OFF depending on the rung's TRUE or FALSE logic state. The output bit that is set or cleared by the rung logic is in the output image table in memory and *not* the real output.

When the last rung is scanned, the PLC copies the contents of the output image table to the real outputs. It then copies the current status of the inputs into the input image table and begins the ladder scan process again. For those who have experience with sequential programming languages such

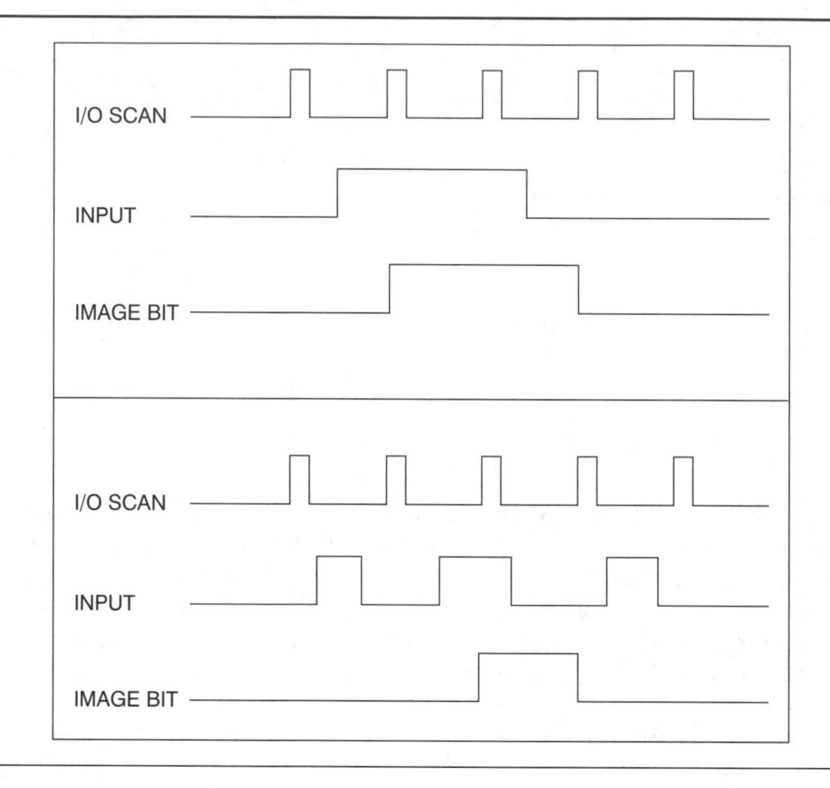

Figure 5–8
Input and PLC scan cycle timing diagram

as Basic or C++, it may be difficult to get used to the fact that the order of the rungs generally has no effect on their operation. Because all of the rungs are scanned all of the time, the order does not matter in most cases.

One PLC scan time is the time in milliseconds between one I/O update and the next. This scan time will vary between different PLC types and will also be affected by the length of the ladder program. One PLC scan time is the minimum amount of time that an input must stay in one state. If an input changes state twice within this time period there is a good chance the PLC will not see it.

Looking at Figure 5–8, the top half shows how a normal input lasting more than one scan time is handled during the I/O update scan. In the bottom half, we see an input that is changing too rapidly for the PLC to read correctly. Note how only one of the three input changes is recognized by the PLC. It is important to understand that while the PLC is scanning the ladder program, the input states being checked by the rung logic statements are those stored in the input image table and not the real inputs. For this reason, the designer never has to worry about an input changing state in the middle of a ladder scan. Note in the top half of Figure 5–8 that the input changes during the ladder scan part of the PLC cycle but the input image table is only changed during the I/O scan.

As long as the PLC remains in run mode, this scan process continues. Because all of the rungs are scanned all of the time, the programmer must write his logic statements so that they each control one step in the control sequence. The mechanical control sequence is determined by our ladder logic functions and not by the order in which the rungs are written. A good programmer will write his program rungs in sequence to make the program easier to read, but it is not necessary for proper PLC operation.

Internal File Structure

In the SLC-500 processor, most of the operations you will be performing will involve program files inside a major processor file. A *processor file* is a collection of programs and data files that pertains to a user project. Processor files are transferable files. They can be up- and downloaded between the PLC and a host computer. As mentioned earlier, the PLC must be in program mode for this to occur. Note that address descriptions are not part of the program file and as such they are not downloaded to the PLC controller.

Program Files

The three kinds of program files associated with a processor file are as follows:

1. *System program files:* Files 0 and 1 are always included in any project. They contain various system-related information and user-programmed information such as processor type, I/O configuration, processor file name, and password if one was used. The user has very limited access to these files.
2. *Ladder program files:* File 2 is always included in any project because it contains your main ladder program. This is the main ladder program that will always be scanned in run mode.
3. *Subroutine program files:* Files 3 through 255 are user-created subroutine files. These files, if they are created, are only scanned in the run mode if they are called by the main ladder program. It is possible to have only a main program file and no subroutine files at all in a small program. It is up to the programmer to decide how to partition her program and how many of these additional files, if any, she will use.

Data Files

Data files contain all of the data associated with the ladder instructions used in your program. They include input, output, status, bit, timer, counter, control, and integer types. Figure 5–9 is a chart that shows the internal memory layout of the SLC-500 processor. In future chapters some of these files will be described in great detail, notably the timer, counter, and control files, so only a short description will be given here because

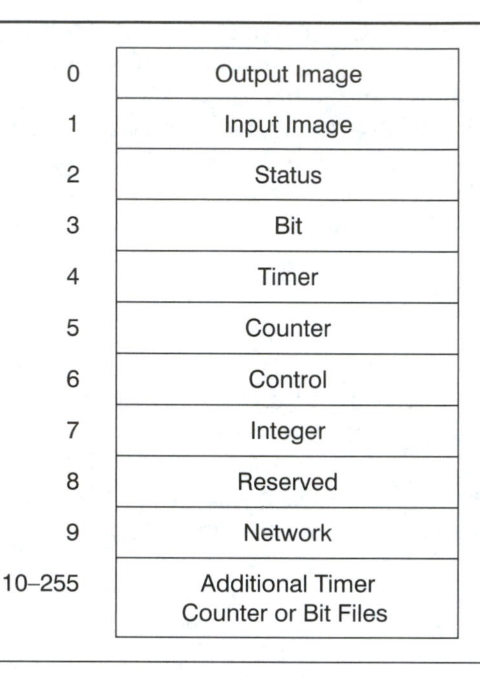

Figure 5–9
Internal memory layout of the SCL-500 processor

the reader needs a better understanding of the various ladder instructions before the data file can be clearly understood.

- *Output image data file (0):* This file contains all of the output data bits generated by the ladder scan and will be transferred to the real outputs on I/O update (see Figure 5–7).
- *Input image data file (1):* This file contains all of the real input data bits. Input data are transferred to this file during I/O update and it is used by the processor during ladder scan (see Figure 5–7).
- *Status file (2):* Various run-time processor status bits are stored in this file. The programmer can monitor the various bits for error or communications problems.
- *Bit file (3):* Locations in the bit file are allocated by the programmer for whatever she needs to keep track of in her program. The maximum length of the file is 256 words of 16 bits each, or 4096 individual bit locations. If additional bit locations are needed, the programmer can create additional bit files in the 10 through 255 file memory space (see bottom of Figure 5–9).
- *Timer file (4):* This is the timer data file. When a timer is created by the programmer, three words of this file are allocated to the timer instruction. The maximum length of this file is 256 words or about 84 timer instructions. If additional timers are needed, the programmer can create additional timer files in the 10 through 255 file memory space.

- *Counter file (5):* This is the counter data file. When a counter instruction is created by the programmer, three words of this file are allocated to the counter instruction. The maximum length of this file is 256 words or about 84 counter instructions. If additional counters are needed, the programmer can create additional counter files in the 10 through 255 file memory space.
- *Control file (6):* This file is used with sequencers and bit-shift instructions. Each time one of these instructions is created, three words are allocated from this file to the instruction. The maximum length of this file is 256 words or about 84 instructions. If additional instructions are needed the programmer can create additional control files in the 10 through 255 file memory space.
- *Integer file (7):* This is a word file that can be used by the programmer to store any word size data he may need to keep track of in his program. These data can be used by the various file instructions, such as copy file or bit-shift instructions. The maximum length of this file is 256 words. If additional file space is needed, the programmer can create additional integer files in the 10 through 255 memory space.
- *Files 8 and 9:* These files are effectively reserved as far as this text is concerned. We will get into some of the networking ideas in the advanced version of this course but not in this basic class.

Addressing Formats

Now that we have described most of the file types, we need to cover the actual addressing language we will be working with to talk to these various files. The three main instruction addressing types are *word, bit,* and *element* addresses.

Word Address

The format of a word address is shown in Figure 5–10. In this example, the timer preset value is in file 4, it is timer 5, and the word we are addressing is its preset value. This addressing scheme can be used to address any word in any timer, counter, or control file. If we wanted to read the accumulated

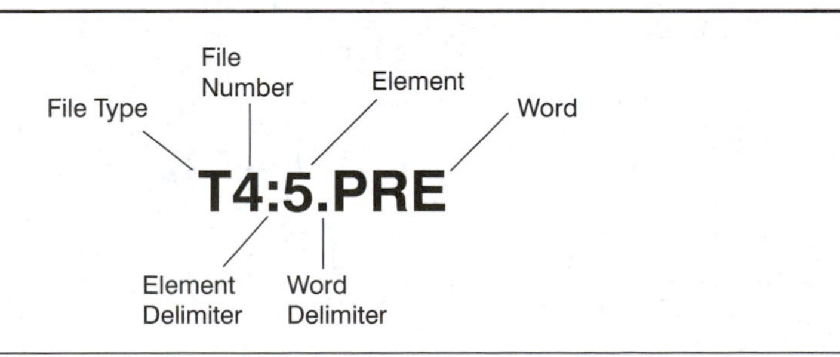

Figure 5–10
Word address format

Figure 5–11
One type of bit address format

TOP OF BIT FILE #3

15	14	13	12	11	10	9	8	7	6	5	4	3	2	1	0	WORD 0
31	30	29	28	27	26	25	24	23	22	21	20	19	18	17	16	WORD 1
47	46	45	44	43	42	41	40	39	38	37	36	35	34	33	32	WORD 2
63															48	WORD 3
79															64	WORD 4
95															80	WORD 5
111															96	WORD 6
127															112	WORD 7

value of counter 25 we would address it as C5:25.ACC, which would read counter file number 5, the 25th counter in the file, accumulated word.

Bit Address

When addressing a bit location in memory, two formats can be used. In Figure 5–11 I have diagrammed the top few words of the B3 bit file. The first addressing method is simply to use the sequential bit number from the start of the file beginning with 0 and ending with 4095. To access the highlighted bit shown in the figure you would enter B3/27. Note that this is the *28th* location in the B3 file because in binary files address 0 is a real place. Most programmers prefer this addressing mode when accessing bit locations.

The second method is shown in Figure 5–12. This address reads file B3, word 01, bit 11. This is the same address as we accessed with the shorter B3/27 addressing format. Either addressing method works, but the sequential number is shorter and therefore it is the method of choice. There are times when you will be required to clear a whole word (16 bits) in

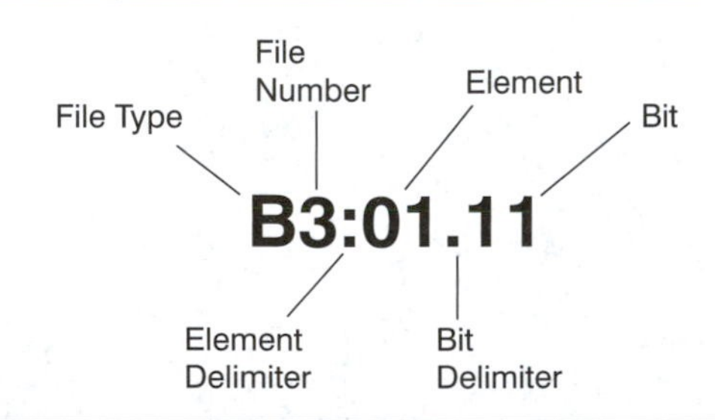

Figure 5–12
Long form of bit address format

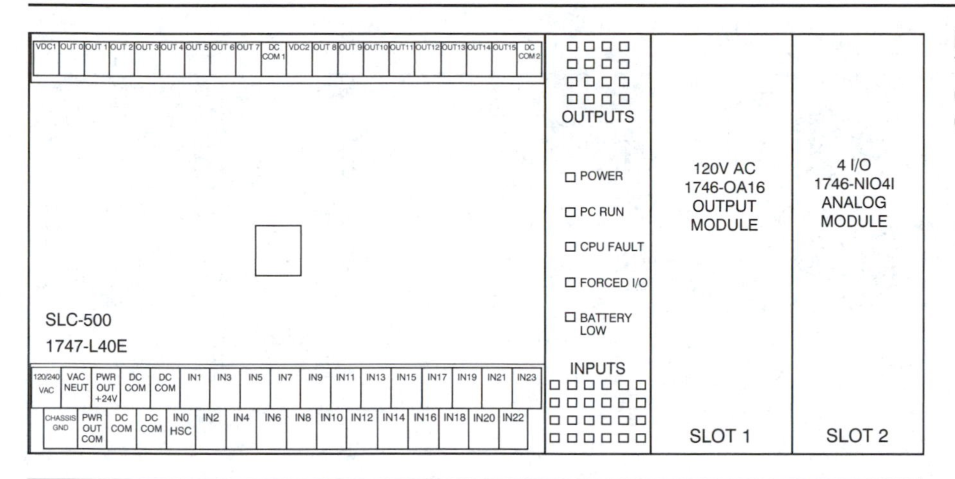

Figure 5–13
Teaching kit 40-I/O fixed controller and rack style controller

the B3 file and this will require you to know in what word the sequential bit number resides. Making a map of the file as shown in Figure 5–11 will help you determine what word holds the sequential bit address you are looking for.

Reading and writing to real I/O addresses is a special case of the bit addressing mode. In our teaching kit we have a 1747-L40E controller. It is a 40-I/O fixed controller with a two-slot expansion rack (see Figure 5–13). When using fixed I/O controllers, the fixed I/O is considered to be in slot 0. In rack-type controllers, the CPU sits in slot 0 and the I/O begins with slot 1.

Our controller has 24 inputs and 16 outputs in the fixed section. We have an additional 16 (120V AC) outputs in the first slot of the expansion rack (slot 1). When addressing real I/O the first letter in the address is an "O" for outputs or an "I" for inputs. (See Figure 5–14). The number following the colon is the rack slot number and then the word number. On I/O modules with more than 16 I/O points the first 16 are addressed as word "0" and the rest as word "1". For example, if

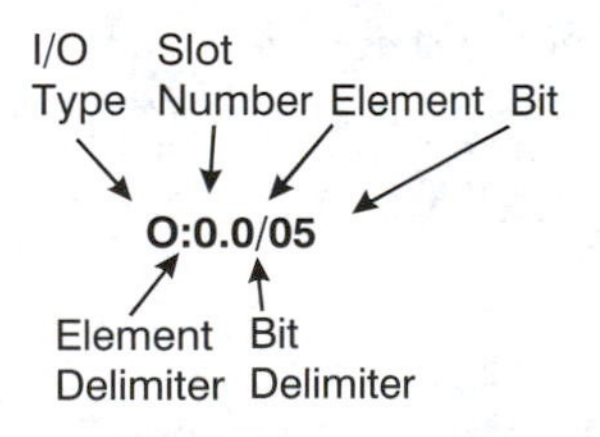

Figure 5–14
Sample of rack and slot I/O addressing format

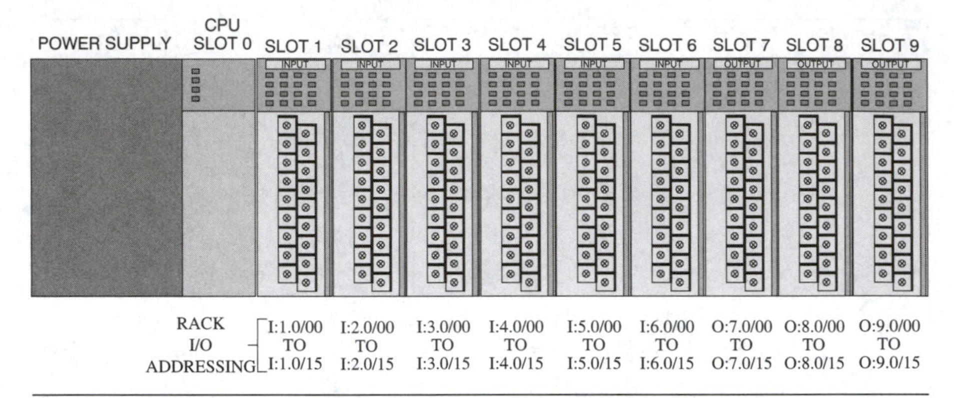

Figure 5–15
A 10-slot rack with six input and three output modules

we had an input module in slot 5 that had 24 I/O points, the first 0 through 15 points would be addressed as I:5.0/00-15, the rest would be I:5.1/00-07. In our kit, the first 16 inputs are addressed as I:0.0/00 to I:0.0/15. The rest of the inputs are addressed as I:0.1/00 to I:0.1/07. The fixed outputs are addressed as O:0.0/00 to O:0.0/15. The 120V AC module in expansion slot 1 is addressed as O:1.0/00 to O:1.0/15. Slot 2 is empty in our kit, but can be used for any additional I/O module you may want to add to the system.

Refer to the electrical wiring diagram included at the end of this book for the kit wiring and I/O assignments. In a multiple-slot rack system the designer decides what modules to place in each slot. In Chapter 6 we will learn how to set up the system file to tell the processor what module types are in each slot.

As stated before, in rack-based systems the CPU module is always inserted in slot 0 or the leftmost slot in the rack. How the other slots are populated will depend on the type and number of I/O modules required to complete the design. For ease of wiring most designers group like modules together. For example, if we have a system consisting of six input modules and three output modules it would make sense to insert all the input modules together in slots 1 through 6 and the output modules together in slots 7 through 9. An example of this configuration is shown in Figure 5–15.

Note the I/O addressing structure shown below each slot position in Figure 5–15. This example assumes we have used 16-port I/O modules in all the slot positions.

Element Address

The format for element address usage is shown in Figure 5–16. Element addressing is used when the programmer needs to read a word in an integer or control type file. The example shown accesses word 15 from file N7.

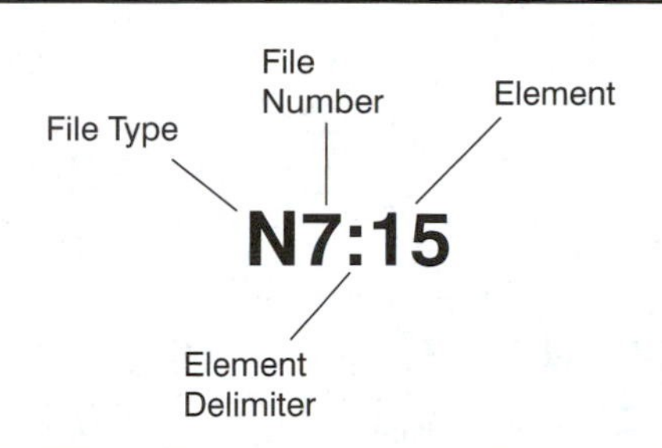

Figure 5–16
Element address format

CONCLUSIONS

To summarize this chapter, the two main modes of operation in our PLC are program mode and run mode. Program mode is used to edit, modify, and document the PLC program. When the run mode is activated, the PLC begins to scan the ladder program and update the I/O. During the ladder scan portion of the PLC cycle the PLC works only on bits stored in the memory and not the real I/O location. The real I/O points are updated during the I/O update portion of the PLC scan cycle. The first 10 files are assigned when we open the program and are listed in Figure 5–9. Files 10 through 255 can be assigned by the programmer to whatever file types she needs.

The three main addressing formats are word, bit, and element.

REVIEW QUESTIONS

1. Why must an input address be stable for one complete ladder and I/O scan for the PLC to read it accurately?
2. Where are the inputs bits stored that are used during the ladder scan part of the cycle?
3. What word in the B3 file contains the bit referenced by the address B3/58?
4. How would you address the ACC value of timer T4:15?
5. During what part of the PLC scan cycle are the real outputs changed?
6. How would you address output 4 in slot 6 if this slot contains an output module?
7. How would you address input 12 in slot 3 if this slot contains an input module?
8. In the 1747-L40E processor we use in the kit, what bit address does I:0.1/3 reference?
9. How would you address counter 8 in file 5 if you wanted to write a new preset value?
10. What address would you use to reference B3:4/12 if you were using the sequential addressing format?

CHAPTER

6

The PLC System Design Process

LEARNING OBJECTIVES

After completing this chapter the reader should understand:

1. The design steps required to complete a control system project.
2. The importance of the relationship between mechanical engineer, electrical engineer, and plant engineer.
3. The functional responsibilities of each member of the design team.
4. The importance of maintaining your project's financial records.
5. How to overcome problems in the debug process.
6. How to finalize your project documentation.

Before we begin designing a control system let us first look at the project from a project management point of view. Many organizational changes have happened in the last 10 years and the engineering discipline has seen its share of new ideas and reorganizations. Although every new design project is different, certain design steps must be completed in the proper order for any project if it is to be successful. The major steps are listed below and each step will be explained in detail following the list. This is a short chapter, but it is very important to understand the process of designing control systems for machines in light of the new team-based engineering design concepts. All the technology you will learn in the rest of this book will be virtually useless if you cannot function well in the new engineering team environment.

1. Interface with mechanical designer and plant MFG engineer to get machine specifications (specs), operator interface concepts, process capabilities, and machine timing.
2. Generate an I/O list from the machine specs and create an electrical bill of material.

3. Generate an overall machine operation flowchart and state diagrams for each major mechanism based on the machine specs and timing charts.
4. Meet with engineering team to finalize charts, parts list, and I/O list.
5. Design electrical system and generate wiring drawings.
6. Generate ladder code and HMI screen layouts.
7. Debug I/O and wiring.
8. Debug program.
9. Generate finalized documentation and standard operating procedures (SOPs).

In most organizations, more than one person is involved with the machine design process. With the accelerated rate of change in today's engineering marketplace, it is hard to find one individual that is good in both mechanical and electrical design technologies. Usually at least one mechanical engineer (ME) and one electrical engineer (EE) will be assigned to a control system project.

The team design approach is one of the biggest organizational changes introduced into corporations today. Most engineering design projects now incorporate a team approach to equipment design. Although you may be the lead person for the electrical and software part of this project, you will have to interface with people in many other disciplines before the project is completed. Figure 6–1 shows a general organizational and responsibility chart for a team-based design project. While your responsibilities will focus on the nine steps we have previously listed, you will be required to keep the other disciplines involved and up to date concerning the project's progress.

One of the biggest areas of concern will be keeping the financial people up to date on spending and budget allocations. Keeping meticulous records of all spending and purchase orders is of the utmost importance. Equally important is keeping track of all material received for each project. A common misconception of entry-level designers is that the purchasing department keeps track of everything and all they need to concentrate on is the design process. It is often a rude awakening when they begin missing scheduled progress milestones due to missing material or late deliveries. Never take a vendor's word that promised delivery dates will be kept. Keeping a list of delivery dates and making some phone calls a few days in advance of the ship date will help prevent having your schedule slip. Also, be aware that the shipping costs of project material are often not

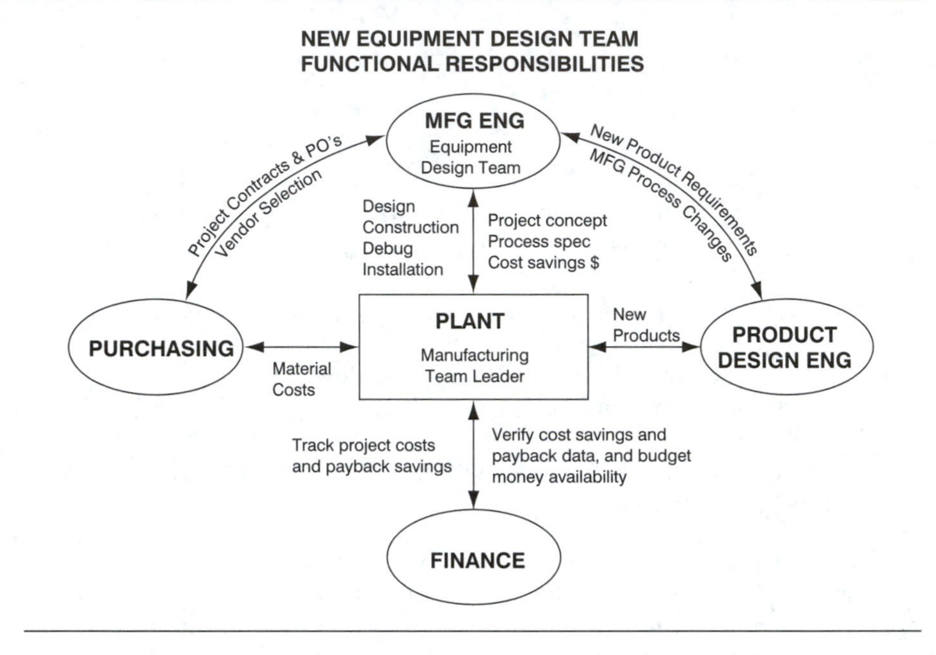

Figure 6–1
Project team responsibilities and interface

included in your price quote, but are charged against the budget for the project. Always anticipate the additional cost for shipping in your project budget numbers.

Keeping a close relationship with the plant manufacturing team leader will pay off big in reducing the inevitable problems that arise during the debug process. Be sure the plant team leader is involved in all operational decisions and in defining the expectations of the equipment. You want the plant people to feel that they have bought into the project. If they feel they have been involved in all the major decisions, they will be more likely to work with you to resolve design problems or when things don't work out quite the way the team expected. If the plant personnel are not involved, an "us versus them" environment with lots of finger pointing and very little cooperation may develop.

INTERFACE WITH ENGINEERS

The first meeting between the mechanical, electrical, and plant engineering disciplines should be scheduled before the mechanical design is started. As the electrical partner in this project, you will be responsible for all of the electrical wiring and drawings, control system elements, and programming.

The three main items you must address at this first meeting are the man-machine interface (HMI) concept, machine cycle timing, and power requirements for the main power feeds. Given that this meeting is very early in the design cycle, you should expect a considerable number of changes in both the I/O list and machine cycle information before the machine's design is complete. Getting the I/O list straight is usually the easier of the two processes.

You may receive the machine timing cycle explanation in any number of forms. Some mechanical engineers will give you a detailed timing diagram of the entire machine, others will only provide a written description of each mechanism and how they function together in the overall machine. Whatever the data transfer method, you must acquire a good understanding of how the machine operates and what is required in both the machine setup mode and under full running conditions. Keep in mind this machine is only in the conceptual stage and the mechanical designer may not have all the mechanism designs completed.

There will be numerous subsequent meetings of the mechanical, electrical, and plant engineers as the mechanical design progresses toward completion. Your concept of the control system design and how it needs to function should become firmer with each meeting.

GENERATE I/O LIST

From the information given to you by the mechanical designer you need to separate the I/Os as to their electrical requirements. From this information you can now determine what type of PLC will be required for this project. Remember, this is a very preliminary list of I/Os. Be sure to design a system that has a lot of room for expansion.

Once a PLC model has been established and the I/O modules determined, you can begin assigning I/Os to actual addresses. During this process you should also keep in mind the machine's operating requirements. You may find there are not enough input functions to accomplish what the mechanical designer requires of the operator who will be running the machine. Consider how you would deal with a machine jam-up and what steps you would take to clear the machine. As these questions come up in the design reviews be sure to share them with the plant team leader and have his group give you feedback on how they want the machine to function. The plant team should verify that you have enough input functions to allow the operator to clear the machine.

I have stated this before, but I again want to emphasize the importance of assigning descriptive names to all I/O points. These names will be used on the I/O address list and should allow the maintenance person to identify the electrical element quickly.

CHART THE MACHINE FUNCTIONS

Now that the I/O devices have been assigned, you must reorganize the machine cycle information given to you by the mechanical designer into a computer flowchart. To understand how your ladder program must operate, it is necessary to chart each mechanical mechanism through its complete cycle. Sometimes it helps to make a chart of all the I/Os assigned to a given mechanism and then walk that mechanism through its cycle noting the I/O states at each step of the sequence. This chart is called a *state diagram.*

Creating a detailed flowchart is necessary for three reasons:

- First, you will be need the flowchart for step 4 to finalize the program design with the mechanical designer. If done properly, the chart will enable him to see the machine process in exacting detail. Many unforeseen problems will show up and be solved during the flowchart process because it forces you to chart each decision the PLC will have to make in the control cycle.
- Second, the flowchart will make generating the ladder program much easier because all decisions have been charted and should now be much easier to code. In most designs the more accurate the flowchart and state diagrams, the easier the ladder logic programming becomes. Each diamond symbol in the flowchart or bubble in the state diagram represents a set of input rung conditions that must be true before the PLC activates the next state change. This flowchart output box or state diagram transition arrow usually represents an output address change.
- The last reason is documentation. One of the worst design assignments you can be given is to make changes to someone else's program after that person has left the company or even your own program 5 years after you completed it. To look at 400 rungs of code and try to figure out what each is doing is a nightmare for anyone. If the code is well documented and has an accurate flowchart and corresponding state diagram, the job becomes much easier.

INTERFACE WITH ENGINEERS AGAIN

This step may actually occur several times, rather than just once. The purpose of this meeting is to discuss with the mechanical designer and plant team leader how you have interpreted their information and see if they agree with your flowchart, state diagrams, I/O list, and man-machine interface concepts. If in your discussions you find problems with the design, now is the time to make the corrections. Often this means the plant team leader takes questions back to the plant manufacturing team for discussion as to how best to solve the problems.

It is not uncommon for the team of designers to move from step 1 to 4 several times on a larger project before all the team members agree they have all the concepts straight and have a workable mechanical, electrical, and operational design. The design team may need to request mechanical changes to the design if they feel they need additional feedback data. They may request the addition of proximity switches if they feel they are needed to operate the machine safely. The plant team may also request additional panel push buttons or a different arrangement of the buttons if they feel they are needed to operate a mechanism manually or to solve a jam-up condition.

At this point a fairly accurate project schedule should be generated and presented to the team. Once the team agrees with the schedule it needs to be circulated to the other groups involved with the project. If funding for this project is not completed, it is a good idea to base the schedule on the number of weeks each element will take rather than providing a project start date. A common problem with engineering schedules that have a fixed start date is the delay in completing the funding paperwork. If the finance department falls six weeks behind in completing the project funding, your schedule is already six weeks behind before you can start. Basing the start date on the day funding is available will go a long way toward having a reasonable schedule.

DESIGN ELECTRICAL SYSTEM AND GENERATE WIRING DRAWINGS

Now that the I/O list is finalized, we can generate machine and panel wiring drawings. A good drawing format to use is to make one drawing or set of drawings of the main control cabinet showing all the main power and PLC hardware power connections. Next, make drawings of all of the I/O connections in order by address. Show unused I/O addresses as open terminals because you may have to add items later and this saves moving things around to add new I/O points.

Be sure the descriptions for each I/O point are accurate and actually describe its machine function because these descriptions will also be used in the program as address descriptions. As stated earlier, input switch descriptions such as SW-1 and SW-35 do not give the maintenance person much to go by.

Next you need to generate an electrical parts list and submit it to the purchasing team member who will send it out for quotation. You may also have to generate a vendor bid package if you are going to use a system integrator or wiring contractor to assemble and wire the equipment.

GENERATE LADDER AND HMI PROGRAMS

Because the rest of this book is devoted to this step, we will not get too involved at this point. Good programmers keep the ladder program in functional modules and try, whenever possible, to keep the ladder rungs in order by machine sequence.

DEBUG THE I/O WIRING

Debugging the I/O wiring is the first step in debugging the electrical system. We first power up the PLC and the DC input power supply and verify that the PLC is not faulted.

The next step is to check that input power is connected to all control panel selector switches and manual push-button control switches. Each switch must be depressed or made to close and the corresponding input address checked to verify that the wiring is correct.

After all control push buttons and selector switches have been checked, we move on to the prox and limit switches. Each mechanical mechanism must be moved through its complete motion and the prox or limit switches adjusted to the proper sensing locations. All of these inputs must then be verified to make sure they activate the correct PLC address.

Once all inputs have been tested, we turn toward the outputs. The same process is repeated on the outputs to verify that the wiring is correct and that each mechanical device moves in the right direction. Even if the I/O wiring connected to an air solenoid is correct, the plumbing may be wrong, which would cause the mechanical movement to be reversed. As you can see, this is often a long tedious process. If any wiring errors are found they must be checked against the wiring drawings. In case the drawings turn out to be the source of the error, a marked-up set of wiring drawings should be kept close by to keep track of changes.

DEBUG THE PROGRAM

Program debugging is very dependent on the individual machine design, but a few generalities can be stated at this point. Programs should be debugged in small pieces. If you have a manual or setup mode in the design, this is the place to start. After you have satisfied the mechanical designer that all the elements can be moved in manual mode, start one station at a time in automatic mode. It is much easier to debug one station at a time than to try to run the whole control program right off the bat.

Keep a set of the program printouts close by to mark up any program changes that are made to the design during the debug process. Many new design engineers attempt to work several weeks without marking up their drawing and program changes. When it comes time to document the machine or when they have a computer problem, they find it is impossible to remember all the small changes that were made. It is far better to keep a set of marked-up electrical prints and program printouts and enter the changes on a daily basis. In this way, nothing gets lost in the rush to complete the project or in the event of a computer malfunction.

PREPARE FINAL DOCUMENTATION

This is the step, from a manager's point of view, that is the hardest to get people to complete. Be sure that all addresses have a description that represents their functions. Describing timers as "timer 1 through timer 55" is of little help to someone who must decipher your ladder program's functions. Each description should be easily understood in connection with the machine's function. The description of a timer as "index start pulse timer" is much easier to decipher than "timer 1." Likewise with counters, a description of "good parts" is much better than "counter 15."

When the design is complete and all the parties are happy with the results, it is extremely important to gather all the last-minute changes and get them into the final documentation. If this is not completed soon after the last debugging is complete, many changes will be lost. Remember, if the design changes are not reflected in the final documentation you will be troubleshooting the same problems the next time around.

■ CONCLUSIONS

In summary, although the exact sequence of events may differ between companies and design departments, this chapter covered the major steps involved in the machine design process. If the number of people working on a particular design project grows larger than three or four, the process becomes even more important.

On large design projects that require the talents of numerous mechanical, electrical, and other technical people, controlling the flow of design data is critical to the success of the project. Team leaders need to meet regularly to pass project data and schedules between teams and to understand the impact design changes will have on the various design disciplines. Some form of process must be instituted that ensures inter-group verification of the mechanical and electrical design changes that occur during the development of the project. Only when all the design groups work together smoothly will the final outcome be cost effective and successful.

As you work through the remaining chapters of this book, keep the team design concepts presented in this chapter in mind. We recommend that the last two lab projects be worked on in a team environment. This is important because a new designer needs to get experience working with a partner or partners. When each person is working on only a piece of the whole design project, the sharing of information becomes imperative if the project is to come together smoothly. The more experience you, the student, can acquire working in a team environment, the easier it will be for you to enter the real corporate design world later on.

CHAPTER

7

PLC Software Development

LEARNING OBJECTIVES

After completing this chapter the reader should understand:

1. How to open and configure a new ladder program file.
2. How to configure the I/O modules in a rack-based PLC system.
3. How to create and edit a ladder program.
4. Entering address descriptions for each ladder element.
5. Entering rung descriptions.
6. Setting up report options and printing a report.

All PLC manufacturers have developed programming software to aid in developing ladder programs. This software will allow you to write, document, and debug ladder programs. You should understand that the PLC contains only the ladder program; all text information is contained in the host computer. When programs are uploaded or downloaded, only the ladder program is affected. When you are monitoring a running program, the host computer is displaying a time slice of the real ladder. Depending on the serial link, there may be a good number of ladder and I/O scans between screen updates. For this reason, fast-changing I/O may not show up on the screen if the input only stays stable for one or two scan times. If you need to verify that a particular condition was present, dummy latch bits can be inserted.

In this chapter we explain how to open a new project file, edit an existing file, enter rung data, download your program to the kit PLC, monitor the program in run mode, and document the ladder program you have created. There are currently two different PLC development programs for the MicroLogix 1000 SLC-500 product. The PLC-500-AI series is a DOS-based program, and the newer RSLogix-500 is a Windows-based program. Although a few companies still use the AI series program we will be showing only the Windows-based RSLogix-500 software in this book.

RSLOGIX-500 SOFTWARE

To open the RSLogix software click on the RSLogix icon or go to *Programs /Rockwell software/Rslogix500 English* and click on the program. The screen printout shown in Figure 7–1 shows the main screen that will come up when the program opens. At this point only the *File Open* and *File New* options are enabled. To open a new program file click on the blank sheet of paper icon shown in the top left corner of the screen or click on *File* and then *New.* This will open the Select Processor Type pop-up window. We will scroll down the list until we find the 1747-L40E processor (see Figure 7–2).

This is the processor used in our teaching kits. If you have a different processor in your lab kit select the correct processor for your kit and then click OK.

The main programming screen will now come up as shown in Figure 7–3. This will be the main screen you will be working on for most of your programming functions. The screen is divided into three main sections.

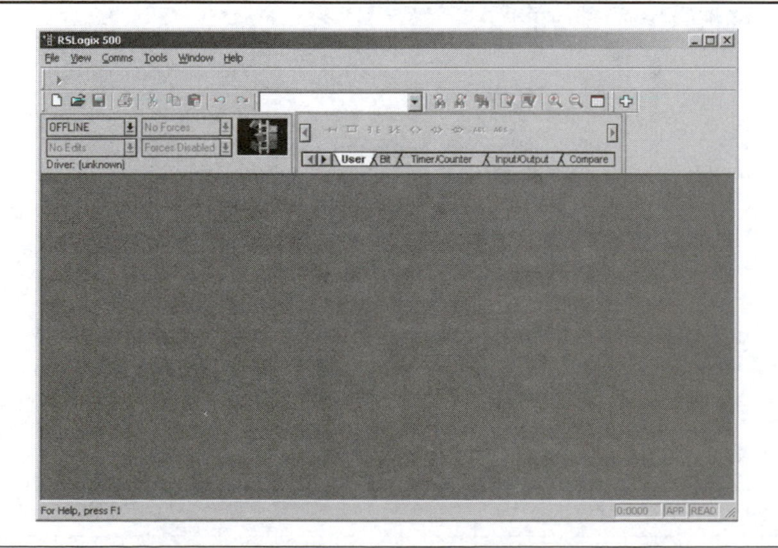

Figure 7–1
Opening screen for RSLogix software

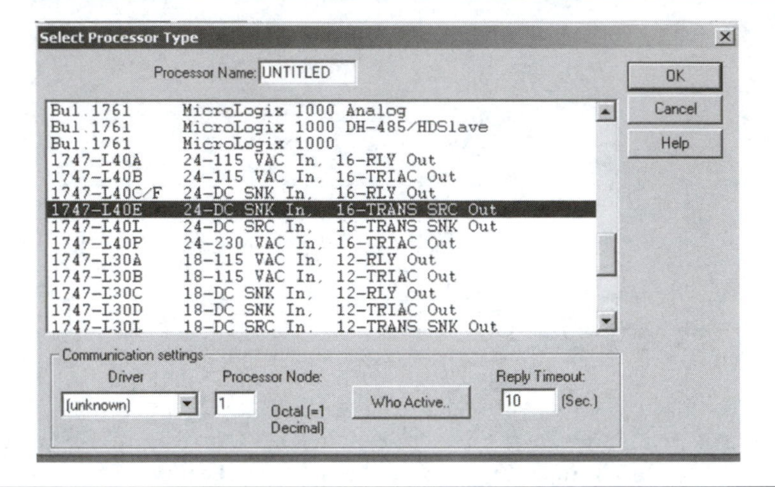

Figure 7–2
Select Processor Type pop-up screen

The top of the screen contains the main toolbars for creating ladder programs, saving and retrieving files, changing between online mode and offline mode, and uploading and downloading processor files. The bottom half of the screen is divided into two parts. On the left is the tree showing all the files in this program file. On the right is a large window where we will enter our ladder program. Once a program is written and downloaded this is also the area where we will monitor a running program as well as edit the ladder program. If you look closely at the tree structure on the left edge of the screen, you will find all the files we discussed back in

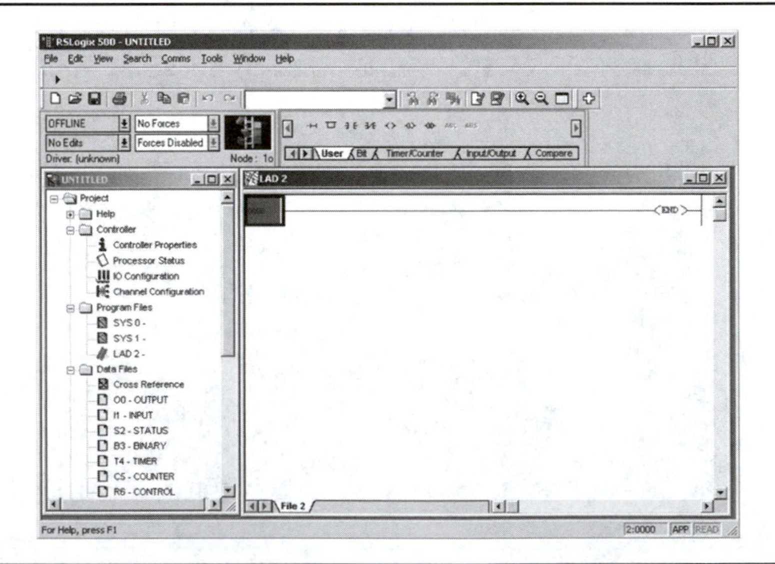

Figure 7–3
Main programming screen

Chapter 5. Note that three program files shown are SYS 0, SYS 1, and LAD 2. The SYS 1 and SYS 2 files are where the program's setup and configuration data are kept. The LAD 2 file is the main ladder program file. Later on we will add files to this list as we create subroutine files.

The next step in generating a new program will be to configure the 2-slot I/O rack that is attached to the PLC in our kits. Click on the I/O configuration icon on the left edge of the screen. This will bring up a pop-up screen for selecting the modules in the two expansion slots. In our training kits we have an 1746-OA16 120V AC output module in slot 1 and nothing in slot 2. With slot 1 highlighted scroll down the I/O module list and pick 1746-OA16 for slot 1 (see Figure 7–4). This is the I/O configuration of our training kit. Be sure slot 2 is empty. If you were writing a program for a 10-slot rack-based PLC you would have to declare the proper I/O module for each slot used in the rack. Once you have slot 1 configured (Figure 7–5) click on the "X" at the top right corner of the pop-up screen to close it and return to the main screen. If you later download your program and get an immediate processor fault condition, be sure to check that you have correctly identified all the modules in the slots. If the processor detects a different I/O configuration than what you entered in its configuration file when it enters run mode, it will not run and will immediately fault.

We are now ready to start entering a few ladder rungs. Directly above the ladder window on the main screen is a set of toolbars that contain all the ladder elements we will need to construct our programs. Most of the symbols we need can be found by clicking on the items in this toolbar. Additional functions can be found by opening the other major toolbar selection lists. This is done by clicking on the tabs below the toolbar for these additional functions.

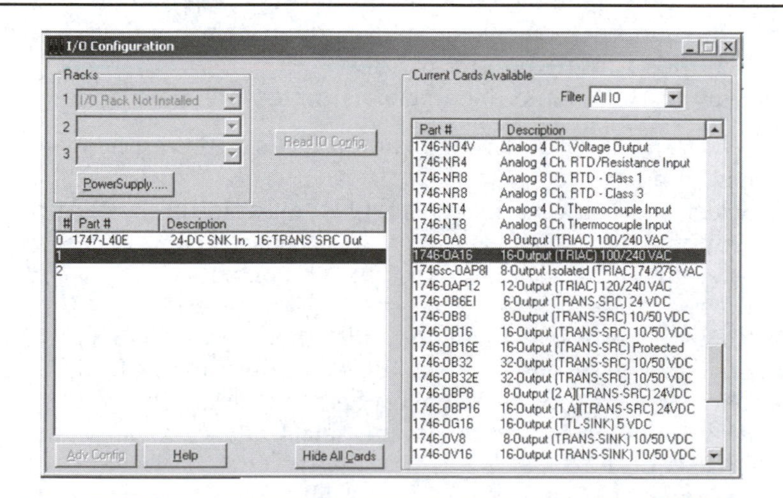

Figure 7–4
I/O configuration pop-up screen with slot 1 highlighted and first I/O module highlighted

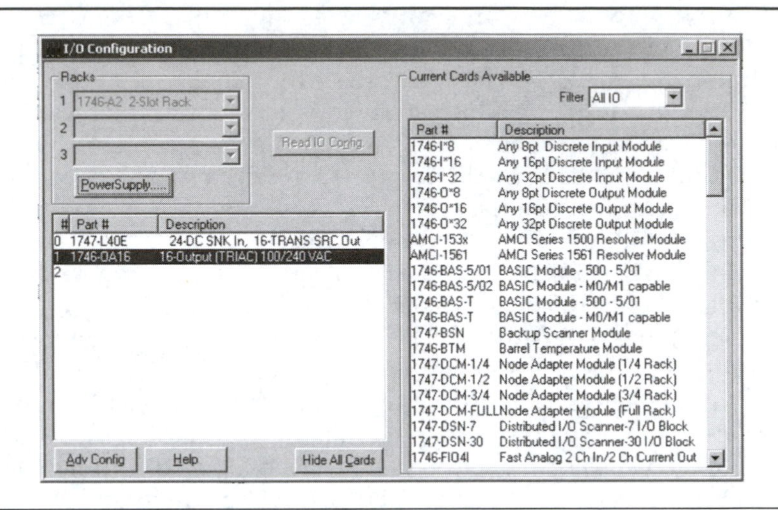

Figure 7–5
I/O configuration pop-up screen with both slots configured

Moving the cursor to the ladder program screen, lower right half, we can now begin entering ladder rungs. Note that there is currently only an empty rung showing with an end statement at the far right. This is the end of the ladder program marker. To insert a new rung right-click on the left side of the current rung and select *Insert Rung* from the pop-up box that appears; or click on the *New Rung* icon at the leftmost position on the element toolbar with the *User* tab highlighted.

To enter a programming element to this new rung you can click on the element you want to add in the element toolbar and drag it down to the rung. Double-clicking on a symbol icon also adds it to the current rung. Once the element is shown on the rung you can assign an address to it by clicking on the element and entering the address. To enter an instruction description right-click on the element and click on the *Edit Description* selection in the resulting pop-up window.

We will now enter and document a simple three-part ladder rung.

1. Select and drag down the Examon ON symbol to the front of the new rung.
2. Select and drag the Examon OFF symbol to the right of the first element.
3. Select and drag a standard output symbol to the far right end of the rung.
4. Click on the first element and enter the address I:0.0/0.
5. Click on the second element and enter the address I:0.0/2.
6. Click on the output element and enter address O:0.0/1.
7. Right-click on the first element, select *Edit Description,* and type "PUSH BUTTON #1".

8. Right-click on the second element, select *Edit Description*, and type "PUSH BUTTON #2".
9. Right-click on the output element, select *Edit Description*, and type "LAMP #2".

Your screen should now look like that shown in Figure 7–6.

The addresses we entered in the first rung come from the address listing in Appendix A at the back of this text. This is the listing of I/O address assignments for the training kit. You will also find the wiring drawings for the training kit in this appendix. If your kit or trainer has a different wiring arrangement you will have to pick addresses that fit the wiring layout of your trainer.

If you entered the wrong address you can fix it by left-clicking on the element, double-clicking on the address, and re-entering the address. Be careful when entering output addresses, as it is easy to enter a zero instead of an "O" as the first character. Note the column of small "e"s on the left edge of the rung. This indicates that the program has not been verified. To verify your single program rung, click on the *EDIT* tab at the top left corner of the main toolbar and select *Verify Project*. If no errors are found the column of small "e"s will disappear. If any of the addresses you entered do not match with the I/O configuration you previously stored, the errors will be shown at the bottom of the ladder screen. You can click on the error listing and find out the exact position in the ladder that contains the error, and the error code will give you a good idea what is wrong with that element.

To change any of the address descriptions you entered, just right-click the element you wish to change, select *Edit Description*, and edit the text in the pop-up box. Once an address has been assigned a description, that description will appear anytime that address is added to another rung. Because the

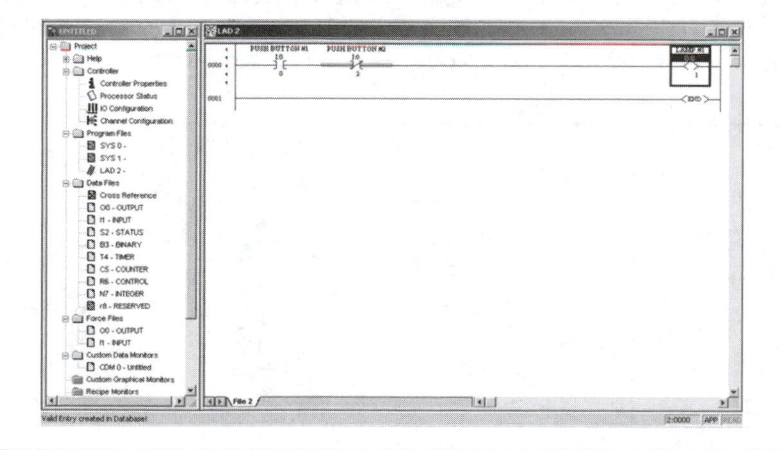

Figure 7–6
A one-rung ladder program

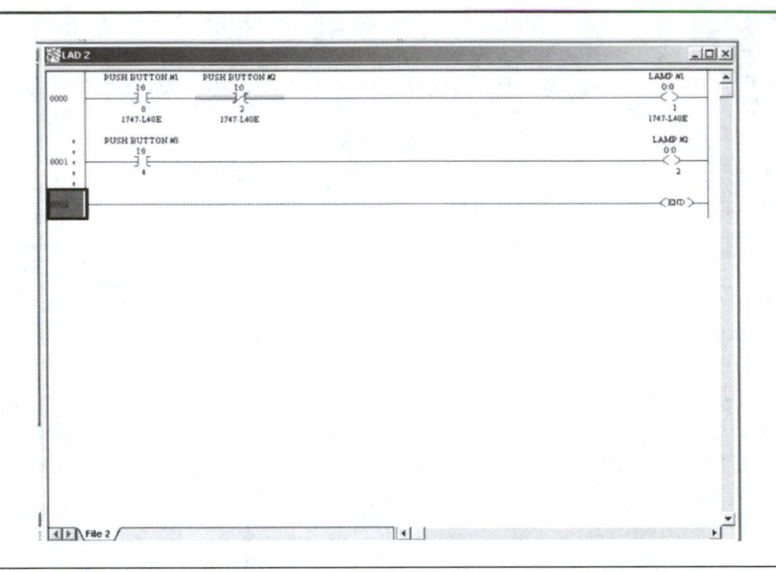

Figure 7–7
Entry of second rung

description is attached to an address you only have to enter a description once and it will appear anyplace that address appears in your ladder.

We will now enter a second rung that will contain a branch rung. Right-click on the left side of the rung with the (end) statement and click on *Insert Rung* or click on the *New Rung* icon on the toolbar. Enter the rung shown in Figure 7–7. The first Examon On element should be I:0.0/4 and the output should be O:0.0/2. You can add the description by right-clicking on the input and selecting *Edit Description.*

We now will add a branch rung to the input element. To do this, click on the *Rung Branch* icon on the *User* toolbar. Drag the element down to the left of the first input element on the second rung. Once in place, drag the red bar to the right side of the input element (see Figure 7–8). We now have a branch rung in place with no elements present on it.

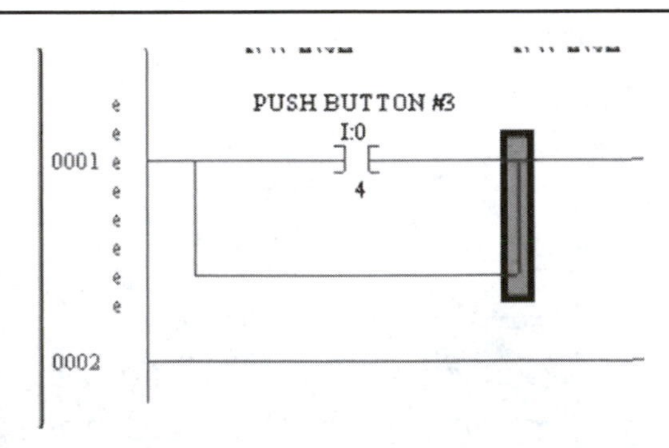

Figure 7–8
Branch rung in place

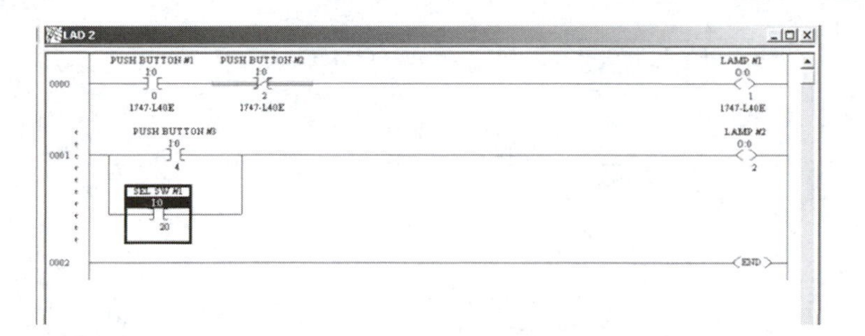

Figure 7–9
Second rung complete

To enter an element on the branch rung click on the *Examon ON* icon and drag it down to the branch rung. Enter the address I:0.1/4 as this is the address of the first selector switch on our training kit. Add the address description by right-clicking again. You may notice that the address changes to I:0/20 once it is entered. The program knows that the address I:0.1/4 is the 20th input of the 24 available on our PLC system and therefore changes it to display in the extended address fashion (see Figure 7–9).

Looking at Figure 7–9 you will notice the small "e"s have again appeared to the left of the second rung. This indicates that the new rung has not yet been verified. Also notice on rung 1 that the program has added some information below each of the elements. This happened when we verified the rung. The program has placed the location of the addresses below each element on the rung. Because all of these addresses are in the PLC's internal I/O or slot 0 the listing is the actual part number for the PLC.

Let's say we need to add an additional branch to rung 2. There are two methods for adding another branch rung. You can click on the same icon you used to add the first branch rung and drag it down to the lower branch. Although this works it is not the recommended method because this would be considered a nested branch and there is a limit of four nested branch instructions you can enter on one rung. You can enter as many nested branches as you like, but when you verify the rung you will get errors on any branch over the limit of four. The recommended way to add another branch is to extend the branch instruction. You do this by clicking on the bottom left corner of the last branch rung. When the small red square appears right-click on the red square and pick the *Extend Branch Down* selection from the pop-up window. Note how the sides of the branch are extended straight down. You can now add whatever element you want to the new branch rung. Adding branch rungs in this manner allows you to add a large number of extended branches.

You can always tell if the branch rungs were extended correctly because nested branches are shown in a stair-step fashion, while the correct way to extend branches is shown with straight sides (see Figure 7–10).

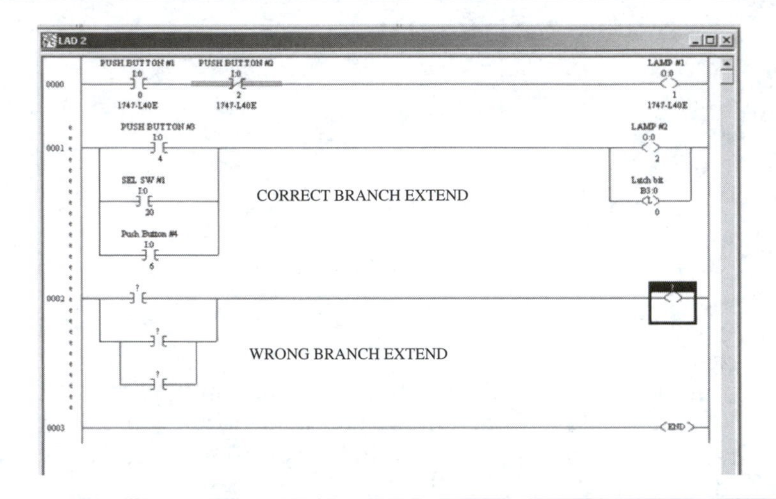

Figure 7–10
The correct and wrong methods of extending branch rungs

Note that if you're only entering three levels of branch rungs as is shown in Figure 7–10, it really doesn't matter which type you use because both will work. But it is good to get into the habit of extending the branch rung down because when you need to enter an eight-level branch rung, you won't remember the limit on nested branches, and will have to go back and re-enter all the elements when your rung fails the verify test.

Figure 7–11 shows the final three-rung program for this example. The program doesn't do anything of value and is only intended to show how to enter and edit a few program rungs. Rung 0 will turn on lamp 1 if push button 1 is pressed and push button 2 is not pressed. Rung 1 turns on lamp 2

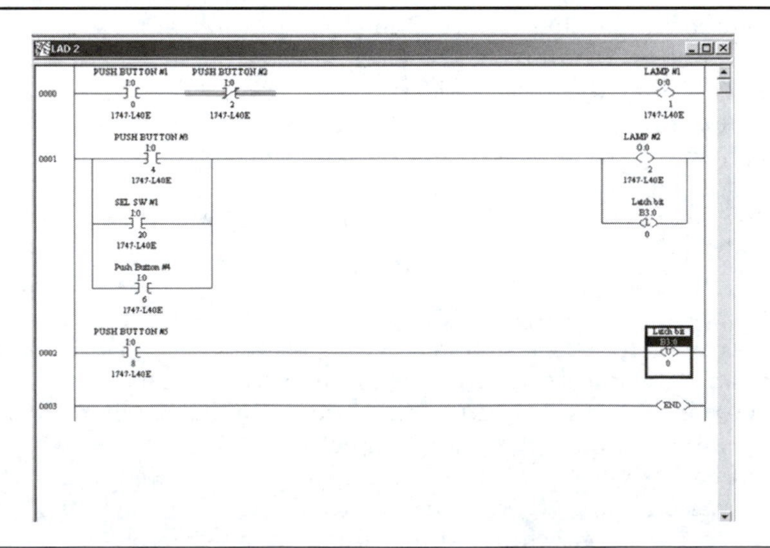

Figure 7–11
Finished three-rung program

and latches internal memory bit B3/0 if push buttons 3 or 4 are pressed or if selector switch 1 is turned on. Rung 2 resets the memory latch bit B3/0 if push button 5 is pressed.

At this point we have a ladder program, but it only exists in our programming computer. To run this program we must download the program to our PLC training kit. Before we download the program we must have the PLC turned on, have the 1746PIC module's flat cable plug connected to the serial port of our computer, and the DH485 plug inserted into the programming port of the PLC. If your training kit has a different communication system you will need to follow the instructions supplied by your instructor.

Click on the *Communications* tab on the main toolbar at the top of the screen. Click on the *Down Load* selection. The first thing the program wants to do is to save the file if you have not already done so (see Figure 7–12). Give the file a name and specify a location to save the file. If you have already saved the file you will be prompted to add revision notes (see Figure 7–13).

As we are just beginning to learn the first steps to programming the PLC you might want to click in the "Do not prompt me for revision notes again" box. Once you are writing a real program this can be a useful tool to keep track of the changes you make to a program as you go through the debug cycle.

Another setting you may want to change is the number of backup files the program keeps. Every time you make a change to your program and save the file, the program keeps a backup copy of the program as it was

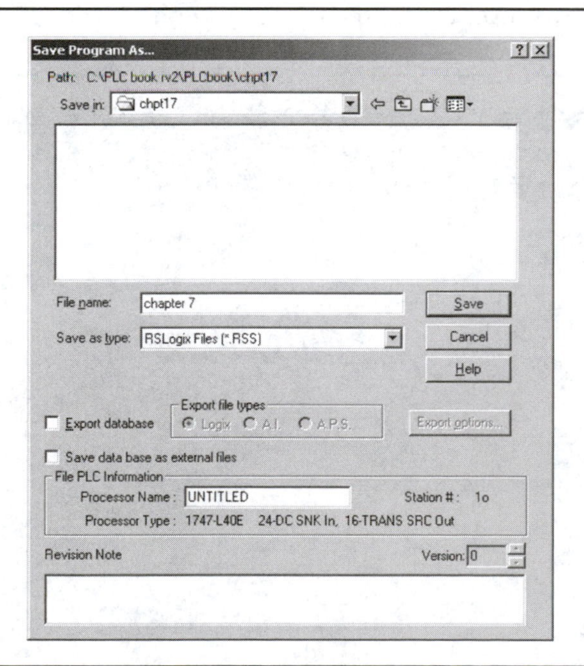

Figure 7–12
Save file screen

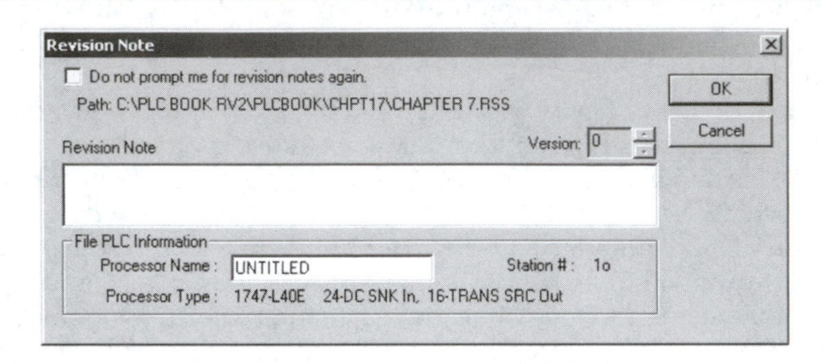

Figure 7–13
Revision note pop-up box

before the changes were made. When editing a large program, this can be a very useful feature, because often the rung changes you make may not work correctly and you may want to return to the previous revision of your program. Keeping more than three backups, is for the most part, a waste of disk space. If you find that your disk's directory is filling up with countless backup files then this setting may be too high. You can change this setting by clicking on the *Tools* tab on the main toolbar and selecting *Options*. The pop-up box shown in Figure 7–14 allows you to change this feature. For class work you may want to change the setting to only one backup file.

Now that we have all our settings correct it is time to download the file to the PLC. Before you actually download your file you should check to be sure your computer and the PLC are communicating. The best way

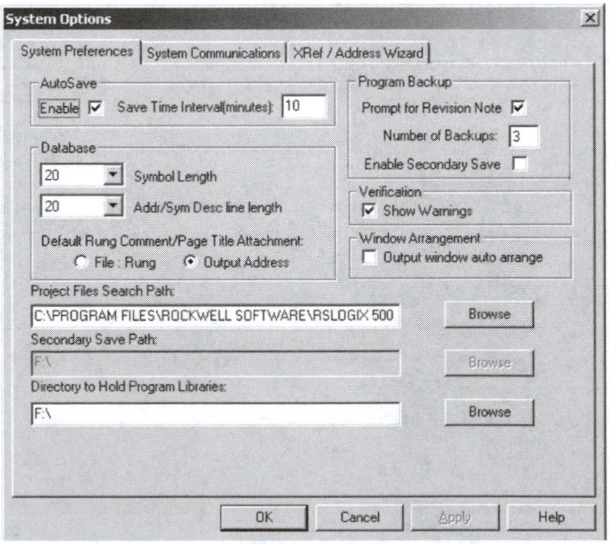

Figure 7–14
Setting the number of backup files

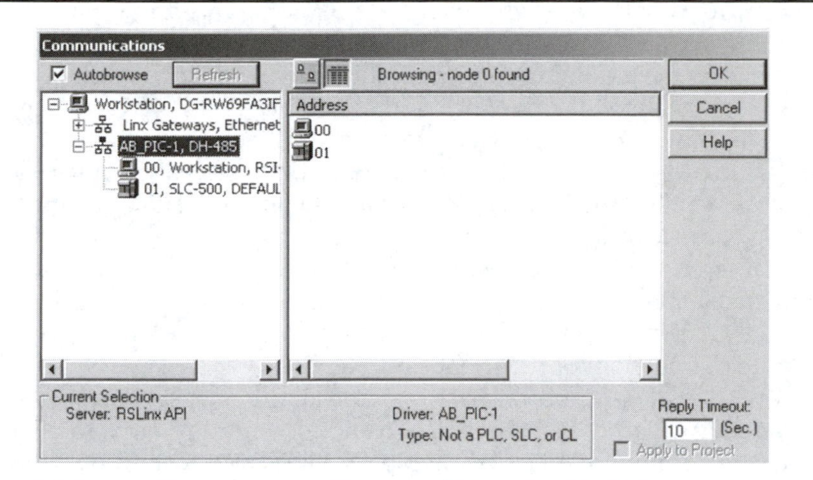

Figure 7–15
RSLinx communication status screen

to do this is to click on the *Communications* tab on the main toolbar and click on the *Who Active* selection. A screen similar to that shown in Figure 7–15 should pop up.

Note under the AB_PIC-1 heading that you can see both the host computer (station 00) and the PLC (station 01). You may have to click on the small sign (+) next to the AB_PIC heading if it is not showing the expanded tree. If your screen looks like that shown in Figure 7–16 then you have no communication between the PLC and your host computer. Note the red (X) over the PLC icon. If your pop-up box looks like Figure 7–15 then you have good communication and can proceed with your program download. If your host cannot communicate you should check the cables

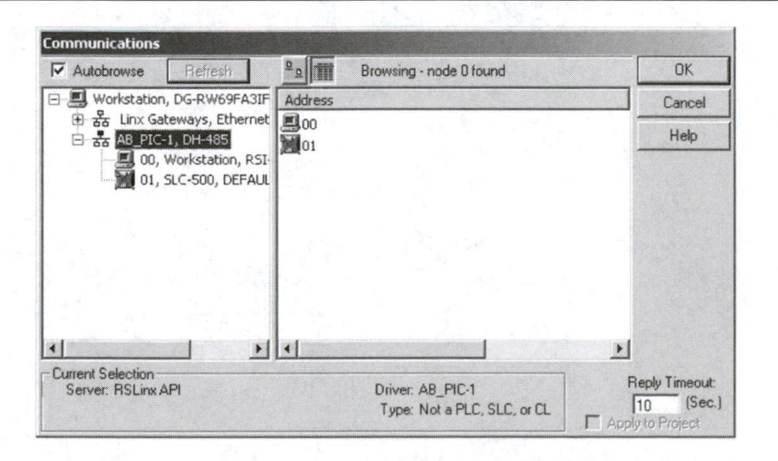

Figure 7–16
RSLinx communication status screen (no communication)

between the two devices. Be sure the PLC is powered up and the programming plug is firmly inserted into the socket.

If you get the message "Communication driver not yet established" click on the arrow next to the driver box and select the AB_PIC driver, then click *Apply*. This may be different if your training kit is set up for some other communication protocol. The next message you should get is a prompt to download the new program file, asking if you want to proceed. This is an important warning because once you download the new program it will overwrite the existing program in the PLC. Click *Yes* and the file will be downloaded. Once the file transfer is complete you will be prompted to change the processor to run mode. Again click *Yes*. The next prompt you will see asks if you want to go into online monitor mode. Click *Yes* again.

You should now be returned to the main edit screen as seen in Figure 7–17. If your PLC is not in run mode, you may have to click the arrow next to the mode box (showing REMOTE RUN) and click on run mode. Once in run mode you should see the rung elements highlight as each button is pressed and outputs come on.

One thing we should explain in Figure 7–17 is that output and latch elements are highlighted if the bit is set on. Note that rung 1 is true, so both the lamp and latch bit are highlighted. If you look at rung 2 you will see that the unlatch function is also highlighted even though rung 2 is not true. This is because bit B3/0 is on and therefore it is highlighted anywhere that address is shown. This can be a little confusing at first, because it appears the unlatch function is activated even though the rung is not true.

As we stated back in Chapter 5 you cannot make ladder rung changes while the PLC is in run mode. To make changes you must switch to offline

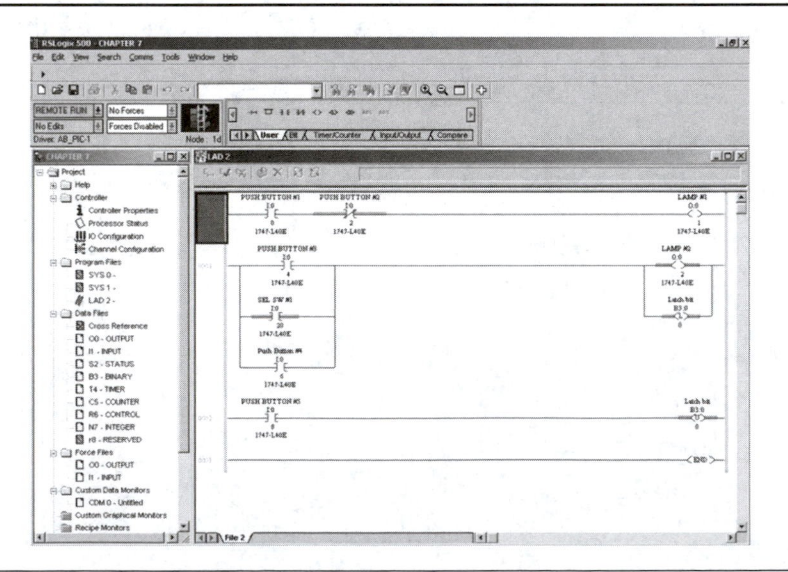

Figure 7–17
Main edit screen showing run mode

mode by clicking the arrow next to the mode box and selecting *Go Offline*. You can now edit your program on your host computer. Note that the ladder program is still running in the PLC and all the buttons and switches are active. All you have done is disconnect from the PLC so you can edit your ladder program. The only way to stop the PLC from running the ladder program would be for you to change it to program mode by clicking on the arrow next to the mode box and clicking on *Program*.

While you were monitoring the running PLC online you should have noticed a slight delay from the time you pressed a push button and that element being highlighted on the screen. This is because the host is looking at a time slice of the PLC program, and it is updated every 100 milliseconds or so over a serial link. Because of this delay you may have trouble seeing inputs that change very quickly.

In later chapters of this text we will want to add several subroutine files to our main program. This is a convenient place to explain how to add files to the system. First let's change the system file LAD 2 name to MAIN. Right-click on the LAD 2 name and then click on the *Rename* selection. Enter the new file name "MAIN".

Next, we want to enter a subroutine file for handling the manual part of our program. To do this we right-click on the program files folder and click on the *NEW* selection. The pop-up box shown in Figure 7–18 will come up with the next file number (3) showing.

Enter a new name, *MANUAL MODE*, in the Name box and the description *Manual mode program* in the Description box. Then click OK. Repeat the procedure for file 4, calling it *AUTO MODE* with a description of *Auto mode program*. Your screen should now look like Figure 7–19 showing the two new files and LAD 2 renamed to MAIN. Note there is a limit of 10 positions on how long the file names can be. Our file 3 name was shortened to MANUAL MOD.

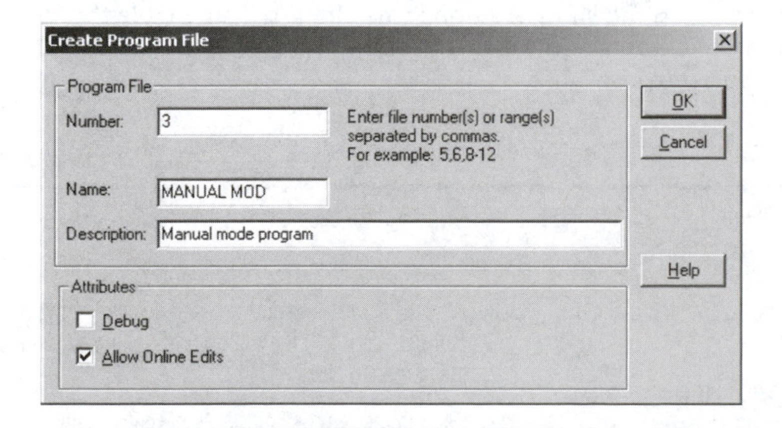

Figure 7–18
Create Program File pop-up box

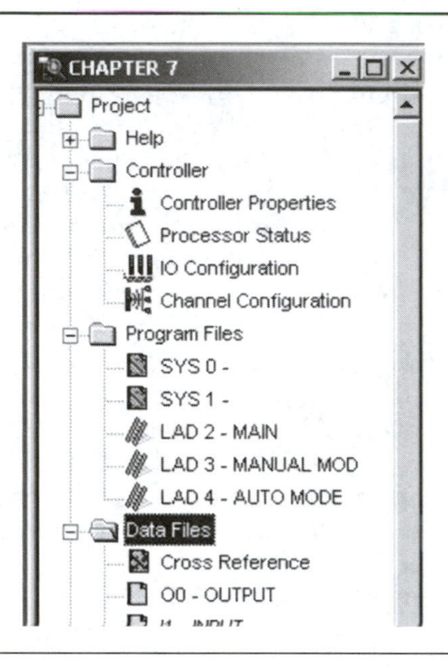

Figure 7–19
Program file list showing two new files

CONCLUSIONS

As we stated earlier, this chapter was not intended as a complete software reference manual for the RSLogix-500 programming packages. Our purpose in showing a sample of how to start and edit a small program is to give the student a starting point to understanding the PLC programming languages. Good manuals are available from the manufacturer for each software package. The problem with most of the software reference manuals is the lack of example programs and examples of how to set up programs from the start. We hope this chapter will enable the first-time student to enter a simple program and download to the PLC for debugging purposes.

REVIEW QUESTIONS

1. When starting a new program what is the first piece of data you must enter?
2. What will happen as you enter a new program if you forget to set up the I/O configuration?
3. Why do I/O changes seem to be delayed between the real I/O event and the highlighting of the element on the screen?
4. What is the purpose of the Who Active utility?

5. What key sequence is used to change an input element description in the RSLogix software package?
6. How do you insert a new rung into an existing program?
7. When a ladder program is uploaded from the PLC to a new host computer how much of the original program will you see?
8. How do you add new subroutine files to the system file list?
9. What should you check if the Who Active report shows a red X on the PLC icon?
10. What is contained in the system files #0 and #1?
11. How do we assign descriptions to element addresses in the RSLogix software?
12. How do we edit element descriptions in the RSLogix software?
13. What are the two ways to add branches to an existing rung?
14. What is the problem with using nested branch rungs?
15. How do you set your preference for the number of backup copies the system saves?

LAB ASSIGNMENT

Objective

The following lab will introduce the student to the process of entering ladder code and downloading, editing, and debugging a ladder program. This lab project was designed to run on the lab kit documented in Appendix A of this text. Although the lab's addresses match the kit addresses this lab is generic in design and can be run on any teaching kit by modifying the addresses to match the kit being used.

1. Start RSLogix-500 and open a new file. Refer to Figures 7–2 through 7–9 at the beginning of this chapter for the step-by-step process of opening a new file. Be sure to set up the processor type and I/O to match the kit that you are using. Save the file as Chpt-7-lab.
2. Highlight rung 0000 by clicking on the left edge, then click on the *New Rung* icon in the user toolbar.
3. Select, drag, and drop the elements from the toolbar to rung 0 as shown in Figure 7–20.
4. Assign the addresses and labels as shown in Figure 7–20.
5. Click on the *Edit* box and then the *Verify Project* selection.
6. Verify that the "e" symbols disappear from the left edge of the rung. If these symbols do not go away, you have an error in your rung and it will be displayed at the bottom of the screen.
7. We now have programmed our first rung, but we need to download the program to the PLC controller. Be sure the interface cable is connected between the computer's COM 1 port and the PLC program port.
8. Click on the *Comms* box on the main title bar and then on the *Who Active Go Online* selection. Your display should show a pop-up window that looks like Figure 7–15 if your connections are good. If your screen looks like Figure 7–16 you do not have a

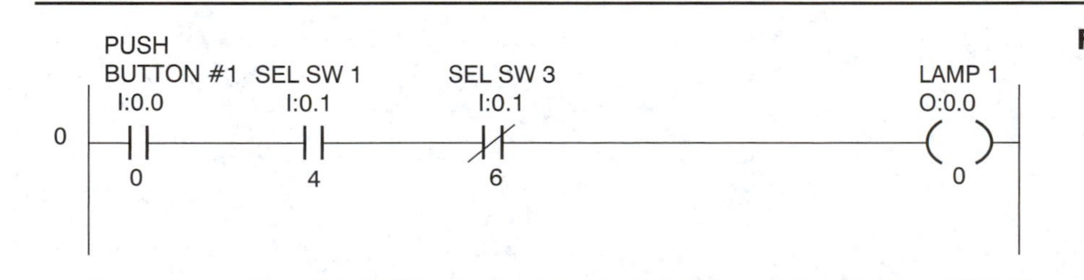

Figure 7–20

Figure 7–21

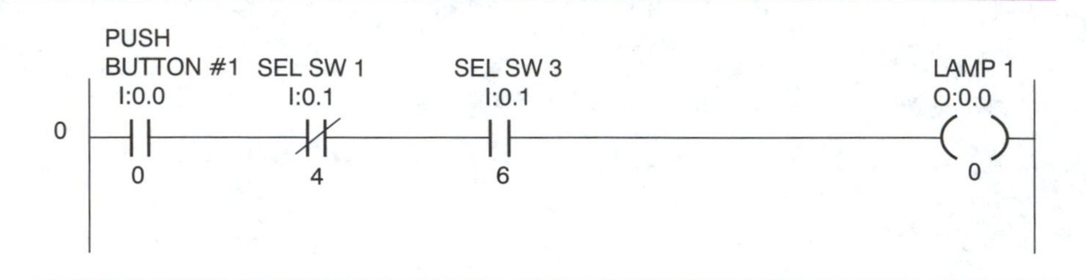

connection to the PLC. Call your instructor to get help with the communication problem.

9. If your Who Active screen shows that you have a good connection you now will download the program to the PLC. Click on the *Comms* box in the main menu bar and then on the *Download* selection. Follow the sequence previously explained in this chapter until your screen looks like that shown in Figure 7–17.
10. You are now in the remote run mode and are ready to test your program. With SEL SW 1 ON and SEL SW 3 OFF press push button 1. Note how the rung elements are highlighted when the element condition is TRUE. Verify that the output lamp 1 turns ON only if the rung condition is true.
11. Click on the arrow next to the box showing REMOTE RUN in the top left corner of the screen. Click on the *Off Line* selection. You have now disconnected from the PLC and can edit the program. Note that if you press push button 1, lamp 1 lights. This indicates the program is still running in the PLC.
12. Edit the program to the form shown in Figure 7–21. Make these changes by right-clicking on the elements and selecting the *Change Instruction* selection.
13. Download the new rung conditions and verify that the changed rung condition performs as shown.
14. Click on *Off Line* and edit the program by adding a second rung as shown in Figure 7–22.
15. Download the new two-rung set and verify that the rung conditions perform as indicated. Note that the state of SEL SW 3 determines which rung is active.
16. Switch to offline mode and add a third rung to the program as shown in Figure 7–23.

Figure 7–22

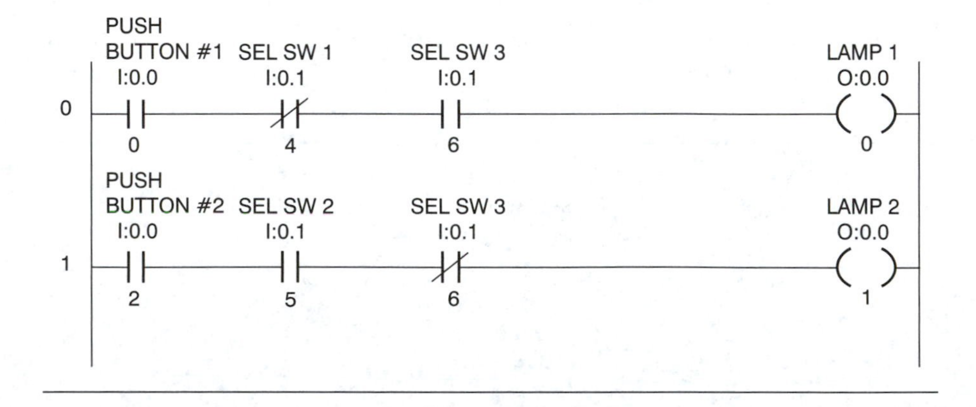

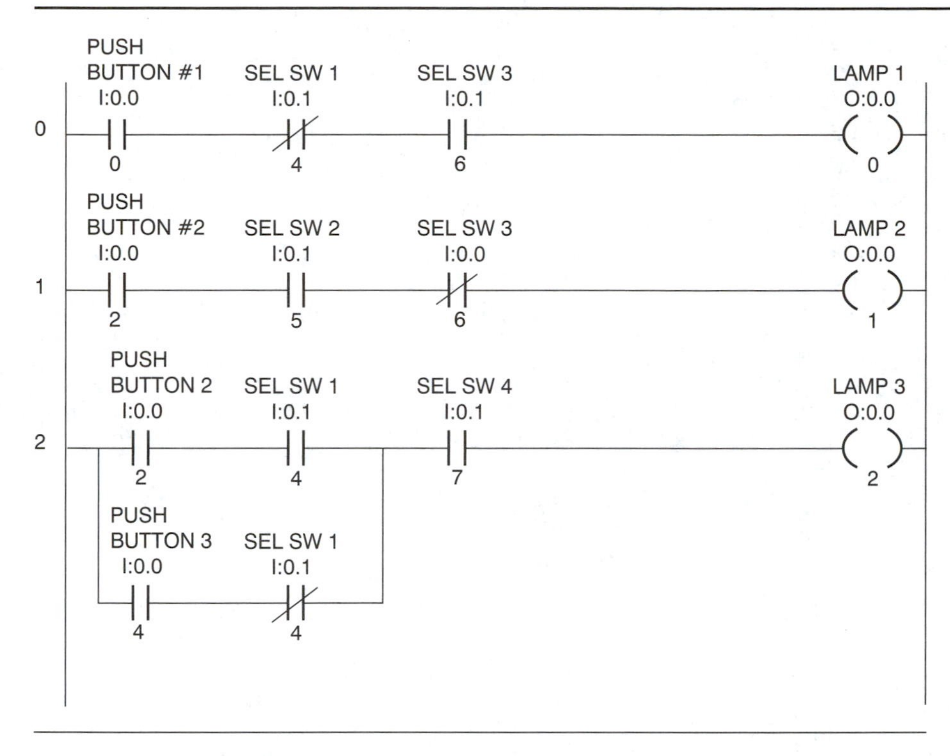

Figure 7–23

17. Download this new three-rung program and verify that all three rungs perform as indicated. Write the Boolean equation that describes rung 2 in the space below.

18. Switch back to offline programming and edit the program to the form shown in Figure 7–24. Be sure to add the third branch to rung 2 by right-clicking on the left corner of the second branch rung, and then clicking on the *Extend Branch Down* selection. Your third branch should look like that shown in Figure 7–24. If your branch rung has a stair-step look to it, you inserted a new branch instead of extending the current branch down.

19. Download the new program and verify that the new rung configuration performs as indicated. Write the Boolean equation for rung 2 in the space below.

20. Add rung 3 to the program to control lamp 4. The Boolean equation for this new rung should be (PB4 + PB5 + PB6) SelSw2 = Lamp 4. Draw your new rung below and enter it into the program. Download the new program and verify that the rung condition follows the Boolean equation.

21. Enter a few new rungs of your own design. Document the rung equations and rung format before you enter each rung. Continue to edit and download program changes until you feel comfortable with the edit-download-debug process.

22. When you are finished, print out a copy of your program and turn it in to the instructor. To print out a copy of the program click on the File box on the main menu and then on the Report Options selection. This will open a window to set up the report options. See Figure 7–25 for the layout.

Figure 7–24

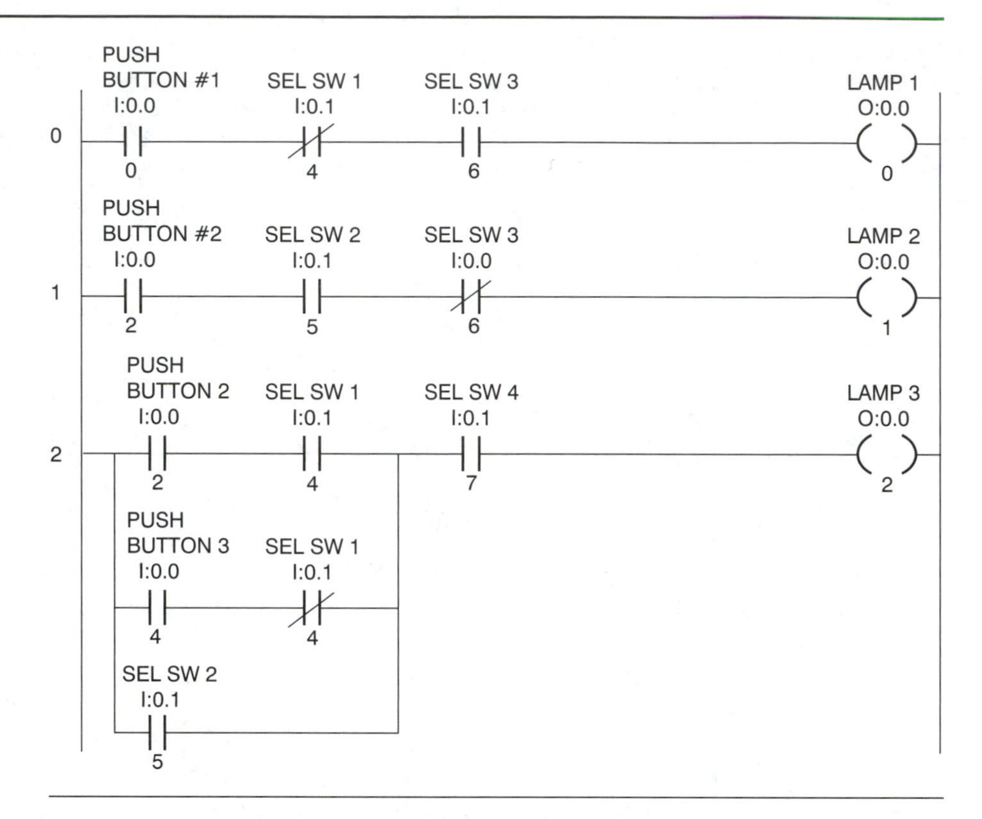

Figure 7–25
Report Options page

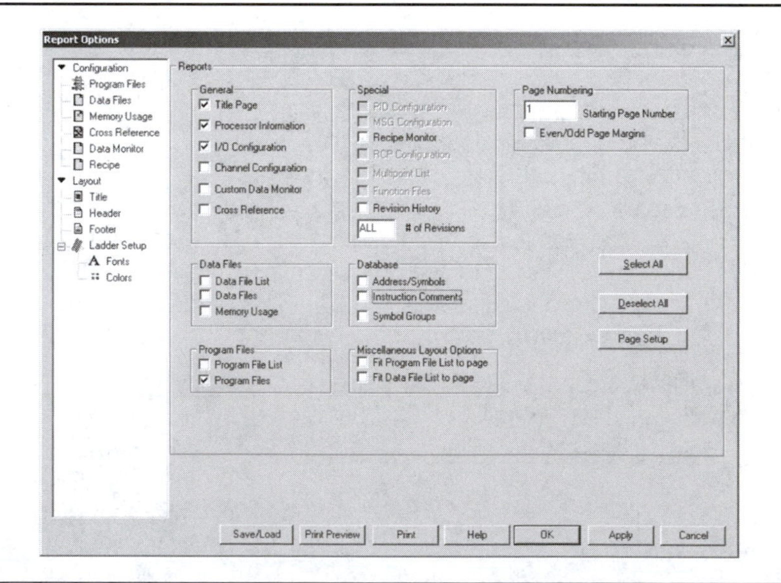

Because we only have a very short program, we will turn off most of the report options. You should have only four check boxes set on as shown in Figure 7–25. After you have set up the options correctly, click on the Title selection listed on the left edge of the window. Click on the default arrow located under the main edit window and select *USER.* Enter your class number and name in the title box. Click *Apply* and then *Print* to print out a copy of your program.

CHAPTER

8

Basic Machine Control Functions

LEARNING OBJECTIVES

After completing this chapter the reader should understand:

1. The basic functions of air valves and air cylinders.
2. Interface concepts for controlling air valves.
3. Position feedback concepts for air cylinders.
4. Indexer types and uses.
5. Controlling indexer functions.
6. Designing a control program for sequencing moving elements.

Up to this point, our discussions have dealt with nonmoving elements. We have covered outputs and inputs connected to lamps and switches, but until now none of the I/Os have driven a moving element. We are now ready to introduce the most two common moving elements in automation machine construction: the air valve with its air cylinder and the cam-driven indexer. We will also design and program our first complete project at the end of this chapter.

AIR CYLINDERS AND VALVES

The most common output element in machine control is the air cylinder with its control valve. This device comes in many different styles and sizes, but they all fall into three main groups: (1) four-way single-solenoid spring return, (2) four-way dual-solenoid two-position, or (3) four-way dual-solenoid three-position, spring return to center.

All of these devices have an air input port and one or two air output ports (Figure 8–1). In the case of the single-solenoid spring return, the air is normally switched to port B, and port A is exhausted with the solenoid off. When power is applied to the solenoid, the air input is switched to the port A side with B exhausted and held there as long as power is applied. When power is removed, the solenoid spring returns to port B. Because of this unpowered and uncontrolled return, this valve type is not recommended for use with any air cylinder. If we use this type of valve for cylinder control and someone presses the E stop push button, all the mechanisms return home uncontrolled. This may cause damage to the machine or harm to the operators. Also, great care must be taken with this valve type when deciding what action will be performed in the unpowered state. For example, if we are clamping some object that is being machined the designer must be sure the unpowered position is the clamp position. Otherwise, if a power failure occurs the clamped object will be released and may cause considerable damage.

The dual-solenoid valve (Figure 8–1, right) has one solenoid on each end of the valve. When power is applied to one of the solenoid coils, the valve switches to that port and stays there even if power is now removed from that solenoid. The valve will only switch ports if the other solenoid is now powered on. This is the recommended valve type for use on all automation cylinders because the cylinder will stay in position even if power to the system is lost.

The third type of air cylinder is used in special cases where the cylinder needs to stay in some position other than fully returned or extended. In this type, the valve is switched to one port or another depending on which solenoid is powered on. When power is removed, the valve moves to a center position. This position can have both ports closed or exhausted depending on which valve style you select. The air cylinder driven by this type of valve is often equipped with an air-driven clamp device to hold it in position when the main air valve is not powered in either position.

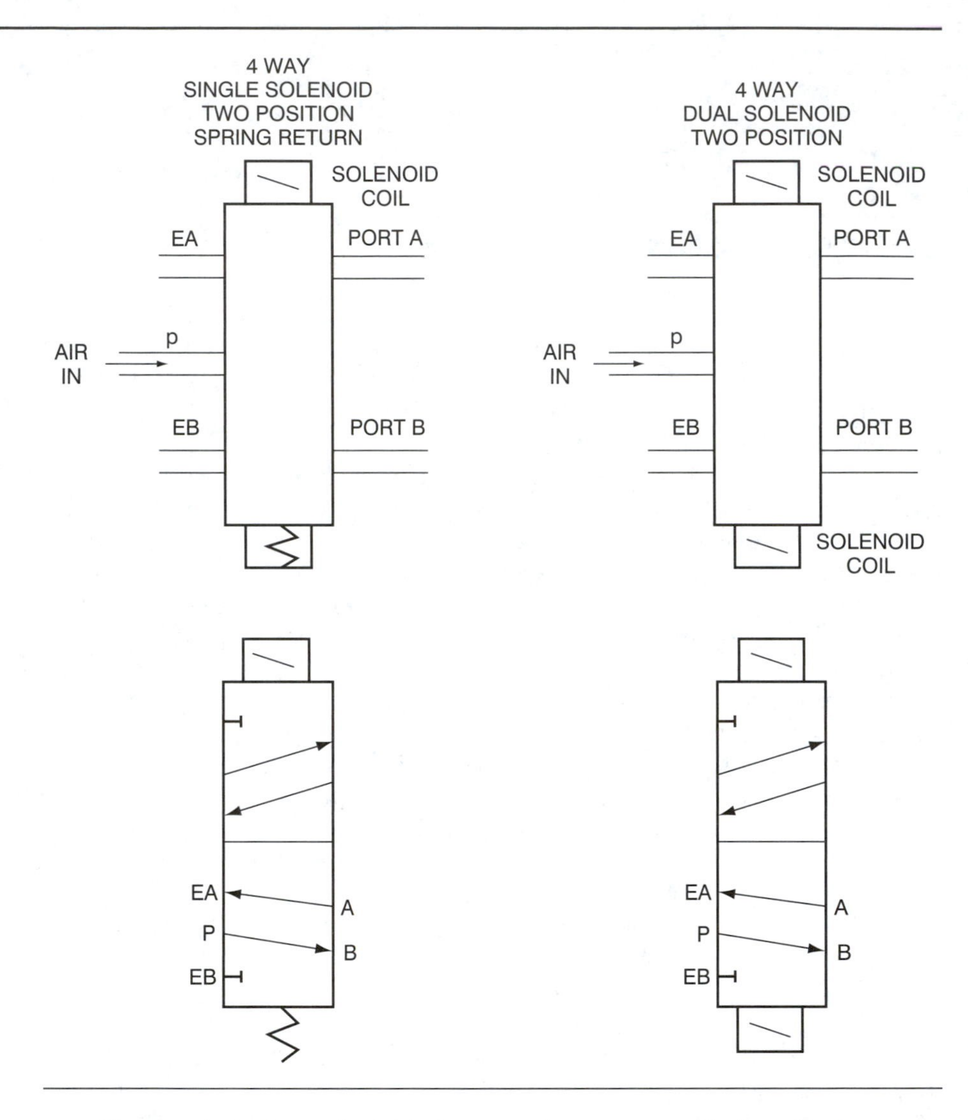

Figure 8–1
Types of air cylinders and control valves

Figures 8–2 through 8–7 show various types of air cylinders and air valves. Note that in Figure 8–3 the air cylinder in the bottom foreground is shown with Hall effect position-sensing switches attached. Also, the air cylinder on the far left is shown with adjustable flow controls attached. It is often easier to use cylinders with the end-of-stroke switches attached than to design brackets and mounting hardware to hold these devices as part of the machine's mechanics. These Hall effect switches can be moved along the cylinder body and then tightened to hold any position. Figure 8–6 shows a cutaway view of a single-acting spring return air valve. When a large number of air valves are needed, it is usually cost efficient to mount them on a common manifold mounting plate as shown in Figures 8–3 and 8–5. This requires only one air inlet pipe and two exhaust ports to be baffled. It also

Figure 8–2
Three single-acting and two double-acting air valves mounted on a manifold. *Courtesy*: Versa Products Inc., Paramus, New Jersey.

allows you to pipe all electrical wiring into one conduit for all the valve solenoids. This makes for a much cleaner layout with less piping and wiring.

For most of this text we will be dealing with the second valve type, the dual-acting valve. Figure 8–8 shows a typical I/O hookup for a dual-solenoid air valve and air cylinder. Note that we need four I/Os for each valve and cylinder combination, two inputs for position sensing of the air cylinder, and two outputs for control of the air valve.

It is the programmer's job to sequence these devices in the proper order to perform the required functions. As we detailed in Chapter 4, the first step is to generate a detailed flowchart or state chart of each mechanism that must be controlled. We can then generate the ladder rungs to control the I/Os as detailed in the flowchart. If we look at any flowchart, we will see that each diamond-shaped symbol defines a set of input conditions that must be made before we can perform the next output statement. On a state chart the input conditions are listed in the box below the circle and the output function is shown above the transition arrow. Writing a ladder program is nothing more than building a set of control rungs that match each one of the decision diamonds in the flowchart. Although this is a simplistic view, it is not far from the truth.

What we are building is a state machine with each rung defining one of the states through which the mechanism will move. Most automation projects contain many cylinders and valves working together to perform the machine's process function. To help explain this process, we have created a simple machine and written a flowchart and a state chart for it. We then

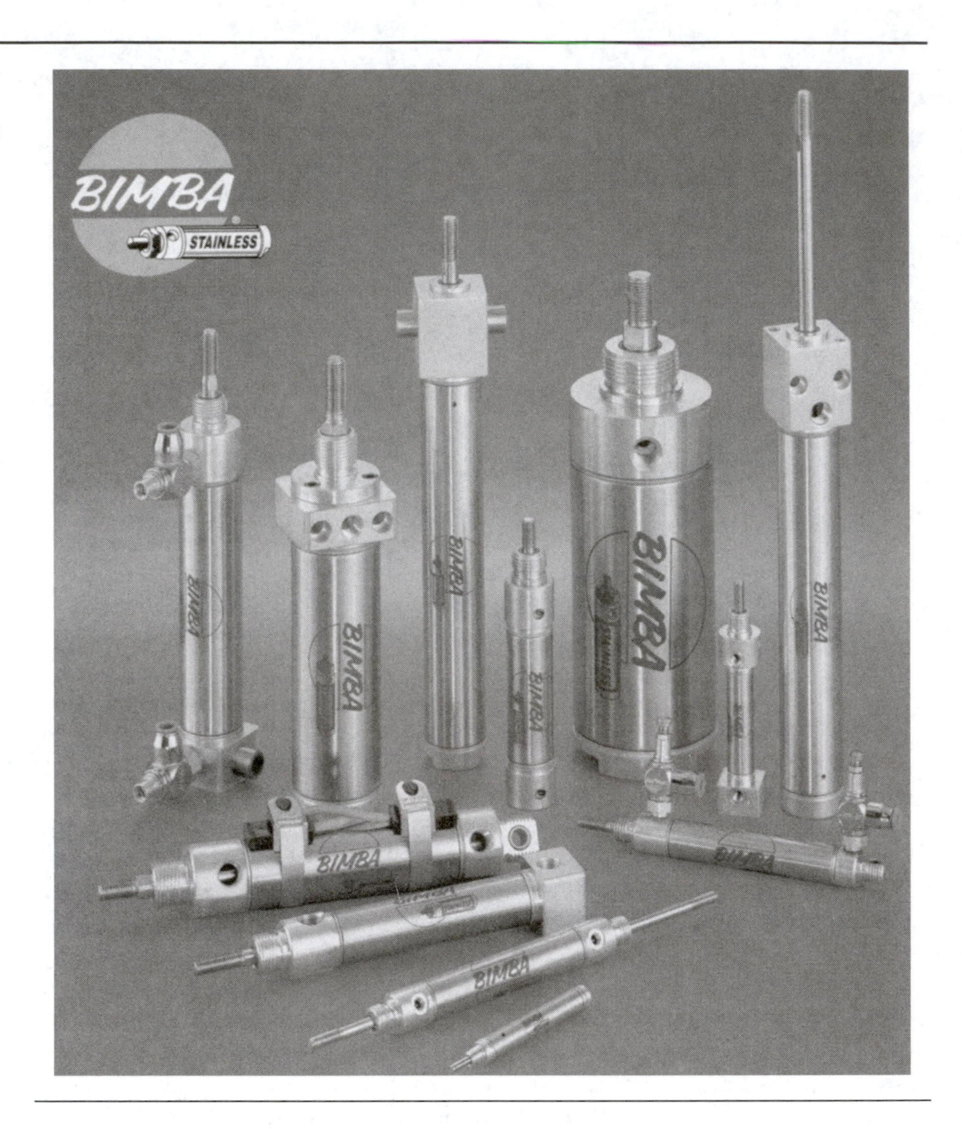

Figure 8–3
Selection of air cylinders.
Courtesy: Bimba Manufacturing Company, Monee, Illinois.

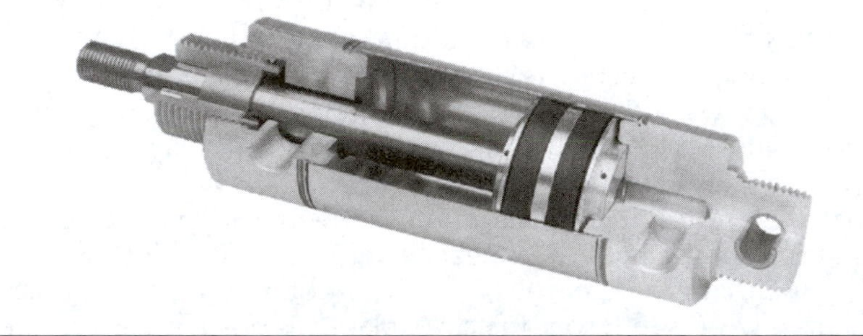

Figure 8–4
Cutaway view of a typical air cylinder.
Courtesy: Bimba Manufacturing Company, Monee, Illinois.

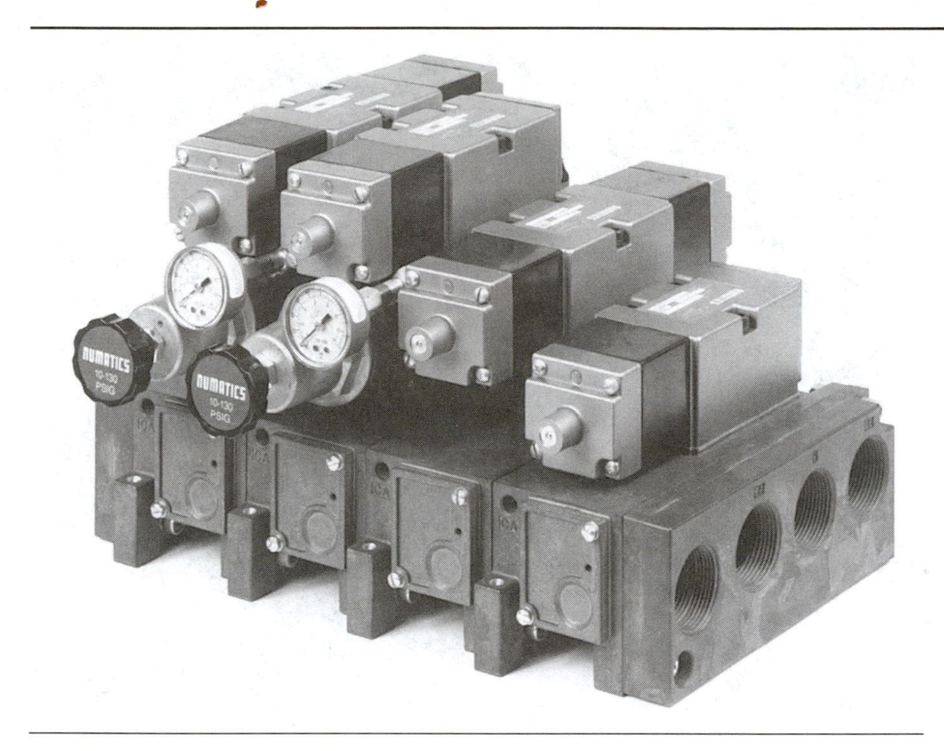

Figure 8–5
Four-valve set on a manifold, two are single-acting and two are double-acting valves.
Courtesy: Numatics, Inc., Highland, Michigan.

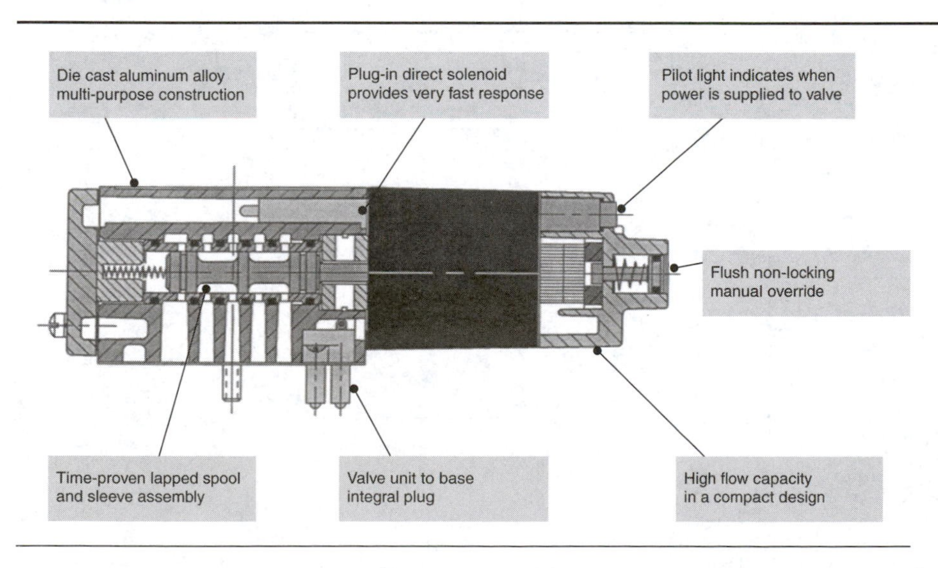

Figure 8–6
Cutaway view of a single-acting direct solenoid valve.
Courtesy: Numatics, Inc., Highland, Michigan.

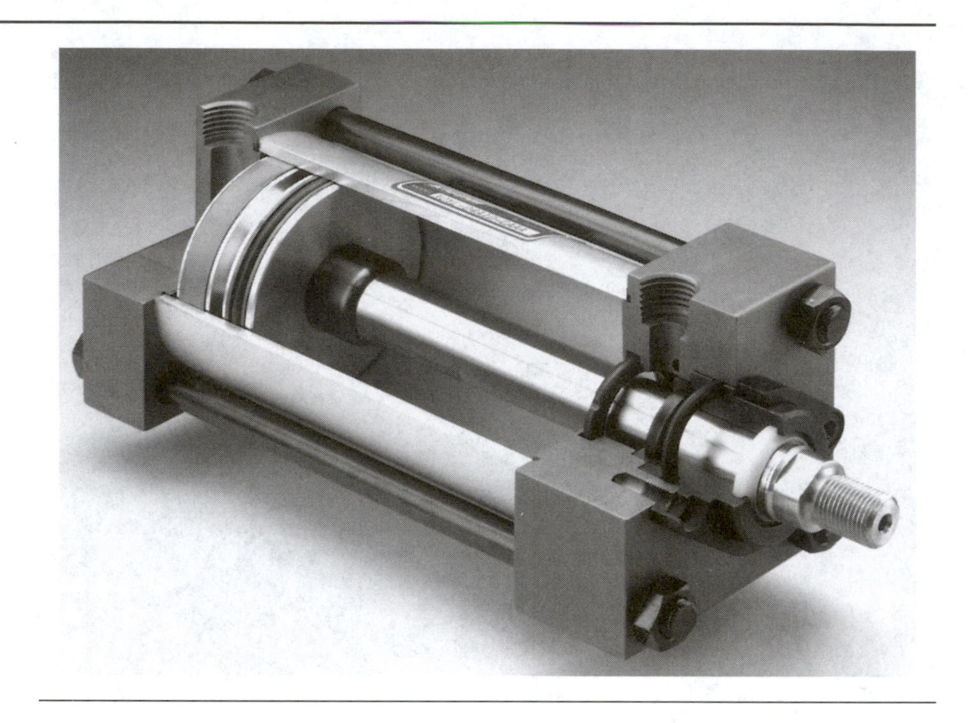

Figure 8–7
Cutaway view of large-bore air cylinder.
Courtesy: Numatics, Inc., Highland, Michigan.

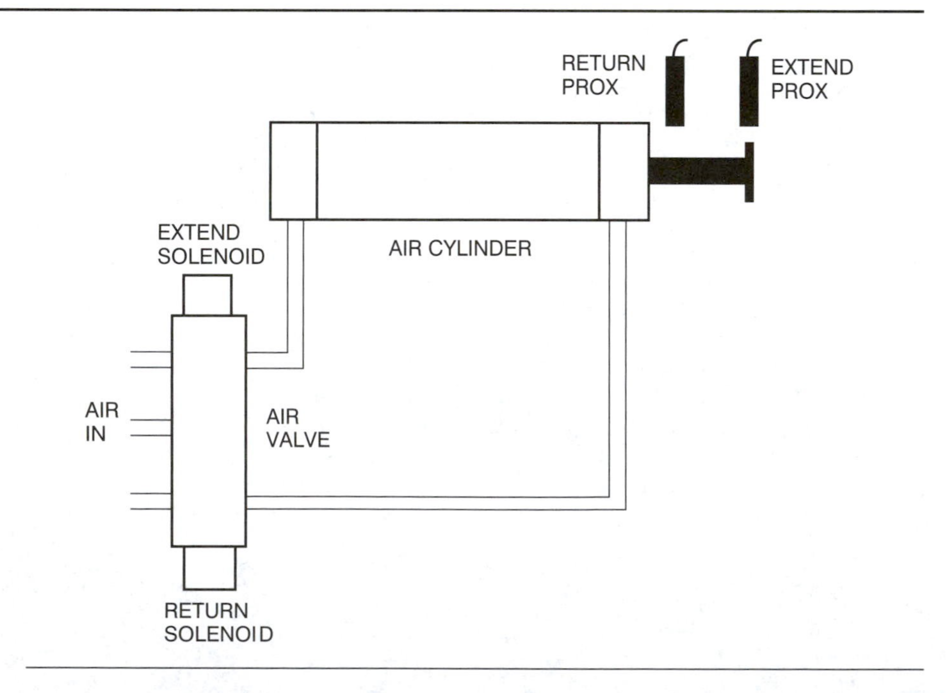

Figure 8–8
Typical I/O hookup for an air valve and cylinder

write a program based on the sequence determined in the charts. The sample project is shown at the end of this chapter. This example must be kept quite simple because we have not yet explained timers, counters, subroutines, or any of the higher math and logic functions that the PLC is capable of performing.

INDEXER DRIVEN EQUIPMENT

The second most common mechanical element found in automation equipment is the indexer. An indexer is a motor-driven device for moving an object from one position to another with a high degree of accuracy. These devices are typically either linear or rotary in design and are usually cam driven. Many companies build these devices but the largest is probably Commercial Cam Company (Camco), Inc. Several examples of indexers are shown in Figures 8–9, 8–10, and 8–11.

Figure 8–9
Large rotary indexer.
Courtesy: Commercial Cam Company, Inc., Wheeling, Illinois.

Figure 8–10
Several small rotary indexers.
Courtesy: Commercial Cam Company, Inc., Wheeling, Illinois.

Cam-driven indexers have a cam attached to the input shaft and a follower wheel attached to the output shaft. As the cam rotates, followers on the output shaft are guided through a path dictated by the cam shape. During part of the input rotation, the cam holds the follower in a rigid position called the *dwell position*. The control engineer must ensure that this dwell position is well defined. On most indexers there is a rotary position cam with a notch or lobe indicating this position. The control engineer must be sure to stop the indexer in this dwell position between index cycles. This is the only position where the output shaft is locked and the mechanism is stable for other devices to perform their mechanical functions. (An example program for cycling an indexer is given in Chapter 9.)

Rotary indexers can vary in size from a foot in diameter to more than 6 feet as you can see by comparing Figures 8–9 and 8–10. It is important to understand that the cam switch shaft rotates once for each table index. This may only be 30, 45, 90, or 180 degrees of table rotation depending on how the indexer was ordered.

Linear-type indexers are normally used to power assembly lines or other assembly equipment that moves in a straight line. Although the drive mechanism is similar to that of a rotary indexer, the output is connected to a precision chain that holds the pallets or fixtures that will be operated on. In any case we still have a dwell position switch cam that indicates the position of the indexer output shaft for sensing the dwell position. This sensor is necessary, because this is the only position in which the chain and pallets are locked and can be operated on. How many inches

Figure 8–11
Combination rotary and linear indexers.
Courtesy: Commercial Cam Company, Inc., Wheeling, Illinois.

of chain movement constitute one rotation of the position switch cam will depend on the gear ratios and chain link spacing.

Some automation equipment can be designed to be almost completely cam controlled. Instead of just one cam, as is used in an indexer, these machines have a timing motor with multiple cams attached to it. As the timing motor rotates, various mechanical elements ride on the outside edges of the cams or in grooves cut into the cams or both. The timing of each movement is dependent on the shape of the cams. We often need to activate additional external equipment at precise times during the timing cycle. One way to do this is to have a series of limit switches also riding on the cams. The problem with this arrangement, in addition to the mechanical switch wear, is that we often need to adjust the timing. This requires the limit switches to be repositioned relative to the cam lobes. This is a time-consuming process and may require many adjustments before the correct settings are found.

A better control system design would be to use an optical encoder attached to the timing motor shaft. We could then feed the output pulses into a high-speed counter module. Because we have a predetermined number of pulses per revolution and a zero position output from the encoder, we can divide the rotary cycle into as many degrees as we have count pulses. We can then activate any output at a precise count number by comparing the counter value with a desired degree of rotation. Adjusting the system timing is now as easy as changing the compare instruction value, and this can be performed much faster than adjusting mechanical switches.

A second advantage of this system is that we can change the activation count based on the motor speed. This is a feature that is very difficult to do mechanically. Because each output is based on a count value from the encoder the machine can be started slowly and gradually brought up to full speed. This ramp up and ramp down of the machine's speed can be a great advantage for complicated mechanisms. We do not discuss this type of machine at length until Chapter 13 because we need to understand timers, counters, and math instructions before we can control this type of equipment.

SAMPLE THEORETICAL MACHINE

Our theoretical machine will be a small clamp fixture and drill press operation (Figure 8–12). In this example we will use two air cylinders located on the kit's motion panel and two push buttons located on the main kit enclosure. For I/O wiring, see the kit wiring diagrams in the back of the lab manual.

Figure 8–12
Sample theoretical machine

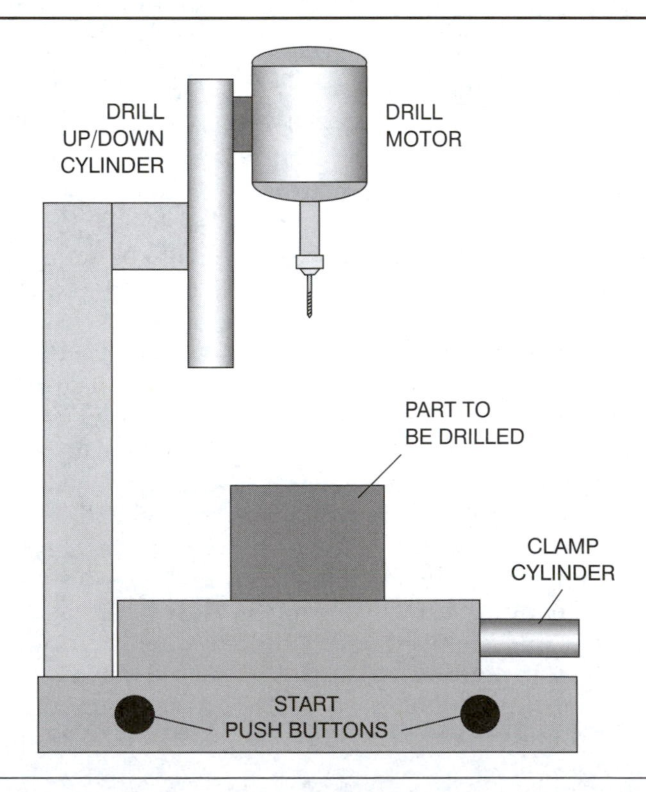

I/O Assignments

The I/O assignments for our theoretical machine are as follows:

Inputs	I	Outputs	O
PB#1	I:0.0/0	Clamp Closed Sol	O:1.0/1
PB#2	I:0.0/2	Clamp Open Sol	O:1.0/3
Clamp Closed Prox	I:0.0/1	Drill Down Sol	O:1.0/5
Clamp Open Prox	I:0.0/3	Drill Up Sol	O:1.0/7
Drill Down Prox	I:0.0/5		
Drill Up Prox	I:0.0/7		

Here is the mechanical sequence:

When **PB#1** and **PB#2** are closed, extend cylinder #1 (close clamp).

When cylinder #1 extend prox **SW** closes, extend cylinder #2 (drill press down).

When cylinder #2 extend prox **SW** closes, return cylinder #2 (drill press up).

When cylinder #2 return prox **SW** closes, return cylinder #1 (clamp open).

When cylinder #1 return prox **SW** closes, and both **PB#1** and **PB #2** are open, the cycle is complete.

The cycle must not repeat if both start push buttons are still down at the end of the cycle. To keep this first example simple, we have made no provisions for a manual setup mode or error detection routines. And although this example does have two push buttons it will not meet OSHA requirements, as the push buttons are not timed. We need to understand timers to create an anti-tie-down circuit.

Figure 8–13 shows the state chart for this program. This is similar to that shown in Chapter 4 when this programming problem was first explained. Note we have now shown only the conditions that are required for each state change and only the outputs or memory bits that are changed with each state change.

Figure 8–14 shows the flowchart created for this sequence. Note that most of the decision diamonds have the NO decision return line connected back to the top of the diamond. This indicates that the cycle waits until the decision condition is positive. We will now construct the ladder rungs to perform the function as specified.

The ladder program is shown in Figure 8–15. Rung #0 consists of three elements. From the flowchart we can see the need for both push-button inputs in this rung, but we also must provide some means of preventing the rung from restarting the sequence if the buttons are held down. To accomplish this, we have added a latch bit that will be set when cylinder

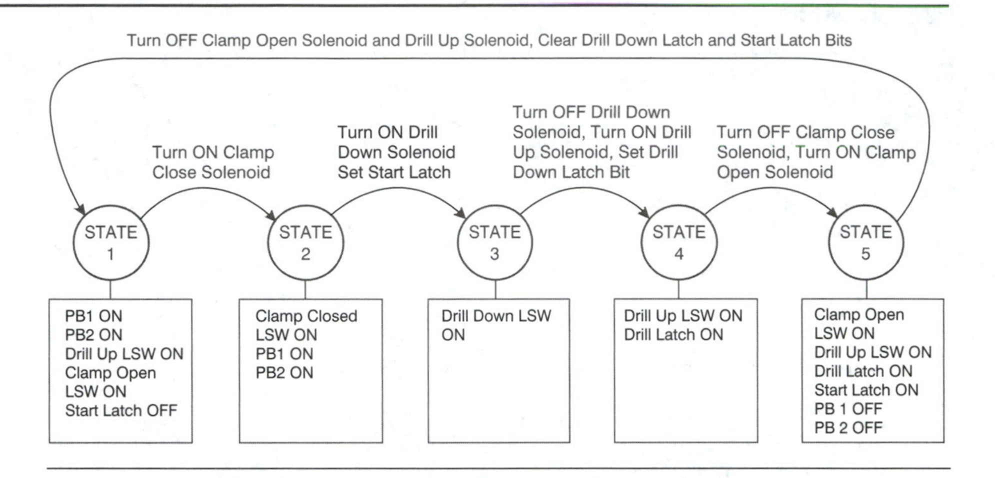

Figure 8–13 State chart for our theoretical machine

#1 (clamp closed) makes the extend prox switch. This will prevent the clamp from closing again if the buttons are held on after the cycle ends. Note we must latch B3/0 to ensure that it stays on until we decide to unlatch it later in the program. Rung #1 watches the clamp extend prox switch and latches B3/0 when it closes. Rung #2 starts the second cylinder (drill) down when the first is extended. We need to add an additional latch bit here to allow the rung to release when the drill reaches the down position. Note that input I:0.0/01 and B3/0 will still be on when we need to return the drill to the up position.

You can use a latch function any time the condition we are sensing will no longer be in effect when we need to act on it. Latching this bit on is very important because we need to hold this bit on even though the drill down prox switch will be open later in the cycle. Without latch bit B3/1 we would have both valves on and the cylinder would not move.

Rung #3 latches the B3/1 bit when the drill reaches the down prox switch. Rung #4 causes the drill cylinder to return to the up position. Note in rung #2 that latch bit B3/1 is set and prevents the output O:1.0/05 from being held on. Rung #5 causes the clamp to open when both the drill up prox switch is made and the latch bit B3/1 is on. Rung #6 ends the cycle by resetting both latches only if both push buttons are open.

In this example we have completed all the steps to program a project. We started with a sequence description, I/O assignments were already made, we then generated a state chart and flowchart, and wrote the program. The student should now download the program and verify that it works.

Note that the latch bits were used to keep track of a condition that had occurred but that had changed again before we needed to use it. This is a typical use for latch functions in ladder programming. Verify that

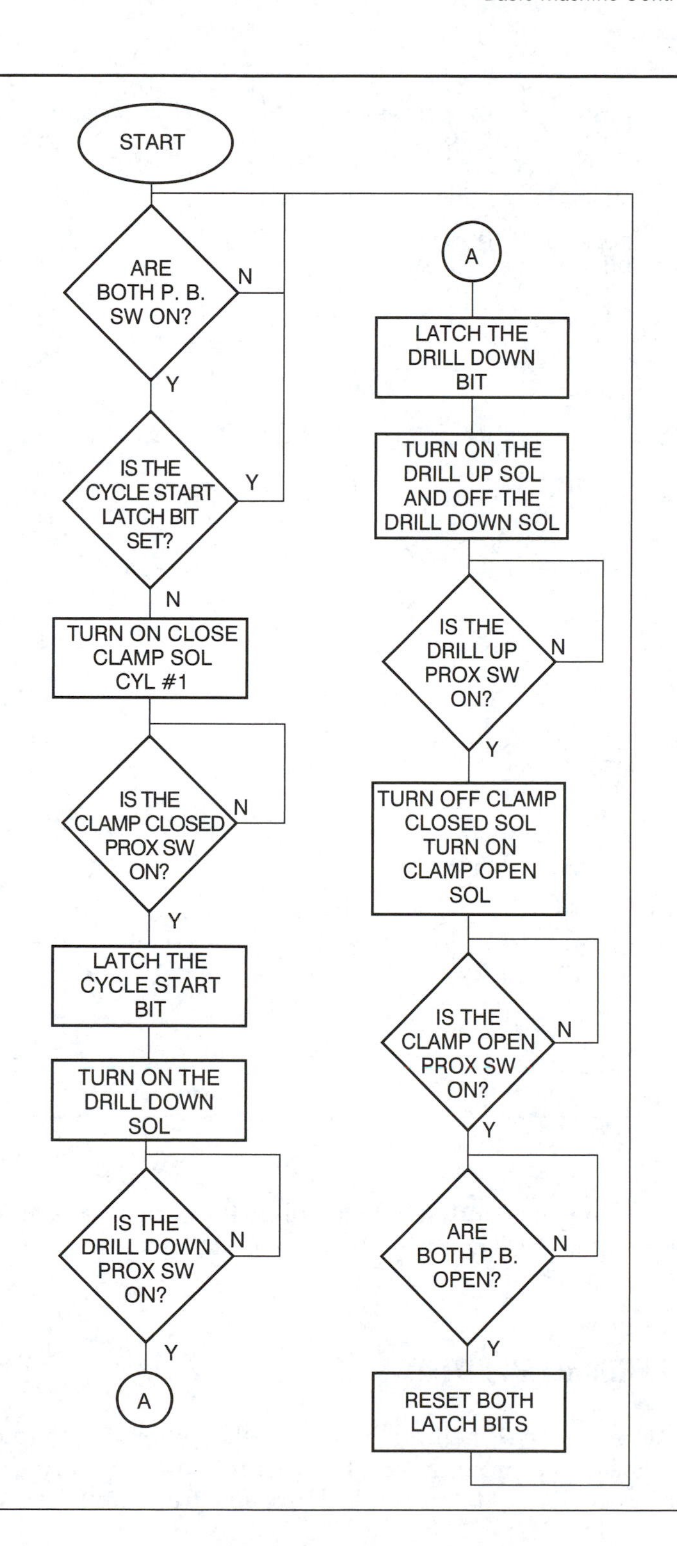

Figure 8–14
Flowchart for our theoretical machine

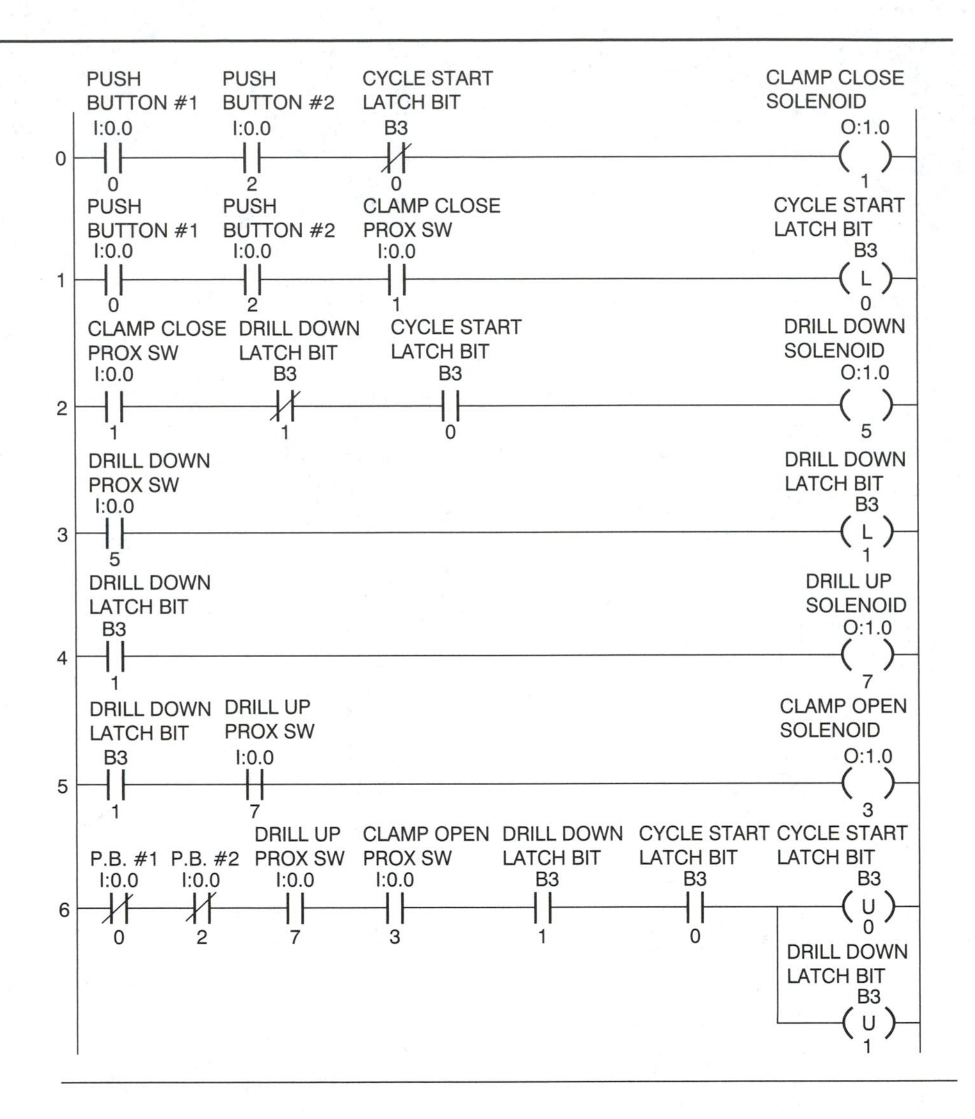

Figure 8–15 Ladder program for our theoretical machine

releasing only one push button will not allow the cycle to be started again. This will be important when we talk about OSHA requirements on manually activated machines.

PROGRAM TROUBLESHOOTING

I would like to restate something here we talked about in Chapter 5. The rungs in this sample program can be rearranged in any order and still work quite well. This is because all the rungs are scanned all the time.

The order in which they are written does not affect the operation of the program. People who are familiar with sequential programming languages always have a hard time with this concept.

I should also mention here that a common troubleshooting problem in PLC programs comes from this same scan process. If you accidentally use the same output address twice in a program, you will see a phenomenon in which the first rung may have all the inputs TRUE, but the output will stay OFF. This is because the second rung using the same output is not TRUE. Because the address bits set during a program scan are in memory, the last rung scanned before the I/O update wins. If this I/O phenomenon shows up in a program you have written, you should perform a search for the output address. In most cases you will more than likely find a second occurrence of the same output that is causing the problem. The only time an output address can be used more than once is if it is used in separate subroutines and only one of those subroutines is active at any given time.

CONCLUSIONS

In this chapter we have introduced the fundamental mechanical elements used in most automation equipment designs. The student should understand that this is only the basic list of devices and that many other mechanical devices exist and are commonly used in control projects. With these basic elements—the air cylinder and its control valve along with the motor-driven indexer—we can create very complex systems.

The example program provided is very limited in scope due to our limited understanding of the PLC's instruction set. It is intended as a first example of the logic necessary to control a mechanical system using a PLC controller. In the following chapters we will increase our understanding of PLC instructions including counters, timers, and various math and logic instructions. As we learn each new instruction type, we will write short example programs to demonstrate how these new instructions are used and what special control abilities they provide us.

REVIEW QUESTIONS

1. Why are single-acting air valves not recommended for controlling moving elements in a control system design?
2. What makes dual-acting air valves a better choice for controlling moving elements in a control system design?
3. Why should both the return and extend positions of an air cylinder be sensed with prox or limit switches?
4. In rotary indexers who decides how many degrees of rotation the unit will move during one position cam rotation?

5. Why must the control systems designer always stop an indexer in its dwell position?
6. What are the main advantages of using an encoder on a mechanism to signal its rotation position?
7. Why would we want to adjust the start position of an external mechanism in accordance with the speed of the controlling timing motor?
8. In the example program, why did we have to latch several status bits?
9. Rework the sample flowchart and state chart to add an additional drill and air cylinder that moves in a horizontal plane and operates after the main drill cycle has completed.
10. Rework the ladder program to sequence the machine with this additional step in the cycle.

LAB ASSIGNMENT

Objective

Introduction to latched output functions and the use of internal bit addresses as substitutes for complete functions.

1. Open a new file in RSLogix and save it as Chpt-8-lab. Although this lab was designed to run on the lab kit described in Appendix A, it is generic in form, and can be run on any lab kit by changing the I/O addresses. Be sure to enter the correct I/O configuration for the lab kit you are using. Enter the ladder program shown in Figure 8–16.
2. Download and run the program. Turn SEL SW 1 on and then press and release PB 1. Note that lamp 1 stays lighted. This is because we have used the lamp output address as a holding branch across the push button. Once the lamp is turned ON it holds itself ON until SEL SW 1 is turned off. This action resets the latch and allows the lamp to turn OFF. This is the typical relay style latch function often used in motor control circuits.

Figure 8–16

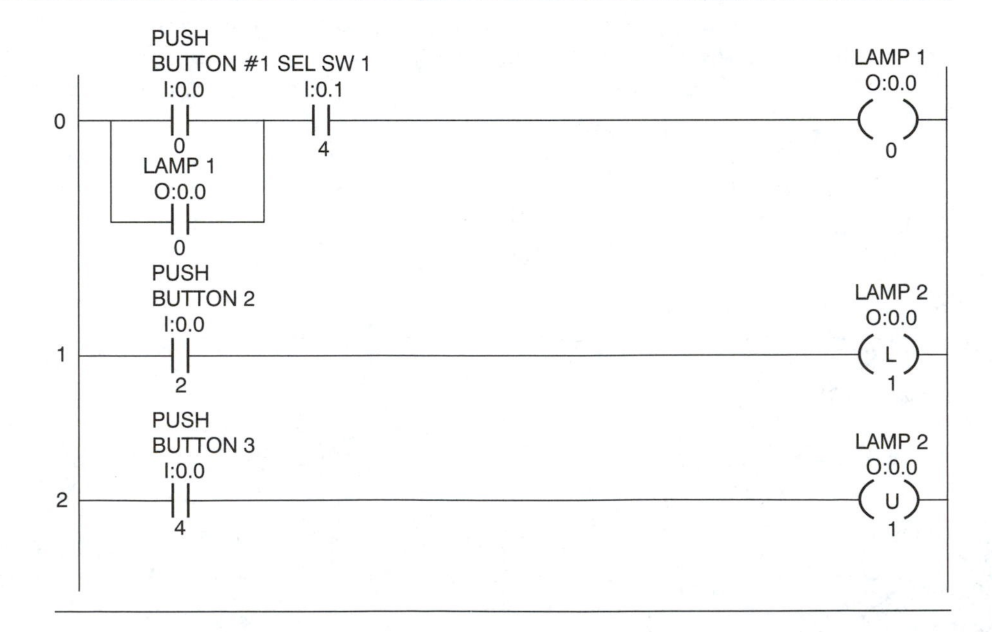

3. Let us now examine rungs 1 and 2. Press PB 2 and lamp 2 is turned ON. When PB 2 is released, note that lamp 2 stays lit. Pressing PB 3 will reset the latch. Note that when the latched output for lamp 2 is ON both the latch and unlatch functions are highlighted. This is because the highlight function is based on the address bit being set ON and not the fact that the function is active. This can be somewhat confusing when debugging a ladder program.
4. Switch to offline mode and edit the program by adding a fourth rung as shown in Figure 8–17.
5. Download the program and run it. Verify that you now have two different functions that will reset the latch. Turn ON SEL SW 1 and then press PB 2. Notice that the latch function will not turn on or highlight even though the rung conditions are TRUE. This is because rung 3 is holding the output unlatched.

Debugging Tip It is good to remember the phenomena we have seen here, because you may see this same problem again when debugging a longer program. Any time you see an output that will not turn ON when the rung conditions are TRUE, look for a rung that is holding that same output OFF farther down in the program.

6. We will now explore function substitution using internal bits. Enter the new rung as shown in Figure 8–18. Download and run the program. Verify that lamp 3 is turned ON only if all three selector switches are ON and one of the three push buttons is pressed.
7. Let us assume for this lab that the three selector switches represent a functional condition that we will need in at least 10 different additional rungs in our program. Let us also say that the three parallel push buttons also represent a functional condition that will be needed in a number of rungs within the program. Rather than

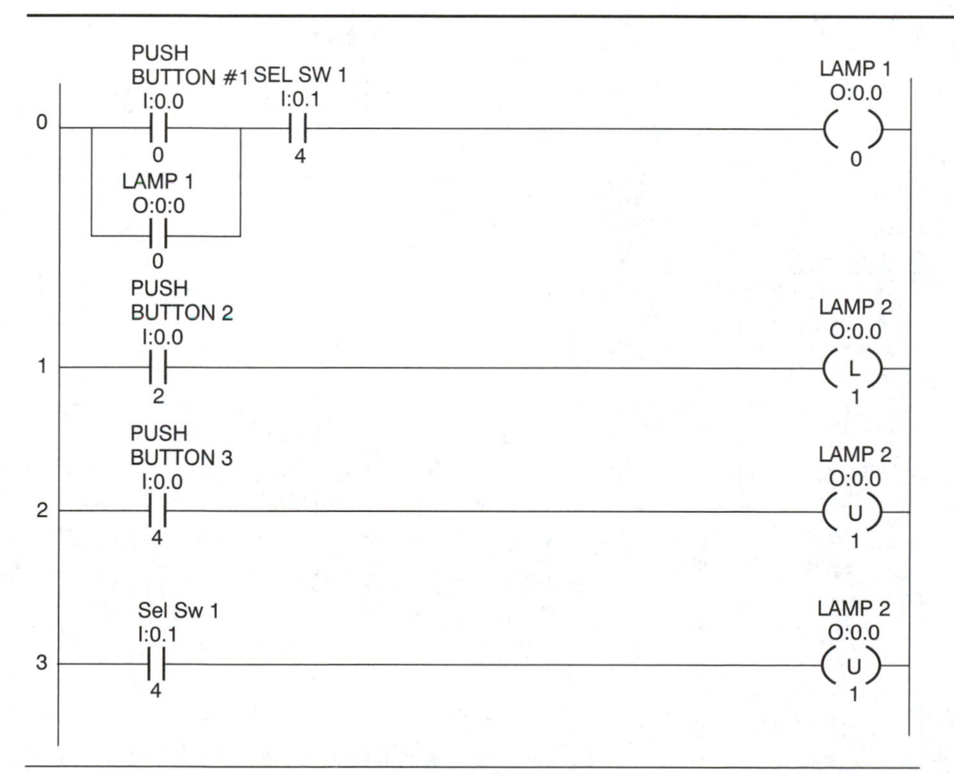

Figure 8–17

Figure 8–18

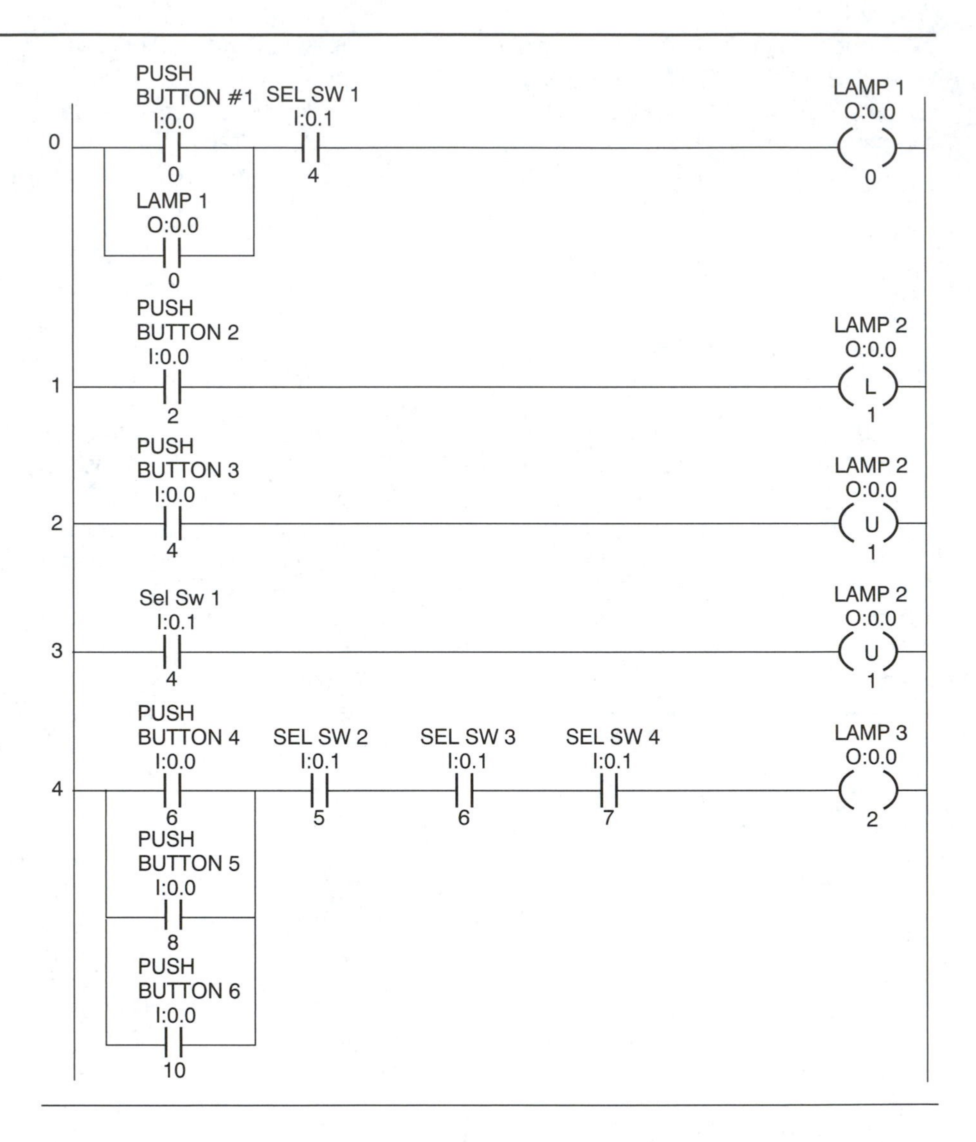

reproducing this same exact three-element function many times within the program, let us assign an internal bit address to this function. We can then substitute the internal address bit in the rungs where the function is needed. Let us rework rung 4 into three separate rungs as shown in Figure 8–19 to demonstrate this idea.

8. Download and run the program. Verify that rung 4 performs exactly the same as it did in the previous version of the program. We now have the advantage of being able to insert B3/0 or B3/1 into any rung where these functions are needed. Because each element in a rung represents one memory location in the PLC's memory, this substitution method can save a lot of memory in a large program.

9. Write a program to extend cylinder 5 when push button 8 is pressed. Hold cylinder 5 extended until push button 7 is pressed to return cylinder 5 to the home position. Cylinder 5 extend output is O:1.0/4, push button 8 input is I:0.0/14, and push button 7 input is I:0.0/12.

10. Set up the report options the same as you did in the lab for Chapter 7 and then print out a copy of your program.

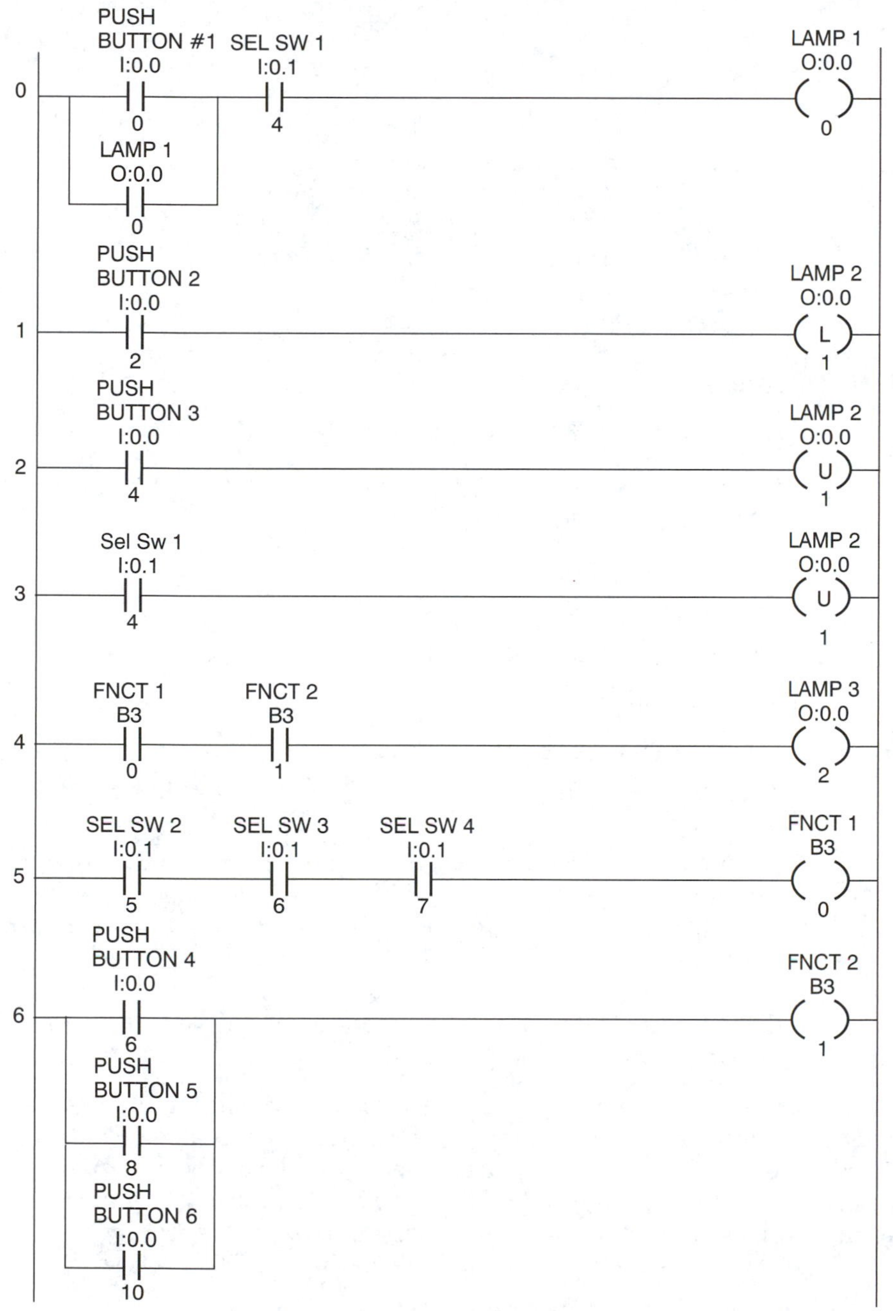

Figure 8–19

PUSH BUTTON #1 I:0.0 0
SEL SW 1 I:0.1 4
LAMP 1 O:0.0 0
LAMP 1 O:0.0 0
PUSH BUTTON 2 I:0.0 2
LAMP 2 O:0.0 L 1
PUSH BUTTON 3 I:0.0 4
LAMP 2 O:0.0 U 1
Sel Sw 1 I:0.1 4
LAMP 2 O:0.0 U 1
FNCT 1 B3 0
FNCT 2 B3 1
LAMP 3 O:0.0 2
SEL SW 2 I:0.1 5
SEL SW 3 I:0.1 6
SEL SW 4 I:0.1 7
FNCT 1 B3 0
PUSH BUTTON 4 I:0.0 6
FNCT 2 B3 1
PUSH BUTTON 5 I:0.0 8
PUSH BUTTON 6 I:0.0 10
0
1
2
3
4
5
6

CHAPTER

9

Timer Instructions

LEARNING OBJECTIVES

After completing this chapter the reader should understand:

1. The basic functions of TON, TOF, and RTO timer instructions.
2. The use of timers to control indexer functions.
3. How to create timing sequences.
4. The use of timers to create OSHA-approved anti-tie-down circuits.
5. Using RTO instructions to provide mechanism cycle times for diagnostics.
6. How to view and edit the timer files.

Timer instructions are used to generate a number of different functions in PLC system programs. Timer instructions can be used to generate the following functions:

1. Generate long, accurate pulse times from short, random-length input signals.
2. Create short pulses on the front or trailing edges of long-duration input signals.
3. Create pulse trains or timing signals.
4. Delay the start of a function, an accurate amount of time, from an input or control bit state change.

As we stated in Chapter 5, file 4 is assigned by default to timers when we open a new project. Additional timer files may be created in file areas 10 and above. Be aware that if you enter a timer instruction and give it an address above file 9, for example, T12:0, you have assigned the entire file 12 to timers. This is TRUE for any file address above 9.

Three timer instructions are available in the SLC-500 processor:

1. TON stands for TIMER DELAY ON.
2. TOF stands for TIMER DELAY OFF.
3. RTO stands for RETENTIVE TIMER DELAY ON.

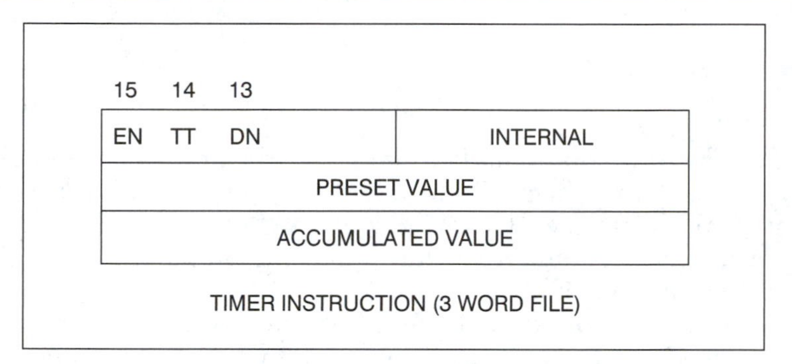

Figure 9–1
Timer instruction file word allocation

Each time you add a new timer instruction it occupies three words of the timer file. The first word in the group contains three status bits, EN, TT, and DN, and they can be used to control other functions in your program (Figure 9–1). The last two words contain the preset time for the timer and the actual accumulated time.

Bit EN is the enable bit and it is on any time the rung conditions controlling the timer are TRUE. Bit TT is the timing bit and it is on any time the timer is on, and the accumulated value is less than the preset value. Bit DN is the done bit and it is on when the accumulated value is equal to or greater than the preset value in TON timers and for the total time in TOF timers.

TON INSTRUCTION

As stated earlier, this is a delay ON timer. The instruction can have one or multiple input conditions. In program mode, when you enter the timer instruction, you will be asked to assign a timer number and a time base. The default time base increment is 0.01 seconds. On rack-based processors you have two additional optional settings of 1.0-second and 0.001-second increments. You will then be asked to enter a preset time. This preset value will be multiplied by the time base increment to give the actual delay time. For example, a preset value of 500 and an increment setting of 0.01 seconds will give a time delay value of 5 seconds (Figure 9–2).

Figure 9–2
TON instruction

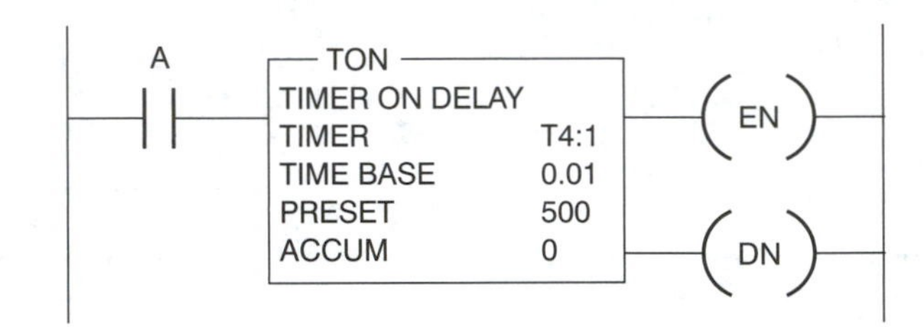

When the rung conditions become TRUE, the timer's EN and TT bits will go on. The timer begins to accumulate time, and will display this accumulated time in the ACCUM position within the timer instruction block. When the accumulated value equals the preset value, the TT bit turns off and the DN bit turns on.

The timer now stops until the rung condition goes FALSE. When that happens, the EN and DN bits turn off and the accumulated value is reset back to 000. From the timing diagram of Figure 9–3, we can see that the time delay period is from the instant the rung goes TRUE until the DN bit is turned on. Note that if the rung goes FALSE before the time-out period is up, the time accumulated is reset to 000 and the DN bit never comes on.

TON timers can be used to create a short pulse at the beginning of a longer input condition or to delay the start of a function for a defined period of time from the start of some other function. Timer reset instructions can be used on TON timers to force a reset even if the rung is still

Figure 9–3
Timing diagram for TON instruction

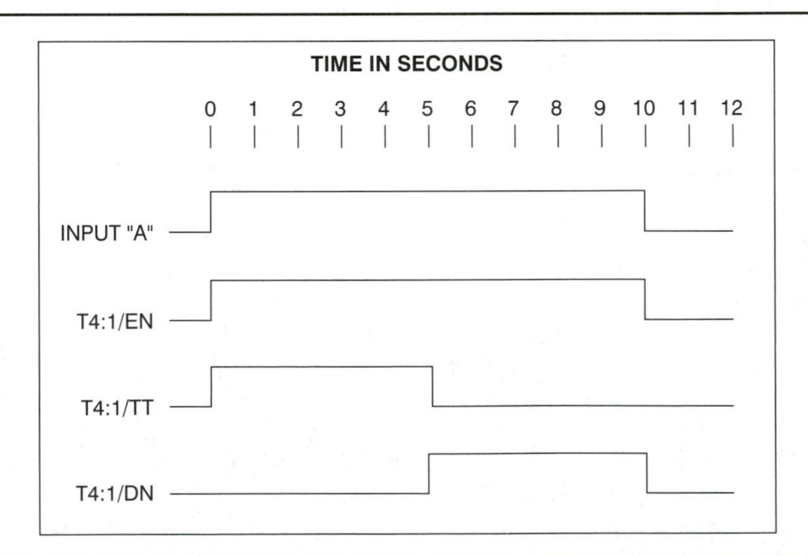

TRUE. Examples of how this TON instruction is used in machine control programs are shown at the end of this chapter.

TOF INSTRUCTION

The TOF timer instruction is a timer delay OFF function. You enter it the same way you do the TON instruction as explained in the preceding section (Figure 9–4). The main difference is that the DN bit turns on as soon as the rung condition goes TRUE. When the rung goes FALSE, the DN bit stays on and the timer begins to accumulate time for the delay period and then turns off the DN bit.

From the timing diagram of Figure 9–5, you can see that the time delay period is measured from the time the rung goes FALSE until the DN bit turns off. The TT bit goes on when the rung goes FALSE and stays on for the duration of the preset time and then resets. While the TT bit is on, the

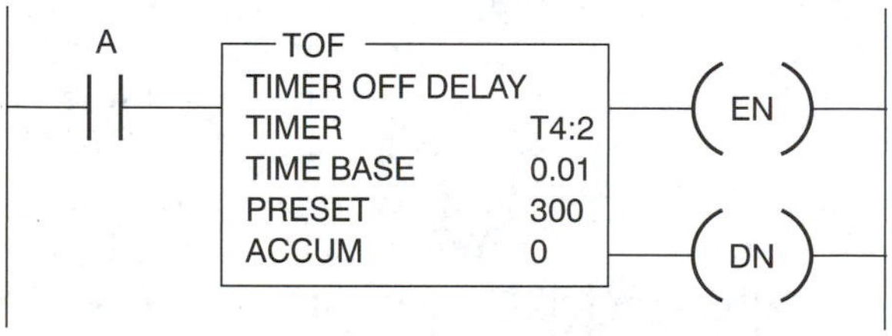

A
TOF
TIMER OFF DELAY
TIMER T4:2
TIME BASE 0.01
PRESET 300
ACCUM 0
EN
DN

Figure 9–4
TOF instruction

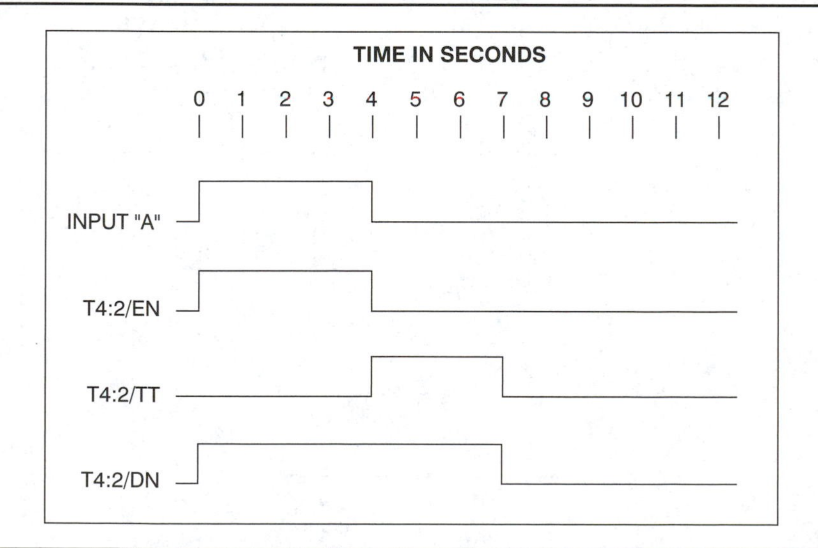

Figure 9–5
Timing diagram for the TOF instruction

accumulated time is displayed in the ACCUM position inside the timer instruction block. This accumulated time is reset when the timer times out and the DN bit turns off. Timer reset instructions (RES) cannot be used with TOF timers.

TOF instructions can be used to create longer output functions derived from short input functions. This is accomplished by using the timer's DN bit as the control bit in another rung. We can also use the TOF timer to generate a short pulse at the end of a long input function by using the TT bit. Both of these functions are explained in the section on timer use in machine control.

RTO INSTRUCTION

RTO instructions are a special case use of the typical TON instruction. The difference arises in that when an RTO-type instruction's input rung conditions go FALSE, the accumulated time is *not* reset. All the status bits work the same as in a TON instruction (Figures 9–6 and 9–7). Because the timer does not reset when its control rung goes FALSE, we need to use a reset instruction (RES) to clear it. When using RTO instructions, you must be careful to always have a corresponding reset instruction for each timer. As you can see in Figure 9–7 the TT bit follows the EN bit until the accumulated time is greater than the preset time. Once that point is reached the TT bit is no longer active even though the timer continues to accumulate time. Note that the DN bit stays on until the reset instruction is used even if the rung condition goes FALSE.

RTO instructions can also be used to accumulate time for maintenance functions and also for diagnostic programs. To use the timer in this fashion you set the time for the maximum delay time 9999. When the RTO timer's input goes TRUE the timer will begin to accumulate time. When the input goes FALSE the timer will contain the total time the input was

Figure 9–6
RTO instruction

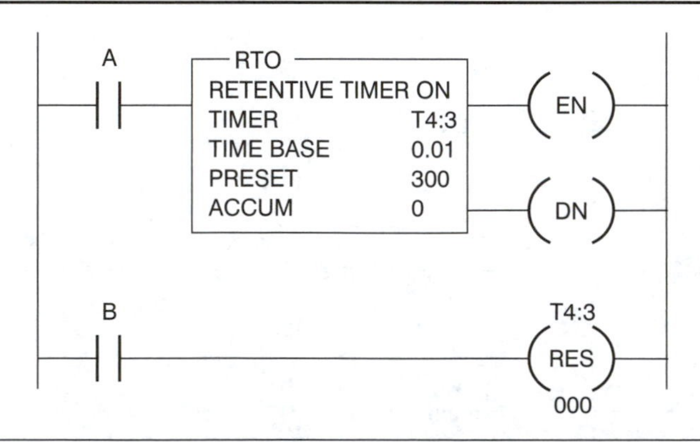

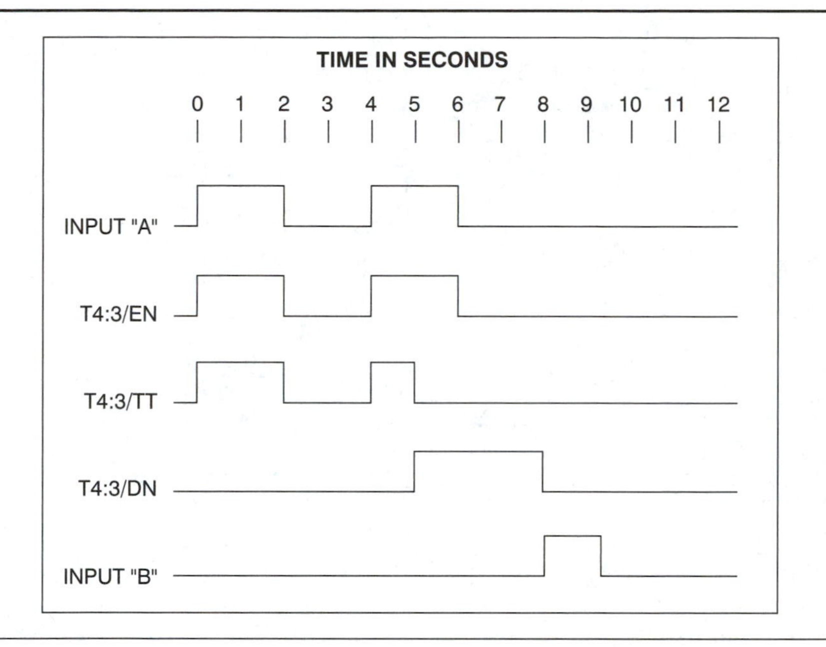

Figure 9–7
Timing diagram for the RTO instruction

TRUE. You then need to transfer this accumulated time value to some other memory location and reset the timer.

USE OF TIMER INSTRUCTIONS IN MACHINE CONTROL

In the electrical world we tend to think of events happening almost instantaneously. In the mechanical world of machine control, we often need to delay events to give the mechanical components time to complete their movements. A good example of this is a prox switch or Hall effect switch often used to sense the end of stroke in air cylinders. This switch is usually set to come on when the cylinder is almost at the end of travel. When this switch activates, the cylinder may still be moving for a hundred milliseconds or so before it hits the physical end stop. If we were to act on the switch input, at computer speeds, we may start the next motion before the first is really completed. This can also be a problem if we are using the air cylinder to clamp an item. Even when we have hit the end of stroke, the air cylinder takes some time to build up to full pressure to hold the clamped item. If this is undesirable, we need to delay for a short time between the two different mechanical motions, and the TON timer is just what we need for the job. Instead of using the actual prox switch input to start the next function, we can use it as an input condition on a TON timer instruction and use its DN bit to start the next function. In this way

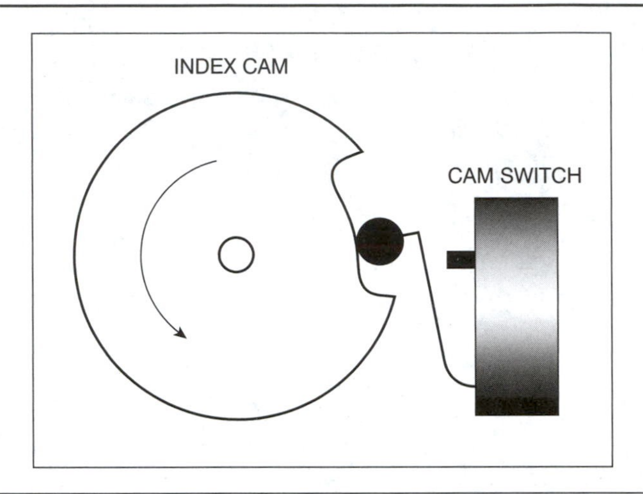

Figure 9–8
Index cam with limit switch

we can tune the delay between motions by adjusting the preset on the timer instruction.

Another example of the use of TON timers is the typical index circuit. Many automation projects contain an indexer of some sort. As discussed in Chapter 8, indexers are rotary devices that move a line or dial plate some defined distance for each turn of the shaft. These devices generally have an index cam with a limit switch riding the cam (Figure 9–8).

The cam switch is in the dwell position until the indexer is energized. It then moves up on the high lobe of the cam and the index is complete when the switch falls back into the dwell position. From a control system designer's point of view, the problem with this arrangement is that the start condition is the same as the complete condition. The problem is even worse if the logic condition used to start the index will still be present after the index is complete. We must be sure we do not start a second index until the input goes off and returns to on again.

The best way to handle this design problem is to use a TON timer to start the index cycle and then let the cam switch stop the cycle. The timer must stay on long enough to get the switch on the high side of the cam lobe. If we are given a total index time of 3.0 seconds, and the time to get out of dwell is .5 seconds, then the timer should stay on for about 1.5 seconds. This is long enough to get the cam switch out of the dwell position, but less than the total index time. (Actually any value between 0.7 seconds and 2.5 seconds will work.) A sample timing diagram and ladder logic rungs for the function are shown in Figures 9–9 and 9–10, respectively. Note how the TT bit starts the cycle and the switch holds the motor on until the dwell position is reached. Because the timer TT bit cannot go on again until the input function has gone off and back on again, we do not have to worry about a repeat index cycle.

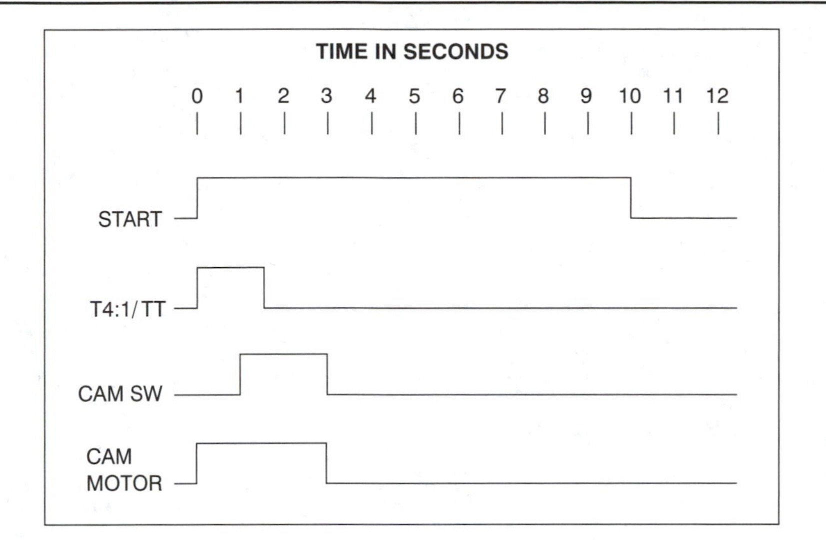

Figure 9–9
Sample timing diagram of an indexer function

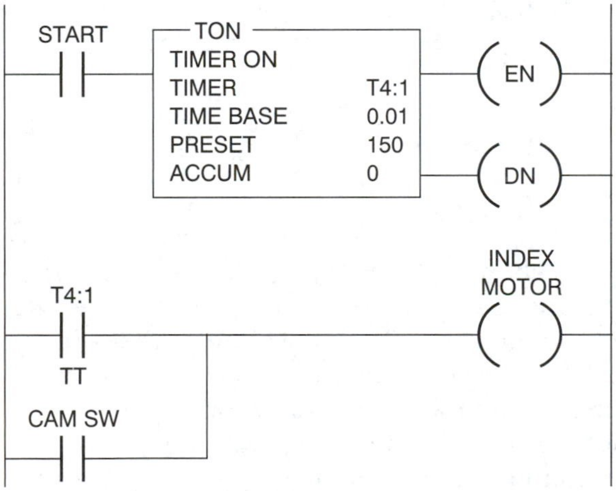

START
TON
TIMER ON
TIMER T4:1
TIME BASE 0.01
PRESET 150
ACCUM 0
EN
DN
T4:1
TT
CAM SW
INDEX
MOTOR

Figure 9–10
Ladder logic rungs to control indexer

In the previous example we used a timer to give a short start pulse at the beginning of a long control function. This concept can be used in a number of ways in other parts of our ladder program. If we use the preceding program to start an index cycle, then the next thing we need to develop is a pulse to tell the rest of the system that the index is complete and it can start the operational functions. We can do this by adding a second timer that is controlled by the cam switch. We will set up this second timer so that it is activated when the cam switch opens at the dwell point. To do this we will use the cam switch NOT function in the control rung (Figure 9–11).

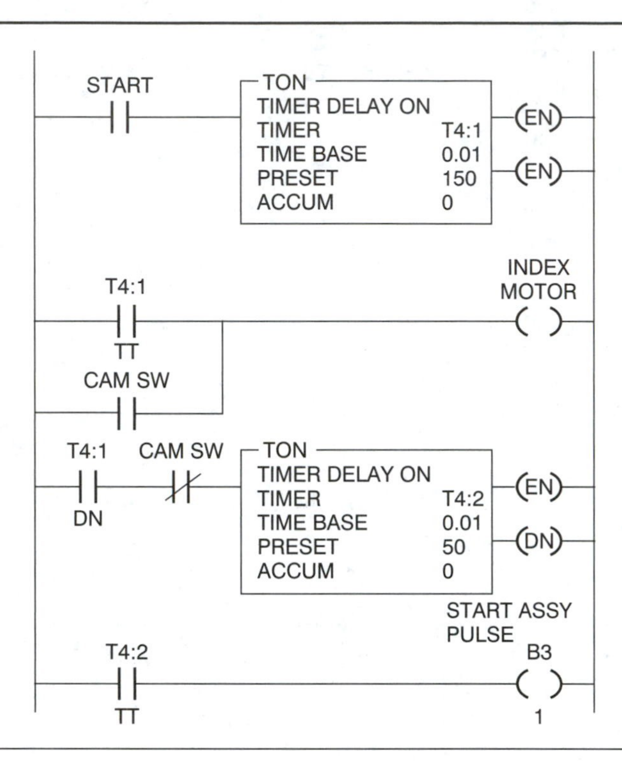

Figure 9–11
Ladder to control indexer and provide start pulse for assembly operation

In this new program the T4:2/TT bit is now used to start the assembly process after the index is complete. We decided a 0.5-second pulse is enough to get the assembly cycle started. When the assembly cycle is complete we no longer have to worry about a repeat cycle because the 0.5-second start pulse will be gone. This start pulse method is one typical way in which index control systems are programmed.

Another use of the TON timer instruction is for diagnostic problem sensing for mechanical elements in our designs. We cover this type of programming in detail in Chapter 16, but I want to explain the timer use here. If we have a mechanism that normally operates in a 5.5-second time window, we need a way of knowing if this mechanism is jammed or no longer working. One way to perform this task is to use an input or group of inputs that represents the home position of the mechanism as a NOT input function to a TON timer instruction. If, as we previously stated, the normal working time of this mechanism is 5.5 seconds, we should set the timer preset at 20% to 30% longer than the normal time. This would work out to be about 7 seconds. Therefore, we will construct a timer rung where the input condition will go TRUE if the mechanism is not in its home position and assign a preset time of 7 seconds.

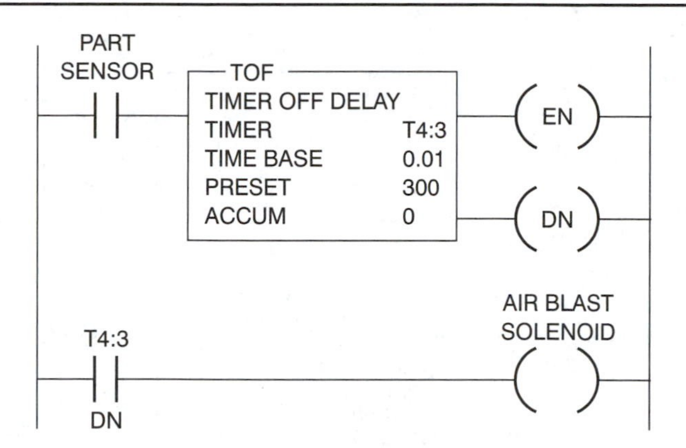

Figure 9–12
Sensor input to start a TOF timer

Under normal operating condition the timer should never time out because the mechanism should return home before the 7-second delay period is up and reset the timer's accumulated value to 000. If for some reason the mechanism jams or becomes inoperative once it starts its cycle, the timer will time out and we can use the DN bit to set off an alarm function to warn the operator of the malfunction.

We now provide an example of a TOF instruction being used in a machine control program. A typical design problem occurs when you must sense a part and then perform a function even after the part is past the sensor. An example would be when you have a part moving down an assembly belt and you must use an air blast to push the part off the belt and down a packaging chute. The sensor that starts the unload function will be cleared as soon as the part begins to move. Our air blast must stay on for additional time to be sure the part moves all the way into the unload chute. To do this we will use the sensor input to start a TOF timer and use its DN bit to control the air blast solenoid (Figure 9–12).

The timer is set for 3 seconds to ensure that the part moves completely off the belt after the part sensor opens. Note that the DN bit will come on as soon as the part sensor activates and stays on for 3 seconds after the sensor opens. This is a typical application of the TOF timer function used in machine control. Note that in this function the TOF timer is stretching the sensor pulse to provide a longer signal time.

The RTO function is commonly used in machine control to generate times for diagnostic functions and reports. Because it has the ability to hold accumulated time until reset, the time value can be copied to a memory location for display on a CRT monitor or it can be printed. In Chapter 16 we will deal with diagnostic programming and display methods, so I do not elaborate on this function here.

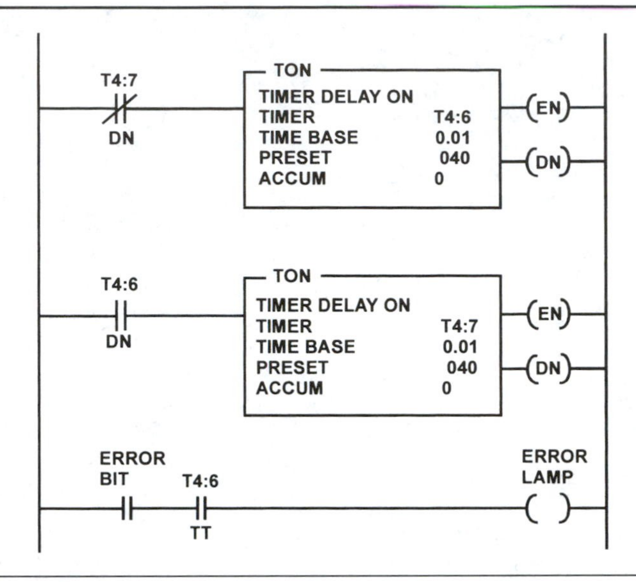

Figure 9–13
Two timers connected to form a pulse generator

OTHER TIMER USES

Another example of timer functions is shown in Figure 9–13. In this circuit we have cascaded two timers to generate a square-wave pattern with a 0.4-second on/off cycle. This circuit is very useful in our programs because the element T4:6/TT can be ANDed with any rung elements to get a flashing lamp or output as shown in the third rung of Figure 9–13. Most system designs should contain the top two rungs in the main section of the program. This allows the programmer to flash any output by inserting the timer's TT function in the control rung.

A timing diagram of the two timer functions is shown in Figure 9–14. Note that the T4:7/DN bit only goes high for one scan time. Because this bit resets timer T4:6, and the DN bit for T4:6 is what is controlling T4:7, it

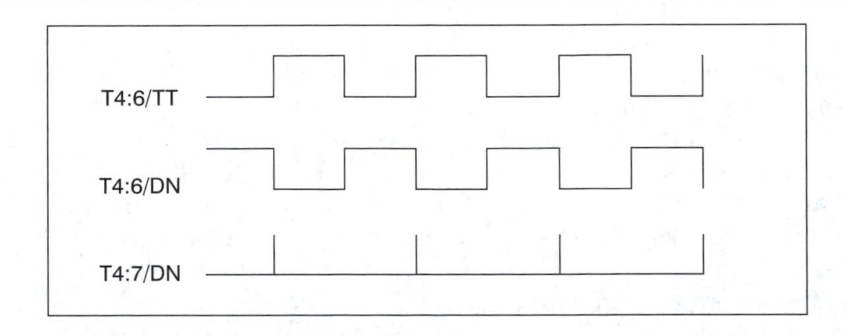

Figure 9–14
Timing diagram for the two timer pulse generator functions

then resets itself. If you are monitoring this function on the computer screen, you will probably never see the T4:7/DN bit light. This is because the computer is displaying a time slice of the PLC's status. Because the computer works over a serial port, the time between screen updates may be on the order of 100 milliseconds or more. The odds of catching the DN bit set on are very small. We can see that the function is working by noting that T4:7 & T4:6 ACCUM amounts are reset to 0000. This computer display update delay time can be a problem any time we are dealing with short pulse times.

The example circuit we look at next is used to time manual cycle start push buttons on a manually activated machine. OSHA states that one way to protect an operator of a manually activated machine is to have two start cycle push buttons and they both must be pressed within a small time window to activate the machine cycle. This prevents the operator from placing a brick or other heavy object on one push button and operating the machine using only one hand. This practice can result in the operator having only one hand attached to work with forever more! In addition, both buttons must be released before a new cycle can be started. This will prevent the machine from double cycling if the operator continues to hold the start buttons down after the cycle is complete.

To accomplish this timing task, we connect each start push button to a separate timer (Figure 9–15). Set the preset of each timer for the allowable time window, in this example, 0.5 seconds. In the third rung we AND the two timers' TT bits and latch an output memory bit B3/12 only if the TT bits are both on at the same time.

The fourth rung then turns on the start cycle bit B3/13, which will begin the machine cycle. In rung 6 the program will pulse "last step complete" when the last mechanical step of the machine has been completed. (Note that this function is not part of this ladder example.) This latches the cycle done bit B3/9, which in turn shuts off the start cycle output bit B3/13. The program stops until both push buttons are released. Rung 5 senses that both push buttons are open and the cycle is done; it then unlatches the cycle start latch bit B3/12. This in turn unlatches the cycle done bit B3/9 in rung 7. Note that in some applications the operator may also have to hold the buttons down until some type of shield moves into place. This program example is designed to show the anti-tie-down function, but the reader should understand that each application must be analyzed and programmed to provide a safe operator environment.

Looking at this program example you can see that the operator must press both start push buttons within the 0.5-second window or the B3/12 start cycle bit will not latch and start the machine. Also the cycle will stop even if the operator is still holding down both buttons at the end of the cycle because the cycle done bit B3/9 shuts down the machine until reset. The only way to reset this bit is to have both buttons released. This is only one example of how timers can be used in safety circuits to prevent injury to an operator.

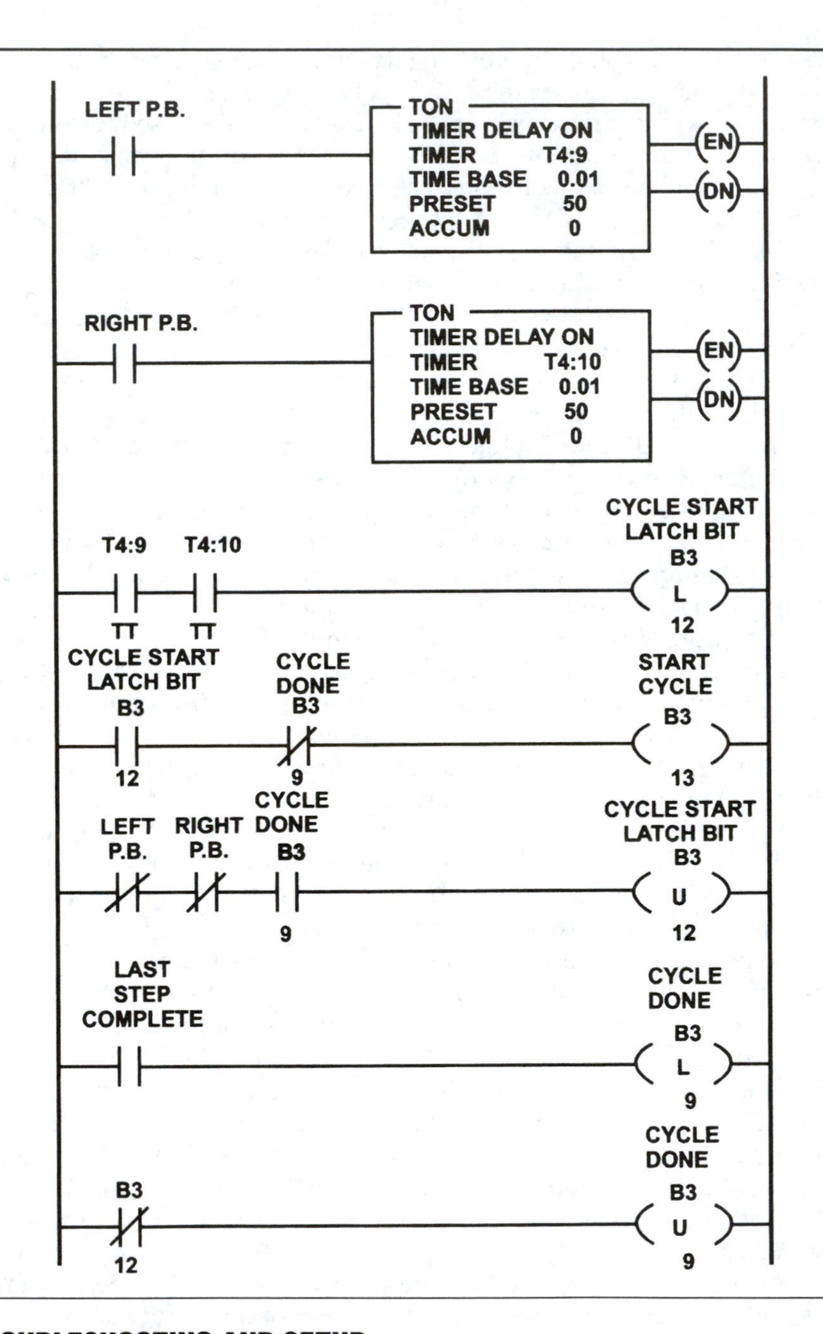

Figure 9–15
Ladder diagram to meet OSHA's anti-tie-down rule

TROUBLESHOOTING AND SETUP

Timers can be monitored in two different ways in the SLC-500 system. While in the online mode you can watch the timer's ACCUM value increment on the screen in the timer instruction box. This is fine if you're

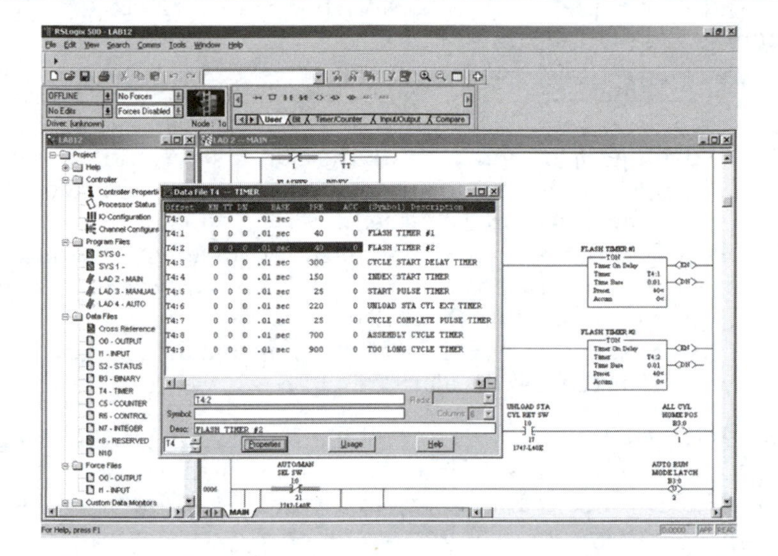

Figure 9–16
Screen printout showing timer file pop-up window

only interested in monitoring one timer at a time. However, if you need to observe multiple timer values or if you wish to change the preset or ACCUM values for any of the timers in your program, you must do so in the file mode. To get to the timer file you need to double-click on the T4 file folder along the left side of the monitor screen. If you have additional timer files you would double-click on the file you wish to view.

Figure 9–16 shows the timer pop-up window that displays all the timers used in this program. You can change any timer value while in run mode by clicking on the value and entering a new value. This is handy if you want to fine-tune your program or when you have a very long time-out period and you want to force the timer to time out early to debug part of the program. You can also monitor multiple timers at the same time by opening this window. Another advantage of this window is realized when you are adding a new timer to your program and need to know what was the last timer used. By opening this pop-up window the programmer can see what timers are used and what names have been assigned to each.

One word of caution: if you change timer values while you are online and in run mode, the changed value is only resident in the PLC. If you exit online mode and make some program changes offline, and then download your changes, the original timer values will be reloaded into the PLC. You can prevent this problem if you save the program while you are still online. This saves the timer value changes you made back to your host computer.

Now that we have a basic understanding of the various timer instructions let us construct a program that uses the new information we have learned. Let us expand on the sample program that we designed at the end of Chapter 8. We will add the following features to that program:

1. Use the timed method to achieve a dual-button safety cycle start.
2. Add an oil pump to the design that will lubricate the piece during the drilling operation. Be sure the pump stays on for 2 seconds after the drill hits the down position to keep it lubricated as the drill is removed from the piece.
3. Check for a damaged drill bit. Time the drill down cycle and cause an alarm if the drill has not finished the down cycle in 10 seconds. If this is the case, abort the drill down cycle and return everything to the start position. Light the alarm lamp and do not allow another cycle to start unless the alarm reset PB is pressed.

Figures 9–17 and 9–18 show the sample program. Rungs 0 and 1 are our start cycle timers and are set for 0.5 seconds. Rung 2 will latch the cycle start bit B3/0 if both buttons are pressed within the 0.5-second time window. Note that if one button is pressed too slowly, both must be released and pressed again to get the cycle started. The alarm bit is included to stop the cycle from being restarted if an alarm condition has occurred.

Rung 3 closes the clamp solenoid if the cycle start latch bit is on and the drill has not been down. The B3/1 not element is necessary to release the clamp close solenoid since the B3/0 bit will still be on when we want to release the clamp. Rung 4 starts the drill down once the clamp is closed. This output bit is shut off by the alarm latch bit B3/3 or the drill down latch bit B3/1.

Rung 5 starts the oil feed motor timer and this TOF timer will keep it on for the required 2 additional seconds. Rung 6 turns on the oil feed motor output any time the T4:2 DN bit is on.

Rung 7 times the length of the drill down cycle. The timer should always be reset before the time-out preset time elapses. Rung 8 latches the alarm bit B3/3 if the T4:3 timer times out. Rung 9 unlatches the alarm latch bit if the alarm reset push button is pressed.

Rung 10 latches the drill down latch bit when the drill down prox switch activates. It is important that this bit be latched because the drill down prox switch will release as soon as the drill head begins to move up. Rung 11 starts the drill up solenoid when the drill down latch has been set or if the alarm bit is set from the too long cycle timer. Rung 12 opens the clamp if the drill down latch bit is set and the drill is in the up position, or if the alarm bit is set.

Rung 13 detects the end of cycle. It unlatches the two cycle latch bits, B3/0 and B3/1, only if the drill is up and the clamp is open and both start push buttons are open. This rung releases the cycle and allows the push buttons to start a new cycle.

Rung 14 turns on the alarm lamp if the alarm latch bit is set.

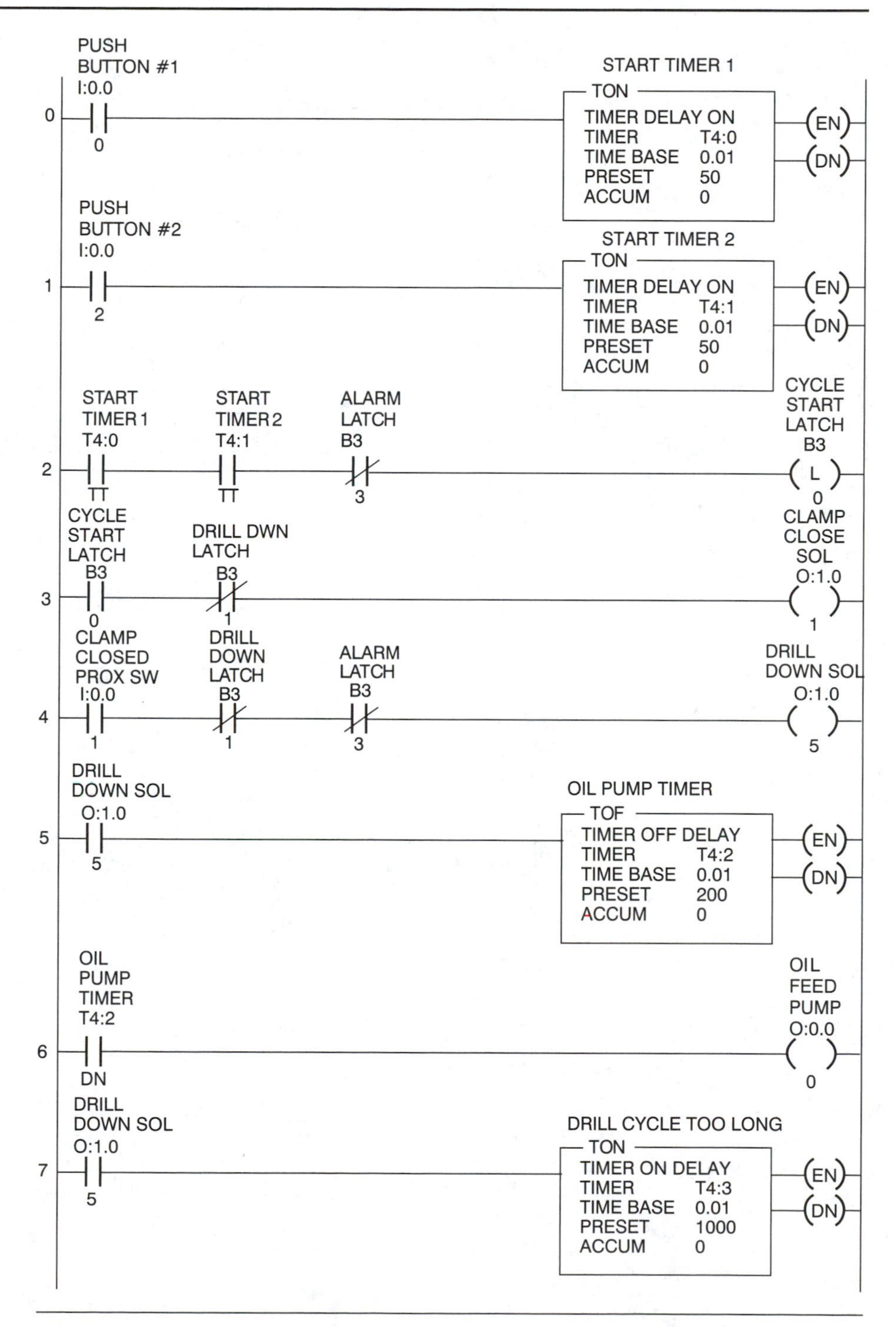

Figure 9–17
Ladder diagram for sample program

Figure 9–18
Ladder diagram continued for sample program

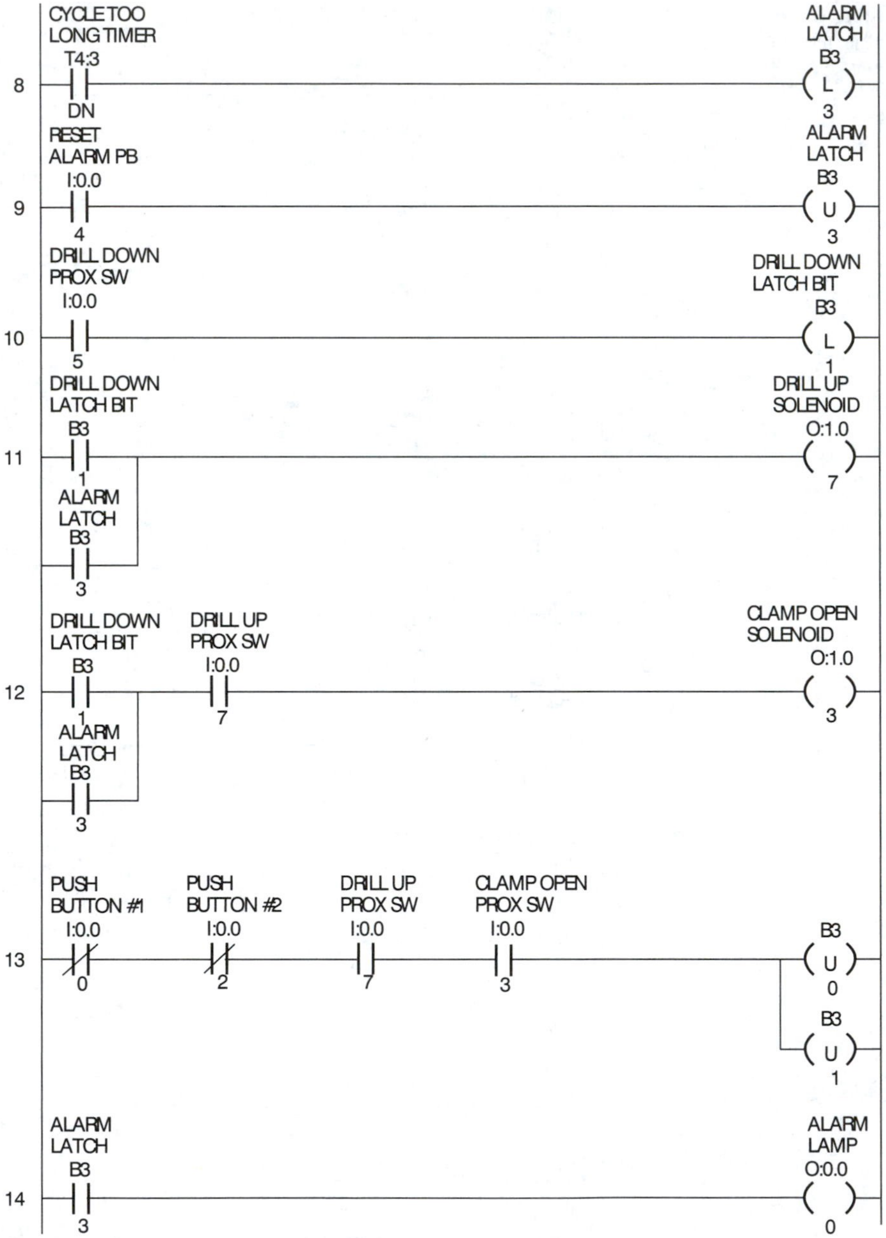

8
CYCLE TOO LONG TIMER T4:3 DN
ALARM LATCH B3 (L) 3

9
RESET ALARM PB I:0.0 4
ALARM LATCH B3 (U) 3

10
DRILL DOWN PROX SW I:0.0 5
DRILL DOWN LATCH BIT B3 (L) 1

11
DRILL DOWN LATCH BIT B3 1
ALARM LATCH B3 3
DRILL UP SOLENOID O:1.0 ( ) 7

12
DRILL DOWN LATCH BIT B3 1
ALARM LATCH B3 3
DRILL UP PROX SW I:0.0 7
CLAMP OPEN SOLENOID O:1.0 ( ) 3

13
PUSH BUTTON #1 I:0.0 0
PUSH BUTTON #2 I:0.0 2
DRILL UP PROX SW I:0.0 7
CLAMP OPEN PROX SW I:0.0 3
B3 (U) 0
B3 (U) 1

14
ALARM LATCH B3 3
ALARM LAMP O:0.0 ( ) 0

CONCLUSIONS

The addition of timer instructions to our bag of programming functions gives us a powerful tool to use in control programming. I have shown only a small sampling of the many uses you will find for these instructions.

A few words of caution to new programmers. It is never a good idea to cascade timers together to build a timing sequence for controlling a mechanism. Programmers new to the field may see this as a quick and easy way to sequence a mechanism through its cycle without requiring much in the way of feedback devices. While it is possible to create such a programming sequence, it will be a constant problem to maintain. A programmer may well find a set of timing values to sequence the mechanism through its cycle for a short period of time. Unfortunately, without feedback devices like cylinder position-sensing switches, any change in air pressure or wear on the mechanical elements will cause timing changes that will result in jam-ups. Once a jam-up occurs the program will go right on sequencing the mechanism and may cause additional damage to the system. Programs written in this manner will need constant tweaking of the timing values to keep the machine running. An additional problem is that the mechanisms will probably run slower than with a feedback-type program because the ladder programmer has to add additional time to each step to ensure that the previous step has completed.

Although there are many good uses for timer instructions, the programmer must be careful not to use them where mechanical variables may cause the time settings to constantly change or drift. The cost to add the additional feedback devices to a mechanism is more than offset by the maintenance headache you will generate trying to control a mechanism with only timed steps.

REVIEW QUESTIONS

1. How many words of memory does each timer instruction require?
2. Explain the function of the TT, DN, and EN status bits on a TON instruction?
3. How does the function of the DN status bit differ between the TON and TOF timer instructions?
4. Why do you have to reset the accumulated time value with a separate reset instruction when using RTO timers?
5. What are the three time base increments available in the rack-style SLC-500 units?
6. Fill in the timing chart (Figure 9–20) for the timer instruction shown in Figure 9–19.
7. What will be the status of the DN bit on a TON timer if the preset is set for 5 seconds and the input condition cycles TRUE for 4 seconds and FALSE for 2 seconds on a continuous basis?

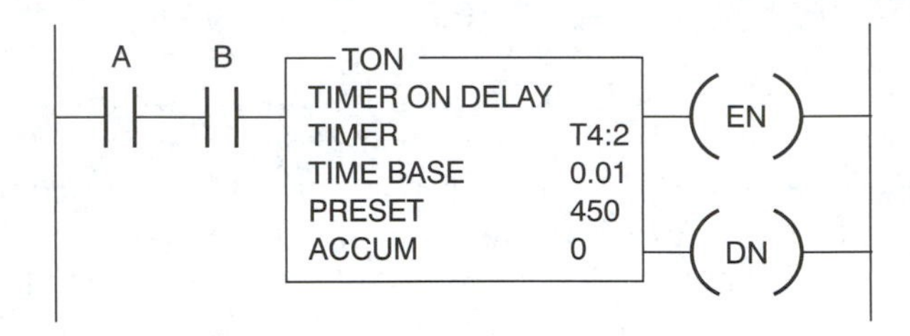

Figure 9–19
Timer instruction for Review Question 6

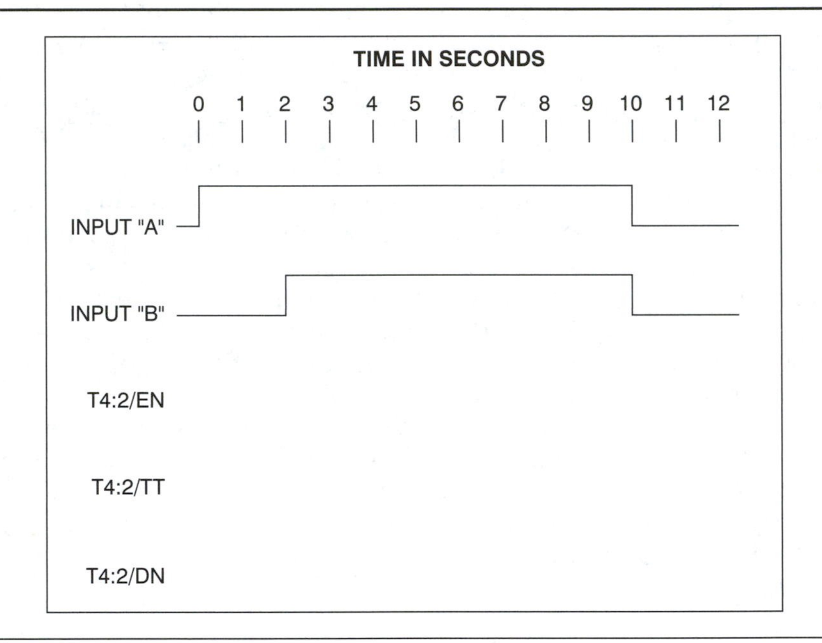

Figure 9–20
Timing chart for Review Question 6

8. Program a set of rungs to detect if the index cam program shown in Figure 9–10 is jammed. You must detect if the cam does not move out of dwell or stops on the high side of the cam. The index time is 2.5 seconds with a dwell time of 0.7 seconds.
9. If you were monitoring the program shown in Figure 9–13, why wouldn't you see the T4:7/DN bit light on the monitor?
10. How can you tell quickly what the last timer number used in your program is?

LAB ASSIGNMENT

Objective

Introduction to the use of timer instructions in control circuits. We will program examples of timed output circuits, index control circuits, flash timer circuits, and methods of accumulating time for use on an HMI display.

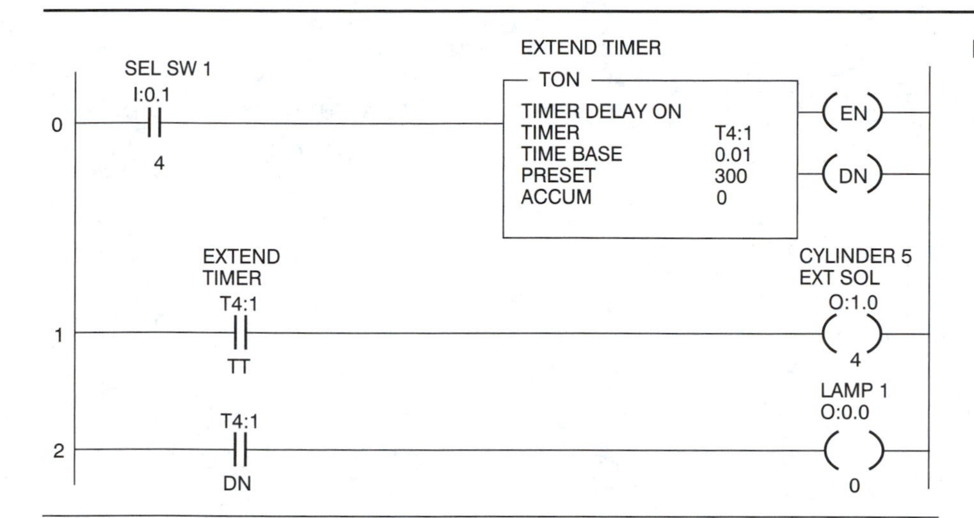

Figure 9–21

1. Open a new file in RSLogix and save it as Chpt-9-lab. Although this lab was designed to run on the lab kit described in Appendix A, it is generic in form and can be run on any lab kit by changing the I/O addresses. Be sure to enter the correct I/O configuration for the lab kit you are using.
2. Enter the three rungs shown in Figure 9–21. Download and run the program.
3. Verify that when selector switch 1 is turned ON cylinder 5 extends and is held extended for 3 seconds before returning. Lamp 1 turns ON when the timer has timed out. Selector switch 1 must be turned OFF and then back ON in order for the cylinder to cycle again.
4. Double-click on the T4 selection on the left side of the screen. When the pop-up box appears change the T4 time to 4 seconds. Note that timer preset changes can be made while the program is running.
5. Switch to offline mode and enter the additional rungs shown in Figure 9–22.
6. These new control rungs will generate a complete index every time selector switch 2 is turned ON. Note that you get only one index cycle, even though selector switch 2 is still on at the end of the cycle. In this example the timer is used to start the index cycle by running the motor just long enough to get the cam switch made. We then let the cam switch hold the motor on until the cycle is complete.
7. In a typical control circuit we normally need to generate a pulse when the index is complete. We will add two more rungs to the program to generate a 1-second pulse at the end of the index cycle. See Figure 9–23.
8. Download the changes and run the program. Verify that lamp 2 pulses on at the end of each index cycle. Note, we determined the index was complete by looking for the cam switch to be off and the T4:2/DN bit to be on.
9. We will now add two more rungs to the program to demonstrate the use of an RTO instruction. Switch to offline mode and add the two rungs shown in Figure 9–24.
10. Download the program and run it. Each time you turn SEL SW 2 ON the index cam will run one cycle. At the end of the index cycle, timer T4:4.ACC will contain the time it took to complete the index cycle. Switching SEL SW 2 OFF resets the timer value in preparation of a new cycle.
11. We have one last example of a timer instruction use to demonstrate. Switch to offline mode and enter the new rungs shown in Figure 9–25. Download and run the program. We have added timers 5 and 6 connected in a free-running oscillator circuit. The circuit puts out a 0.5-second period square wave. Note in rung 11

Figure 9–22

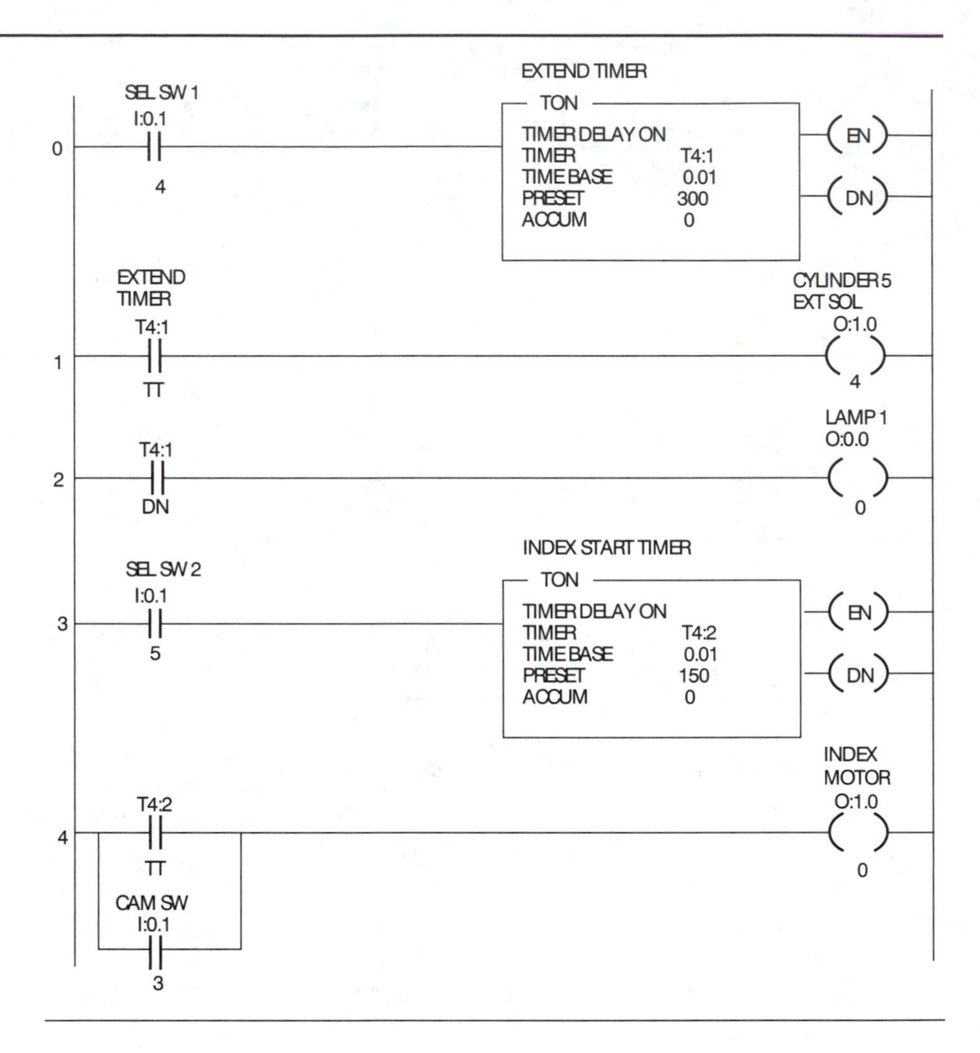

that we use this function to flash lamp 3 any time SEL SW 4 is OFF.

12. Create a circuit of your own design using one latch bit and one timer to perform the following function. When push button 5 is pressed, extend cylinder 1. When cylinder 1's extend limit switch makes, hold the cylinder extended for 5 seconds, then return the cylinder. The cycle time should be the same regardless of how push button 5 is pressed, whether quick push-and-release or held on for 7 seconds.
13. Perform the same operation as the previous program but use cylinder 2 and design it so the cycle is started when the index complete pulse is on after an index operation. In other words, extend cylinder 2 for 5 seconds every time the index cycle controlled by rungs 3 and 4 is complete.
14. Change the preset value to 1 second or 0.25 seconds on only one of timers 5 or 6. What changes do you notice in the flash rate?
15. Design a circuit using a TOF timer so that lamp 7 will flash on for at least 7 seconds when push button 9 is pressed and released quickly.
16. Print out a copy of your completed program.

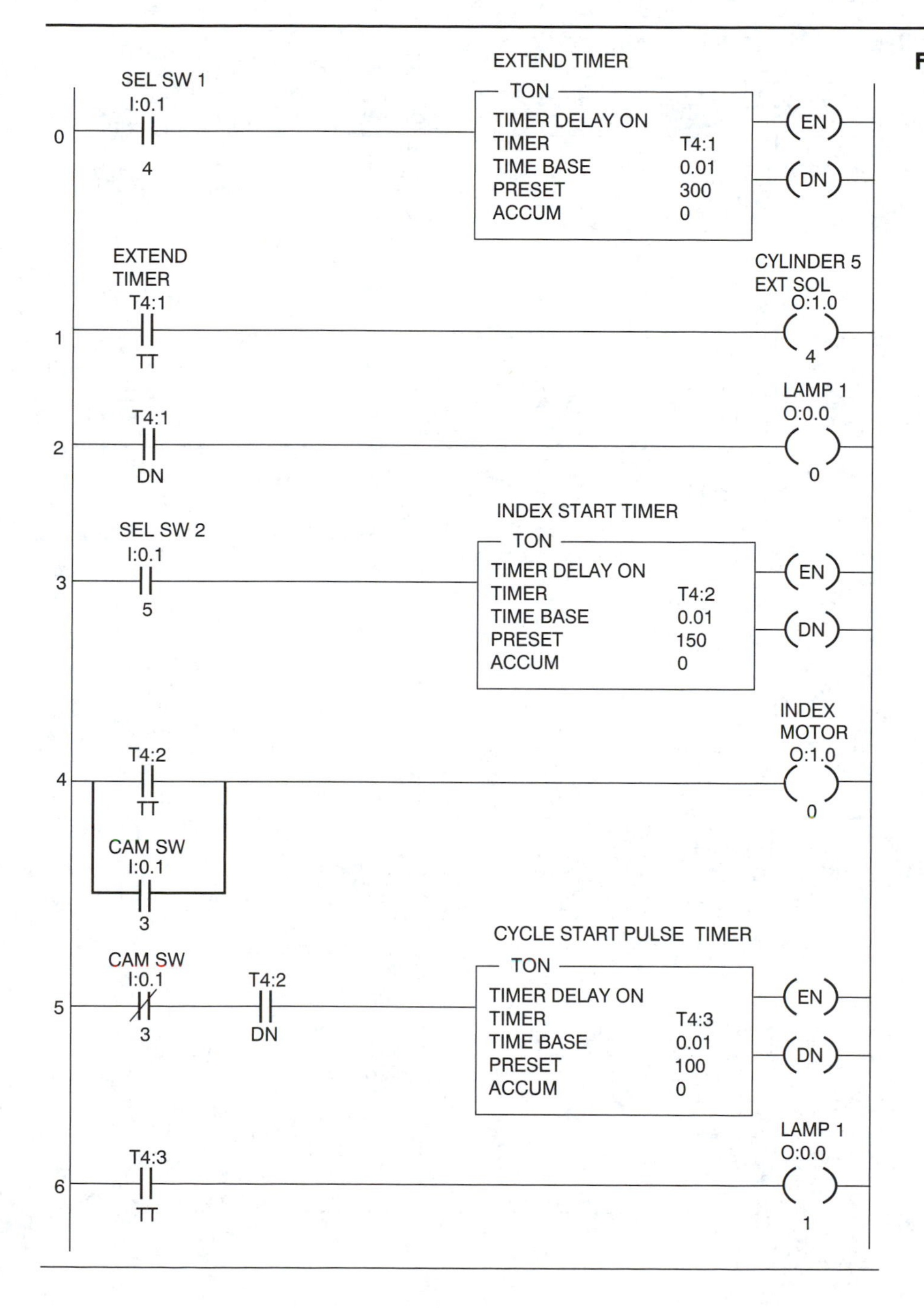
EXTEND TIMER
SEL SW 1
I:0.1
4
TON
TIMER DELAY ON
TIMER T4:1
TIME BASE 0.01
PRESET 300
ACCUM 0
EN
DN
EXTEND
TIMER
T4:1
TT
CYLINDER 5
EXT SOL
O:1.0
4
T4:1
DN
LAMP 1
O:0.0
0
INDEX START TIMER
SEL SW 2
I:0.1
5
TON
TIMER DELAY ON
TIMER T4:2
TIME BASE 0.01
PRESET 150
ACCUM 0
EN
DN
T4:2
TT
CAM SW
I:0.1
3
INDEX
MOTOR
O:1.0
0
CYCLE START PULSE TIMER
CAM SW
I:0.1
3
T4:2
DN
TON
TIMER DELAY ON
TIMER T4:3
TIME BASE 0.01
PRESET 100
ACCUM 0
EN
DN
T4:3
TT
LAMP 1
O:0.0
1
0
1
2
3
4
5
6

Figure 9–23

Figure 9–24

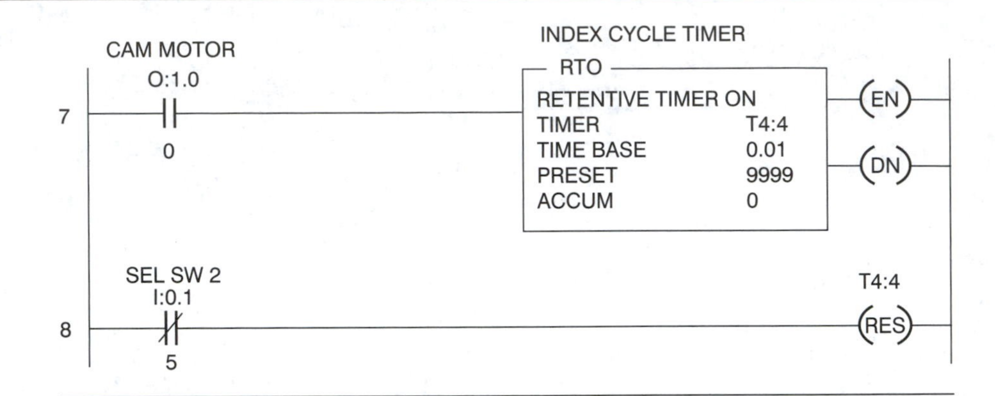

Figure 9–25

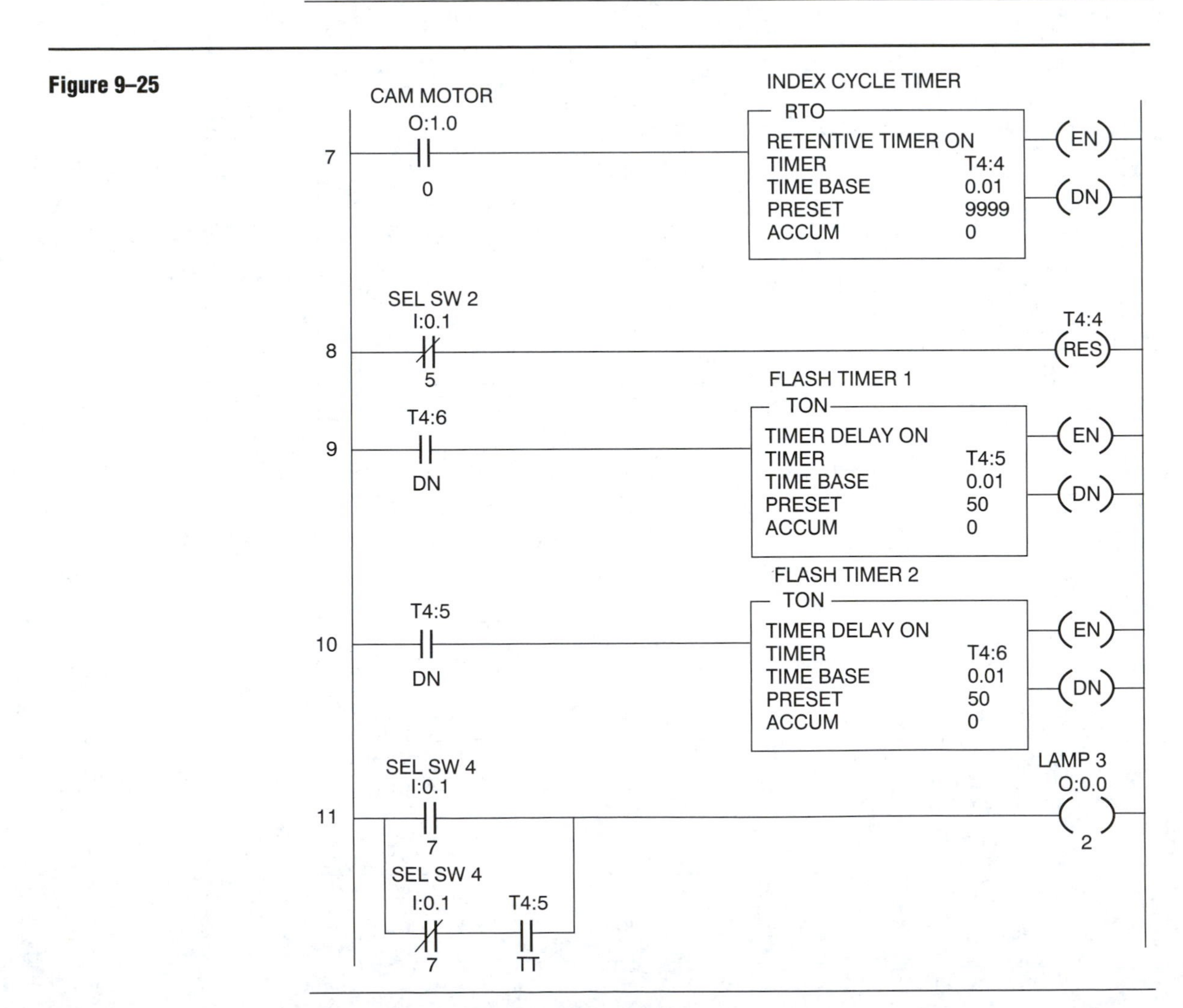

CHAPTER

10

Counter Instructions

LEARNING OBJECTIVES

After completing this chapter the reader should understand:

1. The basic functions of CTU and CTD counter instructions.
2. The use of counters to control asynchronous pallet lines.
3. Methods for changing preset values as part of the ladder program.
4. Methods for determining subcounts using only one counter.
5. How to view and edit the counter files.
6. Time counter combinations for accumulating very long time durations.

Counter instructions are used any time the programmer needs to count actions in a program or accumulate the number of times a function has occurred. Typical counter functions include these:

1. Count the number of items being produced or tested.
2. Maintain the level in a tank by counting the number of gallons pumped in and drained from the tank.
3. Keeping track of very long duration time periods by combining a timer and counter.

Two counter instructions are available in the SLC-500 controller: CTU and CTD. CTU stands for counter up and CTD stands for counter down. As in our previous chapter on timers, each counter instruction you enter occupies three memory locations in the C5 counter file (Figure 10–1). The C5 file is created automatically when you open a new project program file. Additional counter files can be created in the file area above file 10 if needed.

The contents of words 2 and 3 are the same as for timers. They hold the counter preset value and the actual accumulated count. Word 1 is the

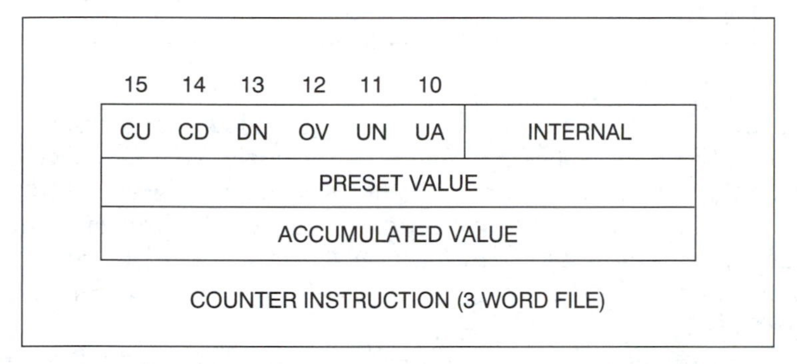

Figure 10–1
Counter instruction word allocation

status word and here we have some new status bits to deal with. Each status bit is described next:

CU is the counter up enable bit and it is on any time the rung to the CTU instruction is TRUE.

CD is the counter down enable bit and it is on any time the rung to the CTD instruction is TRUE.

DN is the counter done bit and it is on any time the accumulated value is equal to or greater than the preset value.

OV is the counter overflow bit and it is on any time the accumulated value has rolled over from the maximum count 32,767. This bit can only be reset with the (RES) instruction.

UN is the counter underflow bit and it is on any time the accumulated value has gone under −32,768. This bit can only be reset with the (RES) instruction.

UA is the update accumulator for HSC instructions only.

Counter instructions are entered in much the same way as timer instructions are. The rung can have a single input or multiple input conditions before entering the counter instruction. When a CTU or CTD instruction is entered, you will be asked to assign a counter number from 0 to 84 and a preset count value. The rung composition is shown in Figure 10–2.

In this example we have entered a count preset of 15. CTU counter instructions work by incrementing the accumulated value once for each

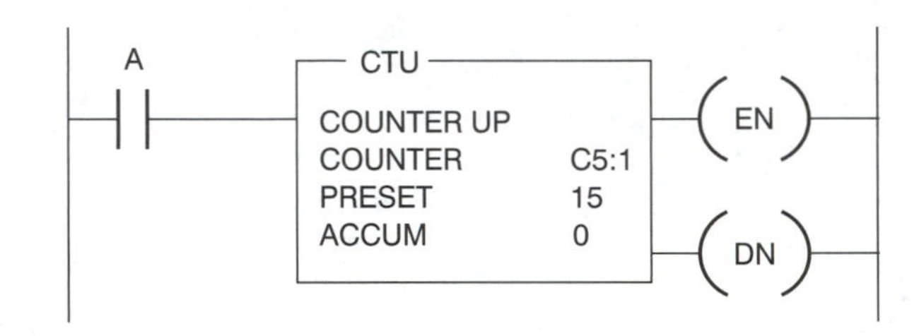

Figure 10–2
Rung composition for counter instructions

FALSE to TRUE transition of the rung logic. CTD counter instructions decrement the accumulated value once for each FALSE to TRUE transition of their rung conditions. When the accumulated count value equals the counter's preset value, the counter's DN bit is turned on. A counter's accumulator value may be reset to all zeros by the reset (RES) instruction. This also resets any over- or underflow bits that may have been set.

It is not uncommon to combine CTU and CTD instructions on one counter address. Because both instructions are working on the same counter address, they both control the status bits. An example of this could be a pallet-style assembly line where there are a limited number of pallet positions between operators (Figure 10–3). Between operators 1 and 2, a maximum of six pallets is allowed. We must be sure that operator 1 cannot release a pallet if six pallets are already between the two gates. To do this, we count up each time operator 1 releases a pallet and we count down each time operator 2 releases a pallet. If we set the counter preset at 6 and use the DN bit to inhibit operator 1 from releasing a pallet, we can maintain a maximum of six pallets between the gates.

A sample program for this control problem is shown in Figure 10–4. We could add a second counter for the pallet limits between operators 2 and 3 and write a duplicate set of rungs to control this section of the line if this was desired. In this way we could control multiple positions on

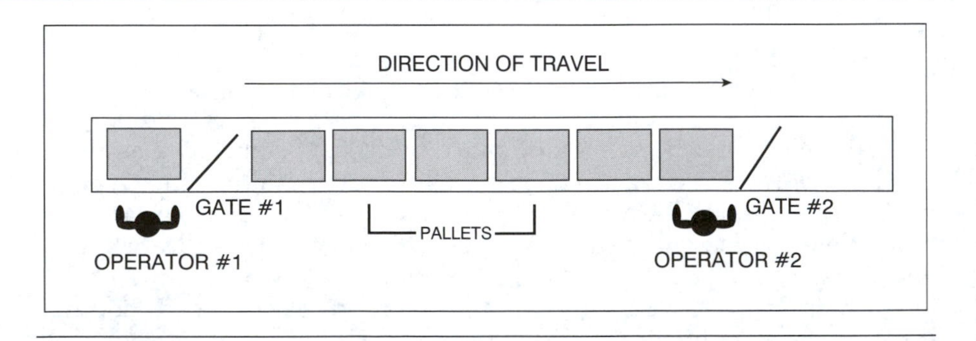

Figure 10–3
Pallet-style assembly line

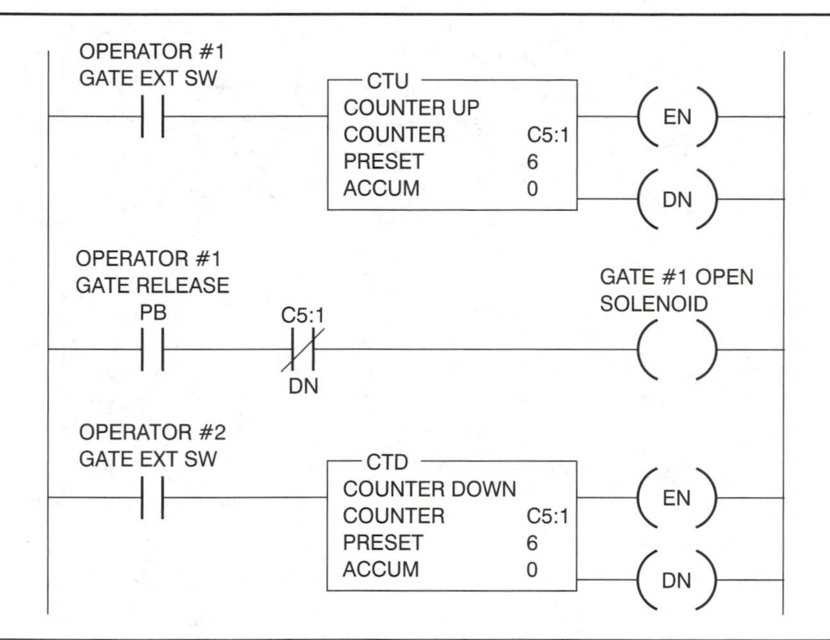

Figure 10–4
Sample program for the pallet-style assembly line control problem

the line that may have different pallet limits between them. This same up/down configuration can be used anyplace the input and output from a location are determined by separate elements. This type of operation is called asynchronous operation because each operator controls his or her discharge gate independently.

In control programs, a programmer often needs the same counter address to supply a number of different counts depending on the setting of an input condition. For example, in a box-filling operation, the control system might need three separate box quantity values based on a control switch that is set for 6, 12, or 24 PCs per box. We could use one counter that has an original program preset of 24, but apply several different MOVE instructions that change the preset value based on the value selected by the control switch. Each MOVE instruction would load a new value—6, 12, or 24—into the counter's preset register. This is one way of providing multiple counter settings that are selectable by moving a selector switch under the program's control (Figure 10–5).

In the example of Figure 10–5 we show a sample program that matches the function we discussed earlier. We have a selector switch that sets the number of parts that are to be placed into a box. The selections are 6, 12, or 24 pieces to a box. When the selector switch is changed, a new value is loaded into the counter's preset register. Each time the part sensor detects a part falling into the box, it adds 1 to CTU counter C5:2. When the counter's ACCUM register equals the preset value, the DN bit is turned on and the full box pusher extend solenoid is activated. When this cylinder reaches its end

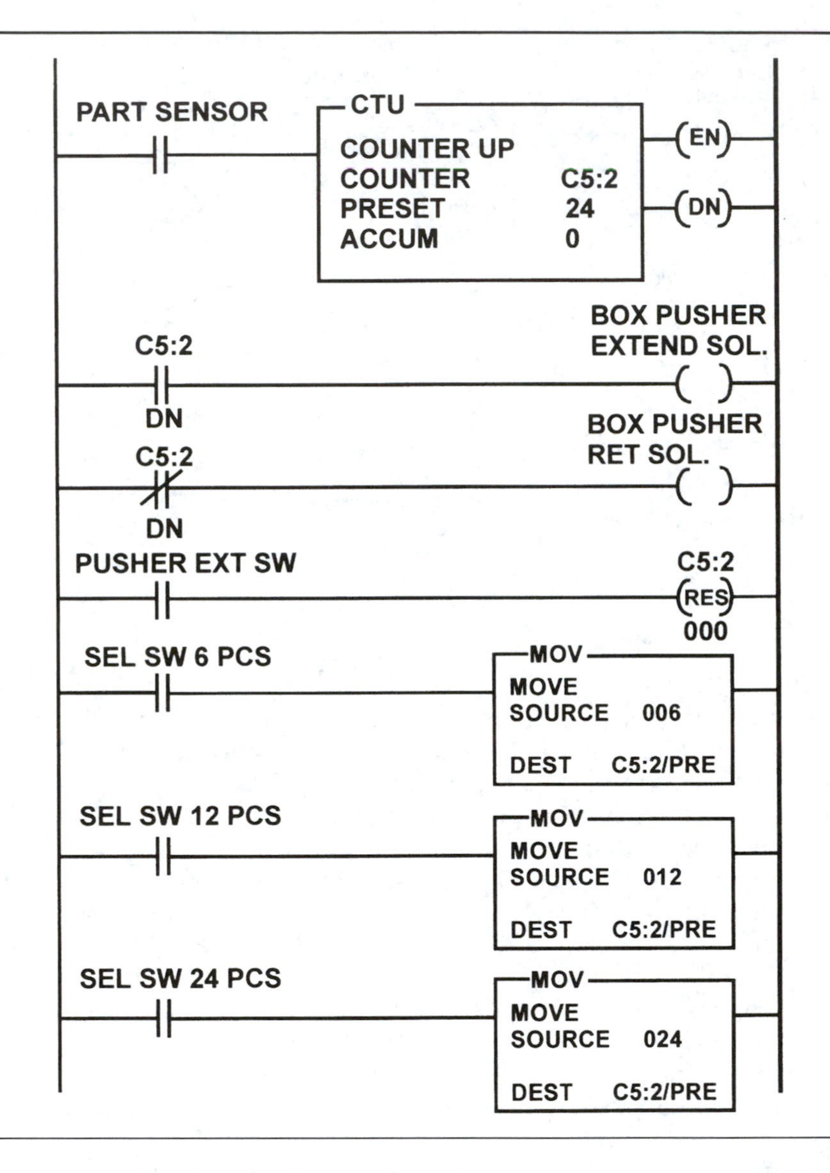

Figure 10–5
Ladder diagram for multiple counter settings

of stroke, the cylinder extend prox switch is activated and the counter is reset to zero. As soon as the counter value falls below the preset value, the DN bit is reset and this allows the box pusher cylinder to return to the home position.

Note that when the MOVE instructions were entered, the programmer had to enter the source value to match the desired count value for that rung's input condition. Because all destinations were the same, the counter's preset value, this address was entered the same for all three instructions.

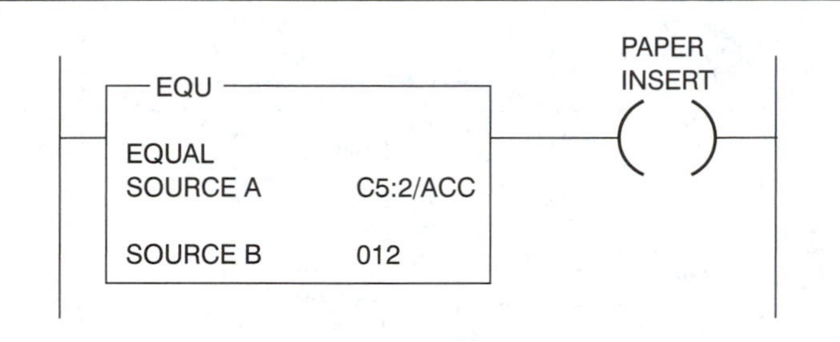

Figure 10–6
Example of the EQU instruction used to detect a specific count value

Both of the previous examples are common programming problems that can be solved using counter instructions. One additional function often required is to start some mechanical action at a count value less than the counter's preset value. Returning to the previous example, if we had the box quantity set for 24 PCs, we might be required to insert a paper divider into the box after the 12th count. To do this we could use the EQU instruction, explained in detail in the next chapter, to test the counter's accumulated value for count 12, and turn on an output when that count is reached (Figure 10–6).

We can use a combination of timer and counter instructions to accumulate long time periods. We often need to keep track of how long a piece of equipment has been running. Maintenance departments are often required to change oil or lubricate equipment every 200 or 500 hours of operation. This can be a problem if the equipment only runs on demand or for short periods of time. A good example is a cooling compressor that may only run a few minutes every hour. Short times can be accumulated with a timer instruction alone, but for very long times the use of a combination of timer and counter instructions is a better choice. We will use an RTO timer because we must accumulate time over many on/off periods.

If we set the timer for a 1.0-second time base and the preset for 3600, we will time out every time an hour has been accumulated. Remember that RTO instructions do not reset their accumulated time when the control rung goes FALSE. We can then use the timer's DN bit to increment the counter instruction that will now accumulate hours of operation. In the example shown in Figure 10–7 we turn on a warning light when we reach 500 hours of operation. The maintenance department can reset the timer and counter by pressing the service reset push button after the service has been performed. This is one case where the order of the control rungs is important. The counter rung must be located between the timer rung and the reset rung. If we were to put the reset rung second and the counter rung third the counter would never be incremented because the T4:7.DN bit would already be reset when we scanned the counter rung.

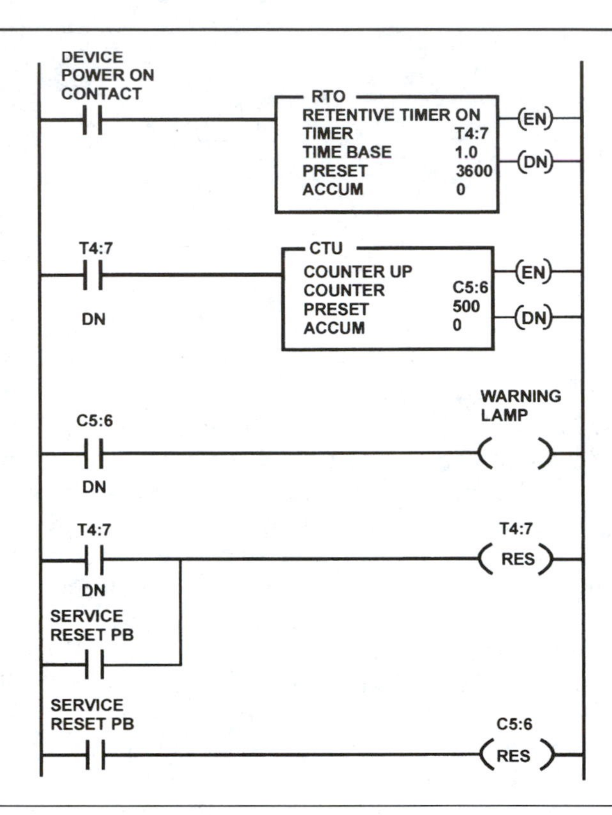

Figure 10–7
Sample program showing how a timer and counter can be used together

Displaying a feed rate is another programming example where the use of timers and counter instructions together fits the requirements. Let's say we are required to display the belt feed rate of a conveyor on an operator's screen. Let's also say we have a prox switch on the drive sprocket that gives us a pulse once per inch of belt travel. We could then set up two timers and a counter to compute the belt speed in feet per minute (Figure 10–8). The two timers are set up to cycle continuously with timer T4:1/TT being ON for 10 seconds and timer T4:2/TT being ON for 0.1 seconds. The T4:1/TT bit is ANDed with the belt sprocket sensor and this allows the counter C5:1 to count sprocket pulses during the 10-second ON time of T4:1. When the T4:1/DN bit comes ON it starts timer T4:2. We use the timer T4:2/TT bit to move the count value contained in C5:1 to a register location. Then we use the T4:2/DN bit to reset the counter. In this way we can update the display screen every 10 seconds with the belt speed of the line. We would need to multiply the stored count value by 6 to get inches per minute and then divide by 12 to get feet per minute. Because we have not covered math and logic instructions, this part of the programming problem will have to wait.

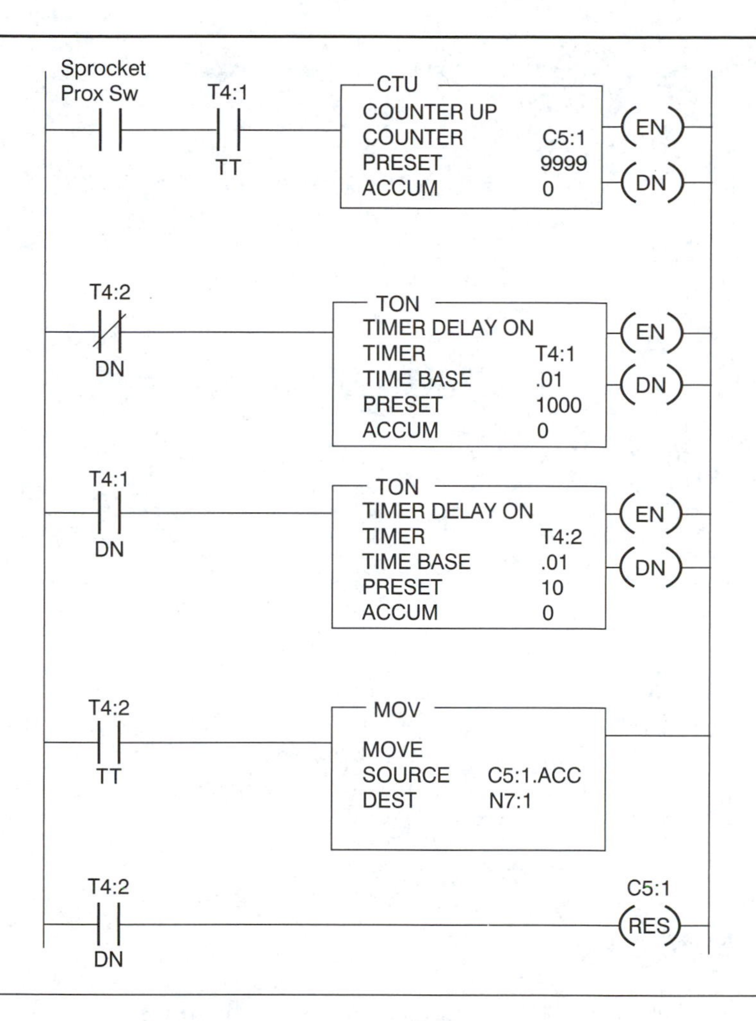

Figure 10–8 Sample program for computing feet per minute

TROUBLESHOOTING AND ADJUSTMENTS

Just as for timer instructions, we have two methods for monitoring counters in our program. If the counter instruction is visible on the screen when we are in online monitor mode, the accumulated value is shown inside the counter instruction box. If we need to observe multiple counter values or wish to change the preset or accumulated value of a counter while in run mode, we must do so in the counter file display screen.

Figure 10–9 shows an RSLogix software screen where the counter file has been opened for display and editing. You open the counter file display by double-clicking on the C5-COUNTER file heading under Data Files on the left edge of the screen. Once this file is opened, you

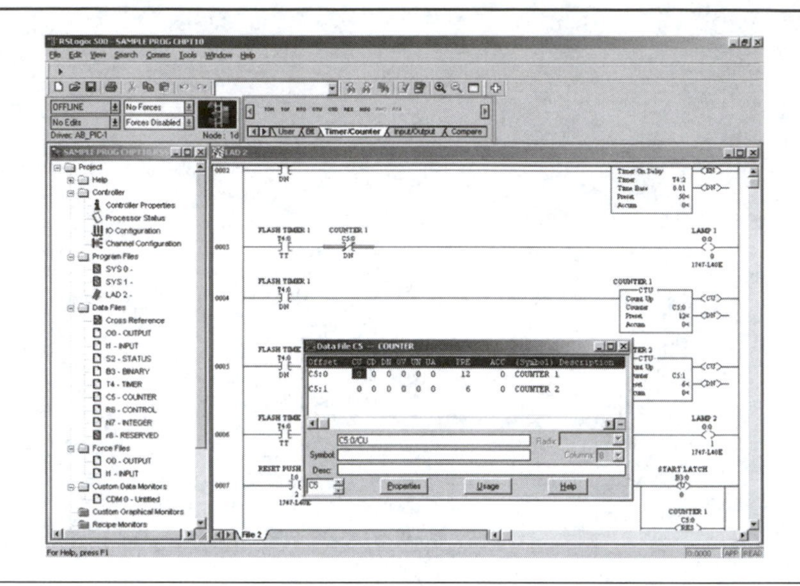

Figure 10–9
RSLogix screen showing counter listing

can change the preset or accumulated values by clicking on or highlighting each value, and typing in a new value. Be aware that this new value is only held in the PLC's memory. If you don't save the file before you exit run mode, the program in the host computer will not be changed, and the next time you download to the PLC the old values will come back.

SAMPLE PROGRAM

We use the sample program given here to demonstrate the use of counters, timers, and latch instructions. The program does not perform any useful purpose other than making two lights flash an exact number of times. The sequence starts when the operator presses the start push button I:0.0/00. This action latches on the cycle start bit B3/0. Rungs 1 and 2 are two timers tied together to make a pulse generator of 0.5-second intervals (Figure 10–10). This pulse generator begins to run as soon as B3/0 is latched on. Rung 3 flashes the first lamp O:0.0/00 each time the timer bit T4:0/TT goes on and the counter C5:0 is at or below its preset count of 12. In other words the lamp will flash 12 times for 0.5 seconds on and off each time.

Rung 4 contains the first counter instruction C5:0. Note that the input to this counter is T4:0/DN. If we had used the timer's TT bit instead, the lamp would have flashed only 11 times due to the fact that the counter would have incremented on the rising edge of T4:0/TT. On the 12th count,

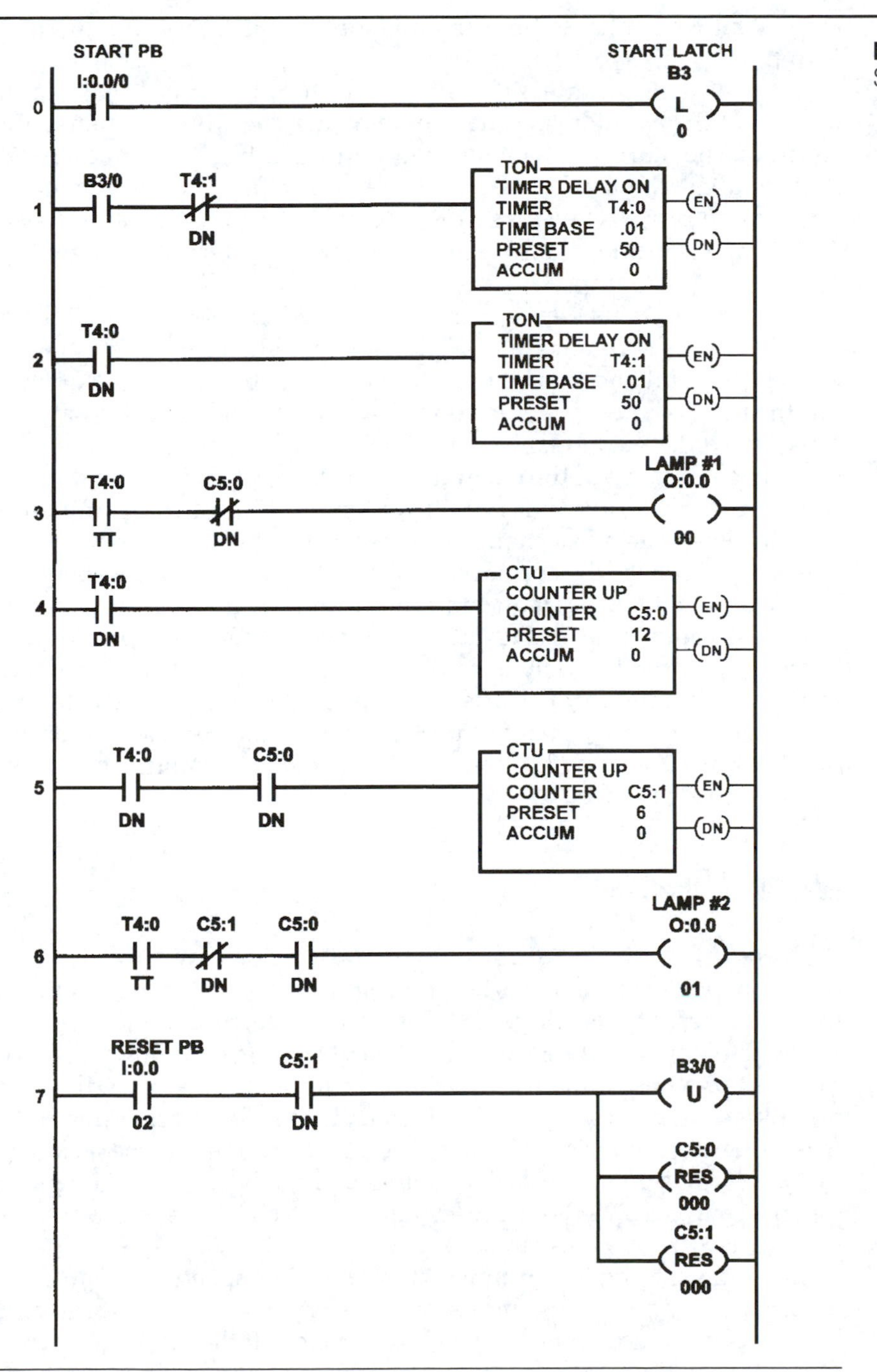

Figure 10–10
Sample program

the DN bit would have come on and stopped the lamp from lighting the 12th time.

When using counters it is important to understand that you must increment the counter *after* the event you are counting is finished. If you count on the start of the event, you will most likely get one less event than you set the counter's preset value for (i.e., one less event than desired). In our case we used the DN bit; therefore, the lamp has completed its flash time when we increment the counter. We will get the full 12 flashes of the lamp.

Rung 5 contains the second counter and here we have a slight problem. Because the first counter C5:0 is incremented on the rising edge of the T4:0/DN bit and this bit will still be on when the C5:0/DN bit comes on, the C5:1 counter is going to increment one count at this same time. This means the second lamp would only flash five times before the cycle is complete instead of the desired six. We can fix this in several ways but the easiest, with our still limited understanding of the complete PLC instruction set, is to just change the preset count to 7.

Rung 6 causes the second lamp to flash only when the first counter is at or above its preset value and the second counter is below its preset value. The last rung resets the whole sequence when the reset push button is pressed but only after the last count is complete. Note that we reset both counters and cleared the start cycle latch bit. As I stated previously, this program does not perform any grand function other than to demonstrate the complexities of using timers, counters, and latch bits together in a program.

CONCLUSIONS

With our new understanding of counter instructions, we have again added two more powerful elements to our programmer's bag of instructions. We must be careful to use these new instructions correctly. As stated before, always increment counters when the event you are counting is completed. This will prevent problems with short counting or unexpectedly shutting down a mechanical sequence in the middle of its last cycle. An additional frustration that new programmers sometimes run into is the lack of a status bit to tell us when we have counted down to 000. The DN bit indicates when we have reached our preset value, but to indicate the 000 status we need to use the EQU instruction.

One final word of caution: The maximum speed you can count is determined by your program's scan time. Any counter input signal must be stable for one scan time to be counted reliably. If the input changes faster than one scan time, the count value will become unreliable because counts will be missed. In this situation you need to use a high-speed counter input or a special counter I/O module suited for high-speed applications.

REVIEW QUESTIONS

1. What file number is created to contain counters every time we open a new project file?
2. What are the status bits contained in the first word of the counter instruction and what do they indicate to the programmer?
3. Why would you assign both a CTU and a CTD instruction to the same address?
4. Is the preset value of the counter instruction adjustable from within the running program and, if so, how can we change it?
5. What compare instructions can be used to give an output at some value less than the counter's preset value?
6. How can counters be used to accumulate long periods of time for diagnostic purposes?
7. Why should the counter value always be incremented *after* the event we are counting has been completed and not as it is starting?
8. How can you see all the counters, along with their functional descriptions, that you have assigned in your program while monitoring the program on your computer?

LAB ASSIGNMENT

Objectives

Introduces the student to using counter instructions in a ladder program. We will create rungs to up and down count values and limit the count to certain values. We also will learn how to change the preset value as a function of the ladder program.

1. Open a new file in RSLogix and save it as Chpt-10-lab. Although this lab was designed to run on the lab kit described in Appendix A, it is somewhat generic in form, and can be run on most lab kits by changing the I/O addresses. Be sure to enter the correct I/O configuration for the lab kit you are using.
2. Enter the rungs shown in Figure 10–11.
3. Download and run the program. Verify that each time push button 1 is pressed the counter adds 1 to its total. Verify also that lamp 1 turns on when the count reaches 15. What happens to the ACCUM value if you continue to press push button 1 after the count has reached 15?
4. Turn ON selector switch 1 and verify that the count resets to zero. With selector switch 1 turned ON press push button 1 several times. Note that the counter stays at zero count. Remember this situation when you are debugging a program. Any time you have a counter that will not increment when the rung conditions are true, look for a rung holding the counter reset.
5. Add rung 3 to the program as shown in Figure 10–12.
6. Press push button 1 until the count is 5. Now press push button 2 until the count reaches zero. Press push button 2 one more time. What happened to the ACCUM value of C5:0?
7. In order to fix this problem we must prevent the counter from counting below zero. To do this we will have to shut off the down count rung if the ACCUM value is at zero. Edit rung 3 and add rung 4 as shown in Figure 10–13, then download and run the program.
8. Repeat the previous count sequence. Notice that the B3/0 bit now disables rung 3 when the count value is at zero, thus preventing the counter from down counting below zero.
9. Next, we will give the operator the ability to change the preset count value. Add the two rungs shown in Figure 10–14, then download and run the program.

Figure 10–11

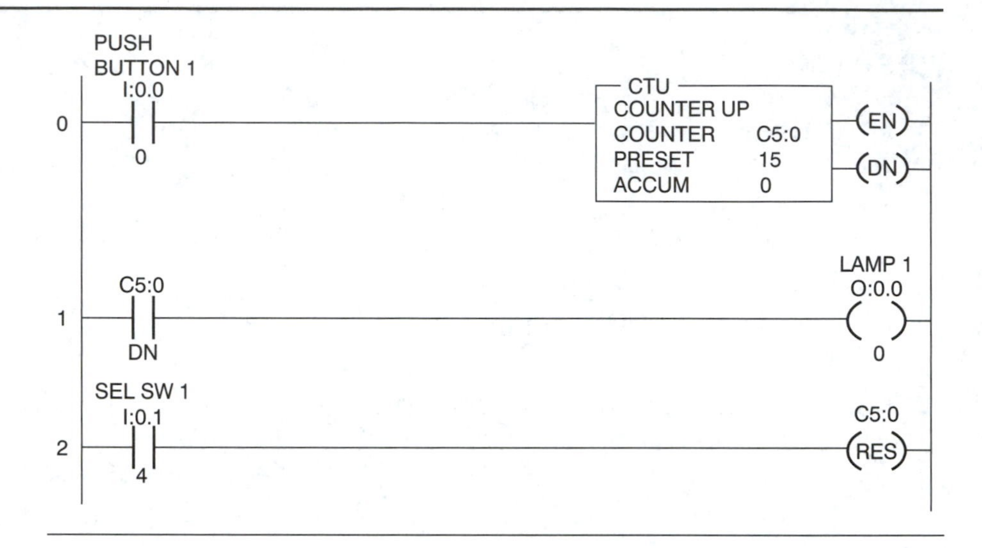

Figure 10–12

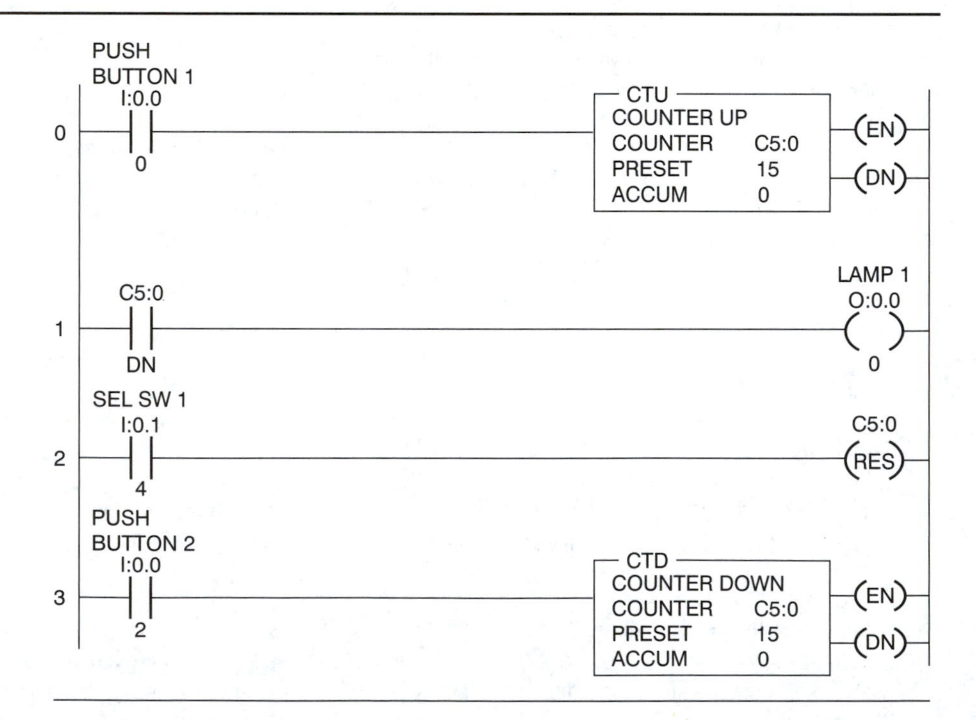

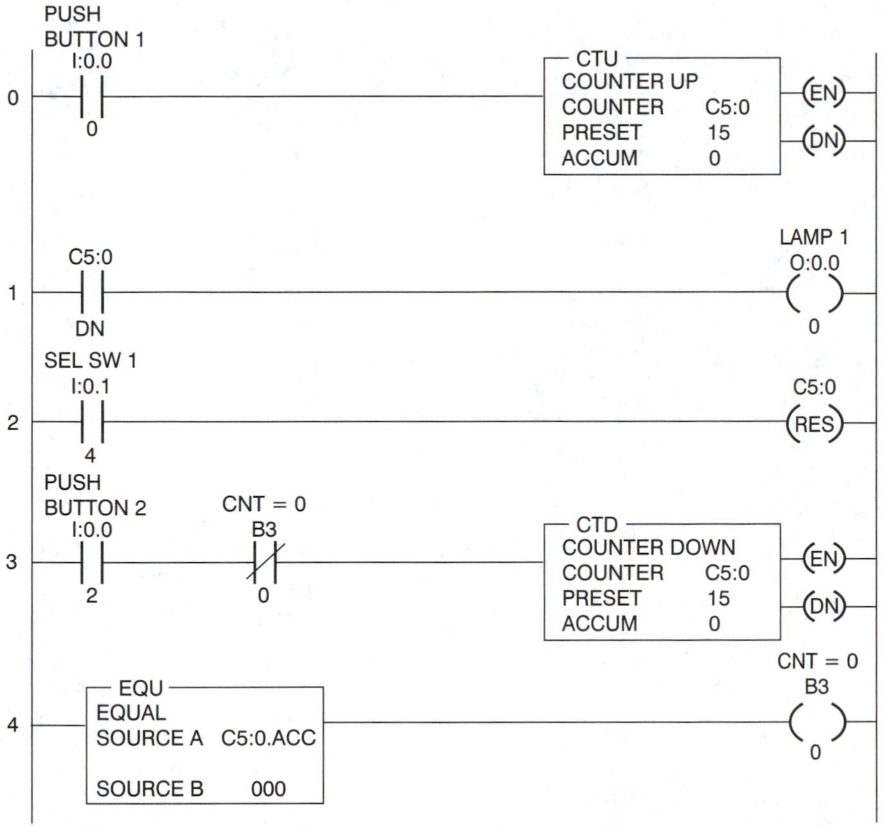

Figure 10–13

10. Verify that you can now change the preset value based on the position of SEL SW 2. If SEL SW 2 is ON the MOVE instruction puts a value of 12 into the counter's preset location. If SEL SW 2 is OFF the bottom MOVE instruction places a value of 24 into the counter's preset location.

11. Add rungs to the program to perform the following function. Use two timers to generate a pulse train that is on for 1 second and off for 2 seconds. When SEL SW 3 is turned ON, use this pulse train to cycle cylinder 3, extend for 1 second, then return for 2 seconds. Use a counter to allow this cycle to repeat 8 times and then stop. SEL SW 3 must be turned OFF and back ON again to restart the cycle. Be sure to count the number of times cylinder 3 cycles to verify that it is 8 times.

12. Print out your program and demonstrate that it works to the instructor.

Figure 10–14

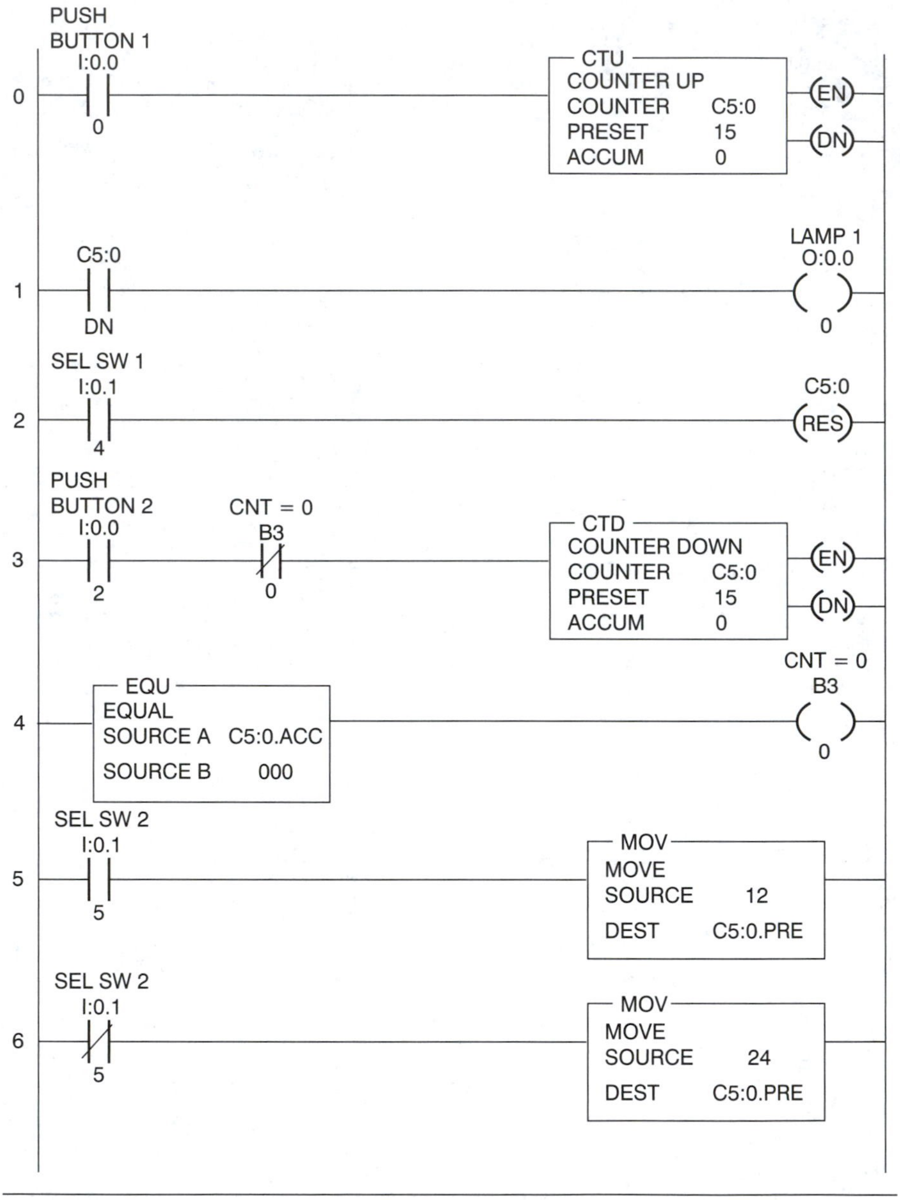

PUSH BUTTON 1
I:0.0
0
0
CTU
COUNTER UP
COUNTER C5:0
PRESET 15
ACCUM 0
EN
DN
1
C5:0
DN
LAMP 1
O:0.0
0
2
SEL SW 1
I:0.1
4
C5:0
RES
3
PUSH BUTTON 2
I:0.0
2
CNT = 0
B3
0
CTD
COUNTER DOWN
COUNTER C5:0
PRESET 15
ACCUM 0
EN
DN
4
EQU
EQUAL
SOURCE A C5:0.ACC
SOURCE B 000
CNT = 0
B3
0
5
SEL SW 2
I:0.1
5
MOV
MOVE
SOURCE 12
DEST C5:0.PRE
6
SEL SW 2
I:0.1
5
MOV
MOVE
SOURCE 24
DEST C5:0.PRE

CHAPTER

11

Logic and Math Instructions

LEARNING OBJECTIVES

After completing this chapter the reader should understand:

1. Program construction using basic comparison instructions.
2. Program construction using the basic math instructions.
3. Program construction using the basic logical instructions.
4. Data move and manipulation instructions.
5. Indexed offset mode of addressing.
6. Indirect addressing mode.

In this chapter we cover all compare, logic, data handling, and math instructions for PLCs. Each instruction is explained in detail as to its function and the resulting status bits that can be changed in the status register.

COMPARISON INSTRUCTIONS

In general, comparison instructions are used any time you need to compare the contents of some address (timer, counter, or memory location) with the contents of some other address or with a constant value. Seven compare instructions can be used with the SLC-500/01 processor and one additional instruction with the SLC-500/02 and higher systems. The instruction mnemonics (a three-character abbreviation for the whole instruction description) and descriptions are listed next. All of these instructions are used as input variables, and if TRUE, will turn on the rungs' output instructions. We have already demonstrated the use of the EQU instruction in the program sample of Chapter 10.

1. EQU: equal
2. LEQ: less than or equal
3. NEQ: not equal
4. GRT: greater than
5. LES: less than
6. GEQ: greater than or equal
7. MEQ: masked compare for equal
8. LIM: limit test (5/02 only)

EQU Instruction

The equal instruction compares the two source location contents and is logically TRUE only if the contents are equal (Figure 11–1). Source A must be a word address. Source B can be a word address or a program constant. There are many reasons to compare a word address with a constant, examples of which are discussed next.

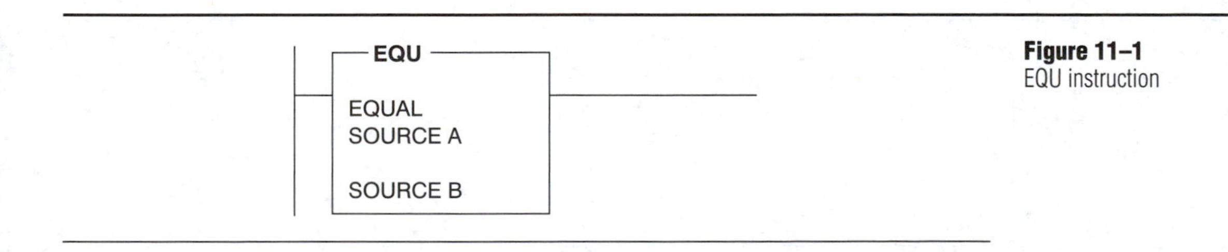

Figure 11–1
EQU instruction

Examples

If our counter is counting objects as they are placed in a 24-count box, we may want to insert a divider at count 12. To accomplish this, we can use an EQU instruction to compare the counter value with the constant value of 12. When the two values are equal the rung is TRUE, and we can use it to fire the output that inserts a divider.

A second example would be if we are using a timer and need to start some action a set period of time before the timer times out. We can use an EQU instruction to compare the timer ACCUM value with a set constant value. When the two values are equal, the rung will go TRUE and the output can be latched on to perform a function.

LEQ Instruction

The less than or equal instruction is similar to the previous instruction except it is TRUE if the source A value is less than or equal to the source B value. It is programmed as shown in Figure 11–2.

Source A must be a word address. Source B can be a word address or a constant value. An example of this instruction's use is seen in a cycle timer circuit.

Example

If we have an assembly line and the index timer is set for 8 seconds, we may want a warning lamp to come on solid after the index and begin to flash 3 seconds before the next index is due. We could use the less than or equal instruction to compare the timer ACCUM value with a constant of 5 seconds. We could then hold the lamp on constant as long as the instruction was TRUE, and start flashing the lamp when the ACCUM value went above the constant.

NEQ Instruction

The not equal instruction is the opposite of the EQU instruction. It compares the source A value with the source B value and is TRUE only if the two values are not equal (Figure 11–3).

Source A must be a word address. Source B can be a word address or a constant value. Because this is the exact opposite of the EQU instruction, we do not provide an example.

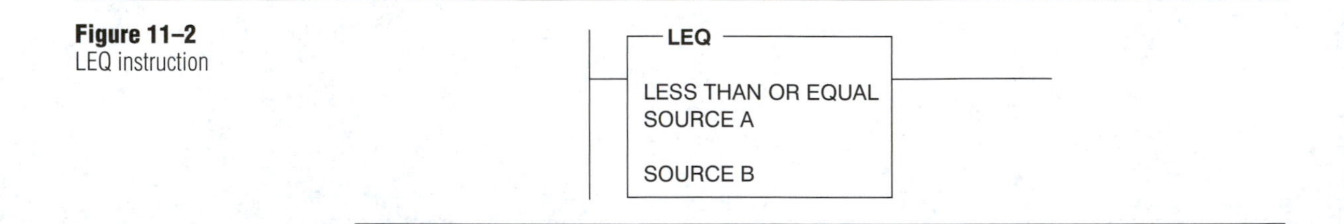

Figure 11–2
LEQ instruction

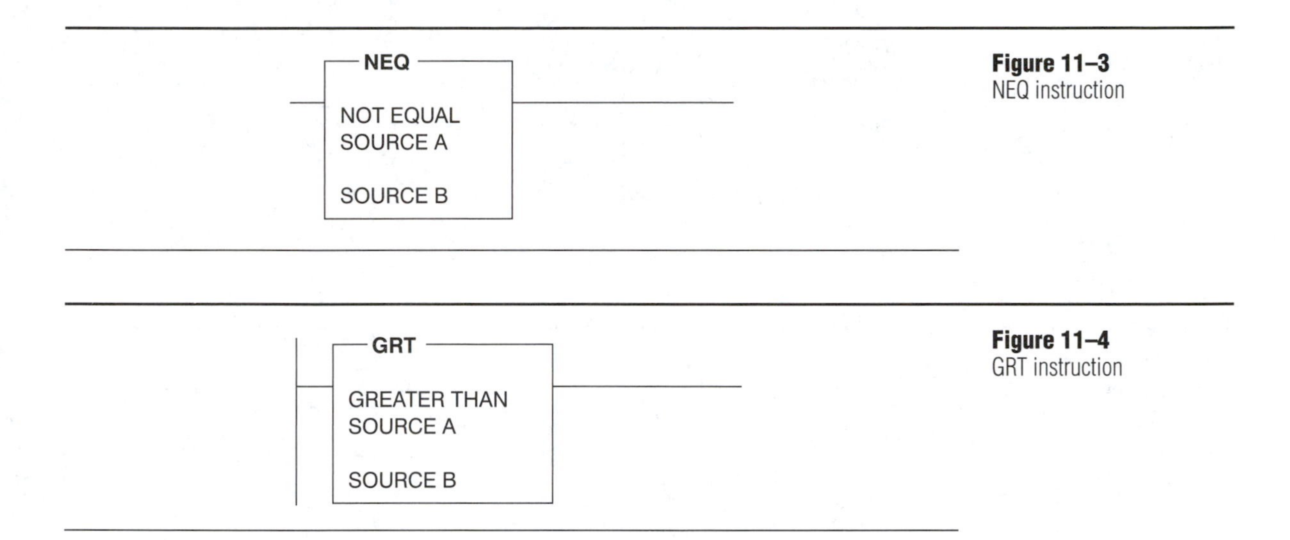

Figure 11–3
NEQ instruction

Figure 11–4
GRT instruction

GRT Instruction

The greater than instruction compares the source A value with the source B value and is TRUE only if source A is greater than source B (Figure 11–4). If source A is equal to or less than source B, the rung is FALSE. Source A must be a word address. Source B can be a word address or a constant value.

Example

As an example we could have an UP/DOWN counter used to control the level in a storage tank. Two separate pumps are used to fill and empty the tank, and each pump has a different flow rate. For each gallon of material we sense being pumped in, we count up one and for each gallon we sense being pumped out we count down one.

The pump control program tries to maintain a tank level of three-quarters full or 175 gallons. We could use the GRT instruction as an alarm monitor by comparing the counter value with a constant of, say, 180. If the counter value is above 180, we have a process control error and we should shut down the pumps and start a warning alarm and light flashing.

LES Instruction

The less than instruction is the opposite of the greater than instruction. This instruction compares source A with source B and is TRUE only if the source A value is less than the source B value (Figure 11–5). If the source A value is equal to or larger than the source B value, the rung is FALSE.

Source A must be a word address. Source B can be a word address or a constant value. Because this is the exact opposite of the GRT instruction, we do not provide an example.

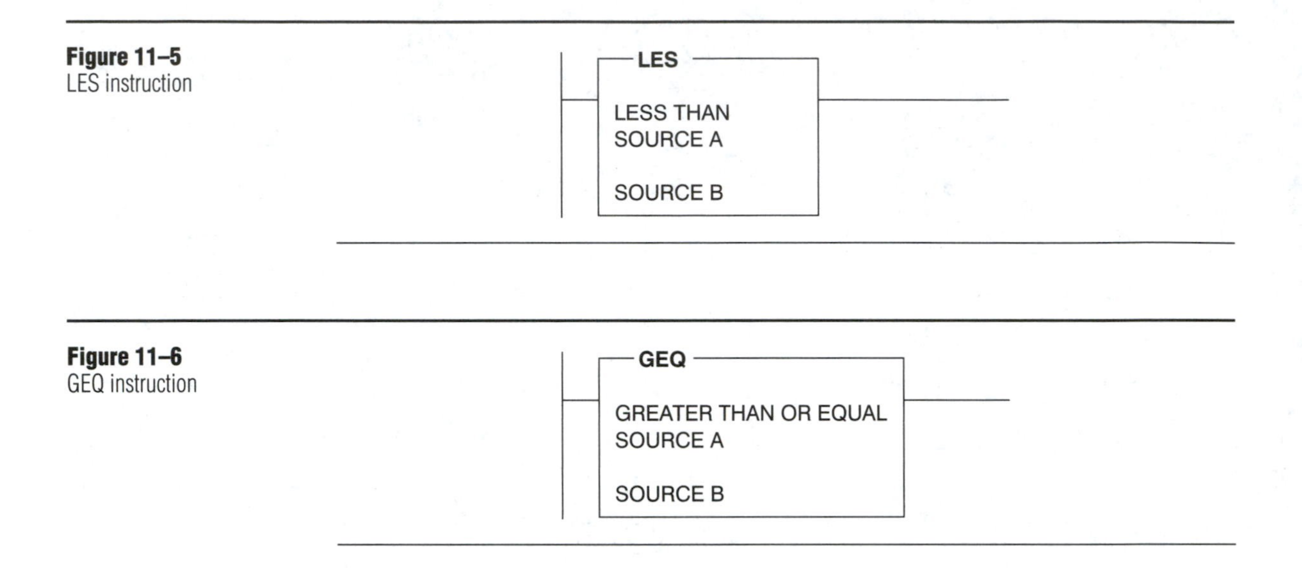

Figure 11–5
LES instruction

Figure 11–6
GEQ instruction

GEQ Instruction

This is the opposite of the LEQ instruction. This instruction compares the source A value with the source B value, and is TRUE only if the source A value is greater than or equal to the source B value (Figure 11–6). If the source A value is less than the source B value, the rung is FALSE.

Source A must be a word address. Source B can be a word address or a constant value. Because this is the exact opposite of the LEQ instruction, we do not provide an example.

MEQ Instruction

This is the masked compare for equal instruction. It compares the source location to the compare location or constant value after the source location has been filtered through a mask (Figure 11–7). The main purpose of this instruction is to allow you to compare only part of an address value with a constant or word address. This is a 16-bit-wide compare instruction. Each bit in the mask that is set to 0 shuts off that bit location; each bit that is set to 1 passes that bit location.

In Figure 11–7, SOURCE is the address of the value you want to compare. MASK is the address of the mask through which the instruction moves data. The mask can also be a hex value. COMPARE can be a word address or a constant value.

Example

In writing program diagnostics we often need to know what state the mechanism was in when it stopped working. If we have an input word with the low 8 bits all being associated with station #1, we could test for certain

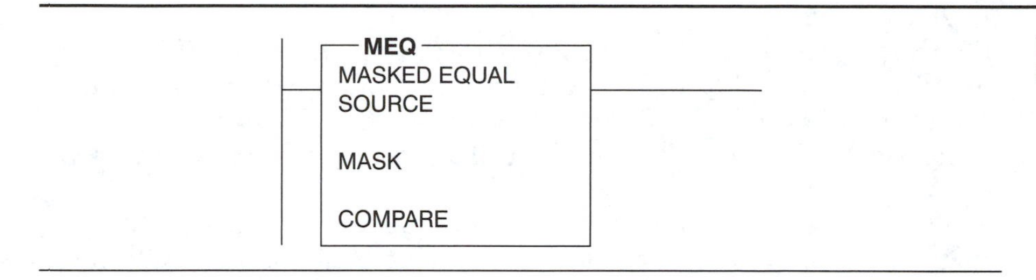

Figure 11–7
MEQ instruction

input patterns that would tell us in what position station #1 stopped. To do this, we need to filter out the high 8 bits since these bits are associated with some other station and have no bearing on our station #1 conditions.

For this example we would set the source to point to the input word for station #1, set the mask value to hex 00FF to mask out the high 8 bits, and set the compare value for the input conditions for which we want to test. If the compare instruction matches, the rung will go TRUE and we can use the output to trigger some diagnostic display message as to the status of station #1.

This same operation can also be used if we need to turn on some device every time the mechanism under control goes through a certain state. This could be necessary for safety because the output would come on only if the mechanism was in the proper state.

LIM Instruction

The limit test instruction is only available in the SLC-500/02 and higher processors (Figure 11–8). Its purpose is to allow the programmer to compare an input value with high and low limit values and give an output if the input value is between the limit values. The low and high limit values you program can be word addresses or program constants subject to the following restrictions:

1. If the test parameter is a program constant, then the limit values must be word addresses.
2. If the test parameter is a word address, then the limit values can be either a word address or a program constant.

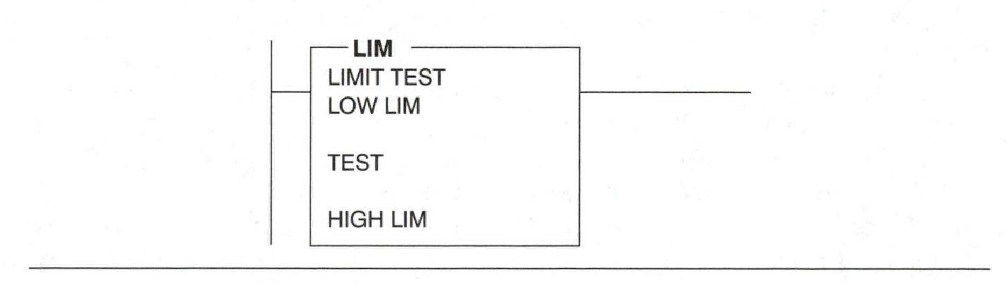

Figure 11–8
LIM instruction

True/False Status

If the lower limit is less than the high limit value, the rung will be TRUE for any value equal to or between the limit values. If the low limit is 10 and the high limit is 18, then the rung will be TRUE for any value from 10 to 18. For any value less than 10 or greater than 18, the rung will be FALSE.

If the lower limit is a larger value than the high limit value, then the rung will be TRUE for all values less than or greater than the limits. If the low limit is 20 and the high limit is 8, then the rung will be FALSE for any value from 8 to 20 and TRUE for any value above 20 or below 8.

Example

A good example of the LIM instruction was described in the section covering the GRT instruction. We said we had a tank of liquid and the program had to attempt to keep it at the three-quarters full level. Let's say we only control the fill pump in this application and the tank is emptied in short dumps of 25 gallons each by some other process. We could use the limit instruction to control a tank full status lamp if we set a low limit of 150 and a high limit of 175. The full lamp would turn on only if the rung goes TRUE, meaning the level is between 150 and 175. We would use a LES instruction to control the fill pump.

LOGICAL AND DATA MOVEMENT INSTRUCTIONS

Four logical instructions and two move instructions are included with the SLC-500 processor. All of these instructions are used as output elements on a control rung. Only if the rung's input conditions are TRUE will the instruction perform its function.

These are the six instructions:

1. AND: logical AND function
2. OR: logical OR instruction
3. XOR: logical XOR instruction
4. NOT: logical NOT instruction
5. MOV: data move instruction
6. MVM: masked data move instruction

AND Instruction

The value at the source A location is ANDed bit by bit with the value at the source B location and the result is stored at the destination location (Figure 11–9). Two arithmetic status bits are set as a result of this instruction. The Z-zero bit is set if the result of the instruction is zero; otherwise it is cleared. The S-negative bit is set if the most significant bit is set; otherwise it is cleared.

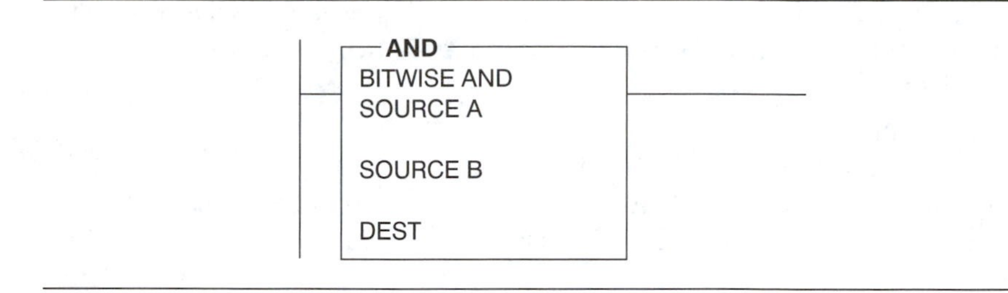

Figure 11–9
AND instruction

AND instructions are used to clear bits or test for bit patterns. An AND instruction can be used as a clear function because any bit ANDed with a 0 is cleared; any bit ANDed with a 1 will remain unchanged. (Refer back to Chapter 1 for math explanations.) To clear the low 4 bits of an output word without affecting the rest of the word, point source A to the word to be acted on, set source B to the hex value FFF0, and set the destination to the same location as source A. When the rung goes TRUE, the source location will have its low 4 bits cleared.

As a test function we can test a single bit to see if it is a 1. Set the source A register to the word to be tested, and set the source B register to a hex value with a 1 in the bit location to be tested. When the rung goes TRUE, check the Z bit to see if it is set. If Z is set, the bit was not set; if Z is cleared, the bit was set.

OR Instruction

The value of the source A location is ORed bit by bit with the source B location and the result is stored in the destination location (Figure 11–10). Two arithmetic status bits are set by this instruction. The Z-zero bit is set if the result of the instruction is zero; otherwise it is cleared. The S-negative bit is set if the most significant bit is set; otherwise it is cleared.

OR instructions are used to set a bit, or group of bits, in a register without affecting the rest of the register's contents. To set the high 4 bits in a register to 1, we first set the source A location to point to the register to be acted on. We then set source B to hex F000 and the destination to the same register address pointed to by source A to store the results. When

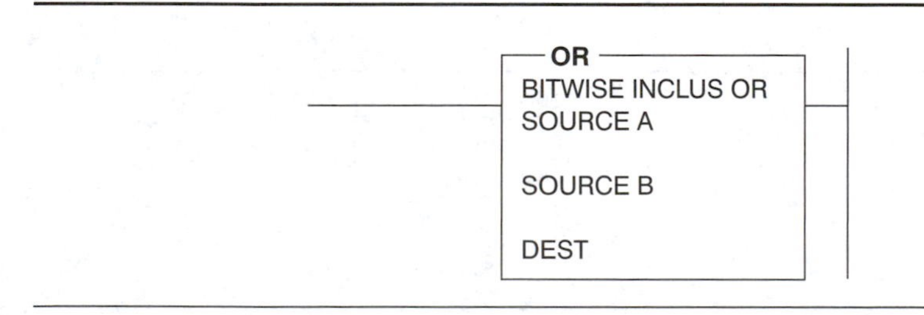

Figure 11–10
OR instruction

the rung goes TRUE, the value at source A will be ORed with the hex value and the result stored back at the source A destination location.

XOR Instruction

This instruction performs a bit-by-bit XOR function between the source A location and the source B location or a program constant and places the result at the destination location (Figure 11–11). Two arithmetic status bits are set by this instruction. The Z-zero bit is set if the result of the instruction is zero; otherwise it is cleared. The S-negative bit is set if the most significant bit is set; otherwise it is cleared.

XOR instructions are used to compare the contents of two locations and will give a zero result if they match exactly. (See Chapter 1 on math and logic functions.)

NOT Instruction

This instruction performs the invert or not logic function on a bit-by-bit action on the source A value and places the inverted result in the destination location (Figure 11–12). Two arithmetic status bits are set by this instruction. The Z-zero bit is set if the result of the instruction is zero; otherwise it is cleared. The S-negative bit is set if the most significant bit is set; otherwise it is cleared.

MOV Instruction

The move instruction allows the programmer to move the contents of one location to another location (Figure 11–13). The contents of the source location are not changed but the contents of the destination location are

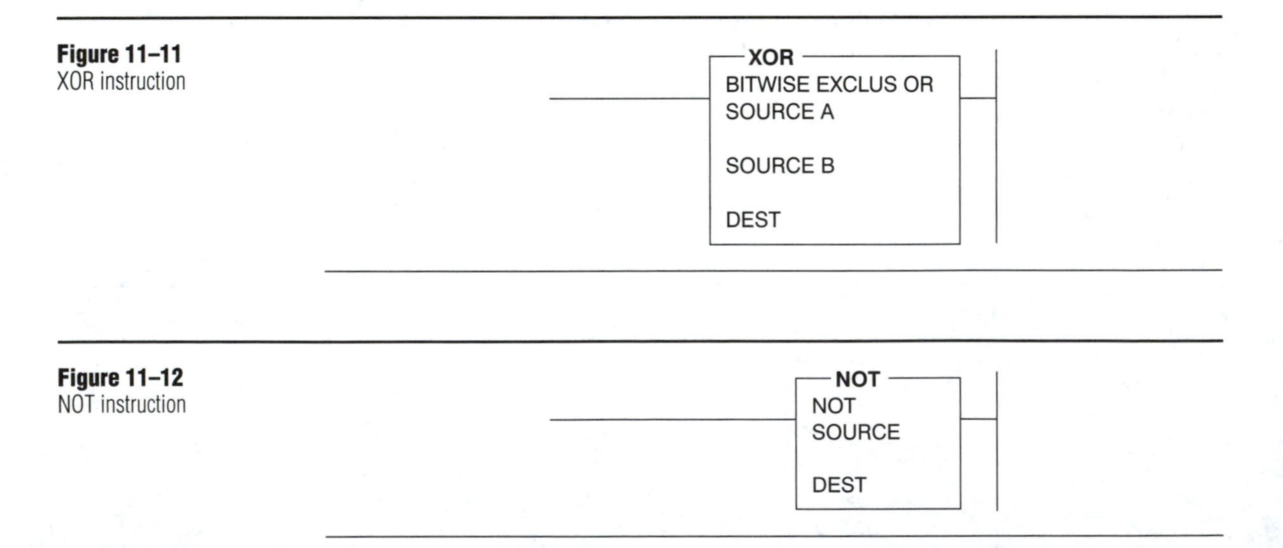

Figure 11–11 XOR instruction

Figure 11–12 NOT instruction

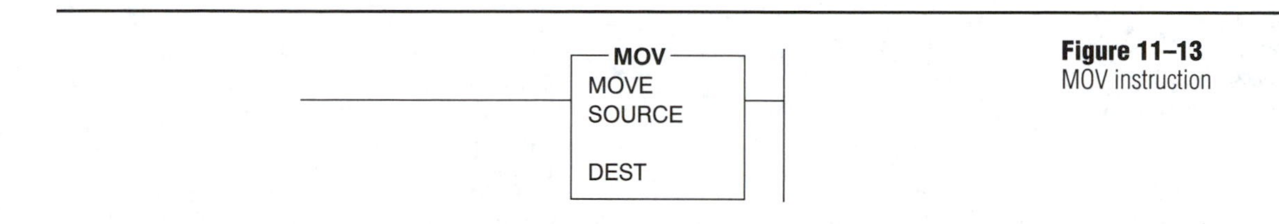

Figure 11–13
MOV instruction

replaced with the moved value. Two arithmetic status bits are set by this instruction. The Z-zero bit is set if the result of the instruction is zero; otherwise it is cleared. The S-negative bit is set if the most significant bit is set; otherwise it is cleared.

We used this instruction in Chapters 9 and 10 to change the preset value of timers or counters. In the chapter on diagnostics, you will also see some examples of how we use this instruction to move cycle time values from RTO timer ACCUM words to a memory address for display on a screen of some type.

MVM Instruction

The masked move instruction is the same as the move instruction except the data are moved through a mask before it is stored in the destination (Figure 11–14). Two arithmetic status bits are set by this instruction. The Z-zero bit is set if the result of the instruction is zero; otherwise it is cleared. The S-negative bit is set if the most significant bit is set; otherwise it is cleared.

The masked move instruction moves data from a source location to a destination, and allows portions of the destination data to be masked by a separate word. In Figure 11–14, SOURCE is the address of the data you want to move. MASK is the address of the mask through which the instruction moves data. The mask can also be a fixed constant hex value. DEST is the address to which the instruction moves the data.

As an example, in Figure 11–15 we move the data from B3:5 to B3:6 and strip off the high 4 bits of each half of the word. As long as the rung remains TRUE the program executes the move for each scan.

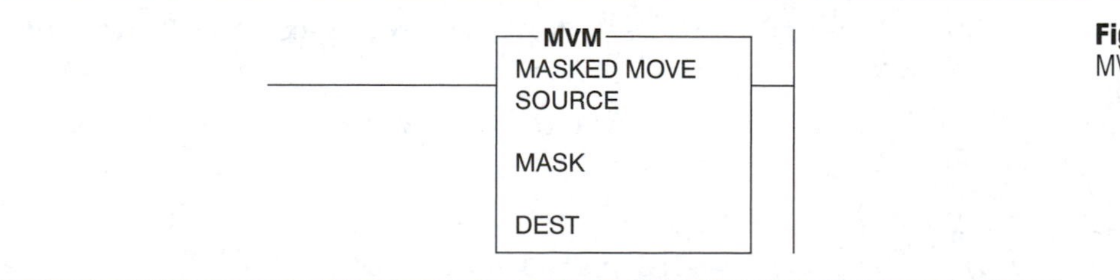

Figure 11–14
MVM instruction

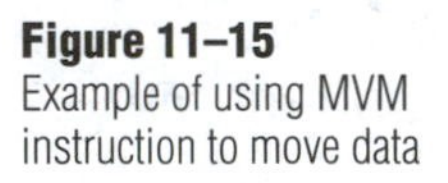

Figure 11–15
Example of using MVM instruction to move data

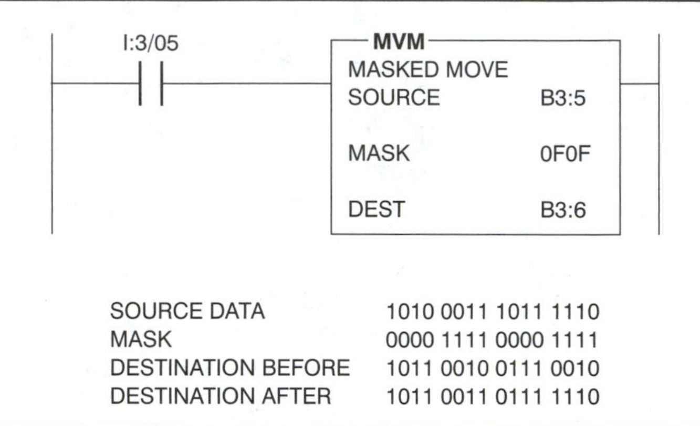

MATH INSTRUCTIONS

Ten math instructions are included with the SLC-500 processor and two additional instructions with the SCL-500/02 and higher processors. All of these instructions are used as output elements on a control rung. Only if the rung's input conditions are TRUE will the instruction perform its function.

These are the 10 math instructions:

1. ADD: add instruction
2. SUB: subtract instruction
3. MUL: multiply instruction
4. DIV: divide instruction
5. DDV: double divide instruction
6. NEG: negate or twos-complement instruction
7. CLR: clear instruction
8. TOD: convert to BCD instruction
9. FRD: convert from BCD instruction
10. DCD: decode instruction

The SQR (square root) and SCL (scale) instructions are for the SLC-500/02 and higher processors and are not discussed further in this text.

After the instruction is executed the arithmetic status bits in the status register are updated.

The S:0/0 carry bit is set if a carry was generated; otherwise cleared.

The S:0/1 overflow bit is set if the result did not fit in the destination register.

The S:0/2 zero bit is set if the result of the math instruction is zero.

The S:0/3 sign bit is set if the result is negative.

The S:13 bit contains the least significant word of the 32-bit values of the MUL and DDV instructions. It contains the remainder for DIV and DDV instructions. It also contains the first 4 BCD digits for the FRD and TOD instructions.

The S:14 bit contains the most significant word of the 32-bit value generated by the MUL or DDV instructions. It contains the unrounded quotient for the DIV and DDV instructions. It also contains the most significant digit for the TOD and FRD instructions.

ADD Instruction

The value at source A is added to the value at source B and the result is stored at the destination address (Figure 11–16). The following arithmetic status bits are affected:

- **C:** The carry bit is set if a carry is generated and cleared otherwise.
- **V:** The overflow bit is set if the result does not fit in the destination location.
- **Z:** The zero bit is set if the result is zero.
- **S:** The negative bit is set if the result is negative.

SUB Instruction

The subtraction instruction subtracts the value at source B from the value at source A and places the result at the destination address (Figure 11–17). The following arithmetic status bits are affected:

- **C:** The carry bit is set if a borrow is generated and cleared otherwise.
- **V:** The overflow bit is set if the result underflows.
- **Z:** The zero bit is set if the result is zero.
- **S:** The negative bit is set if the result is negative.

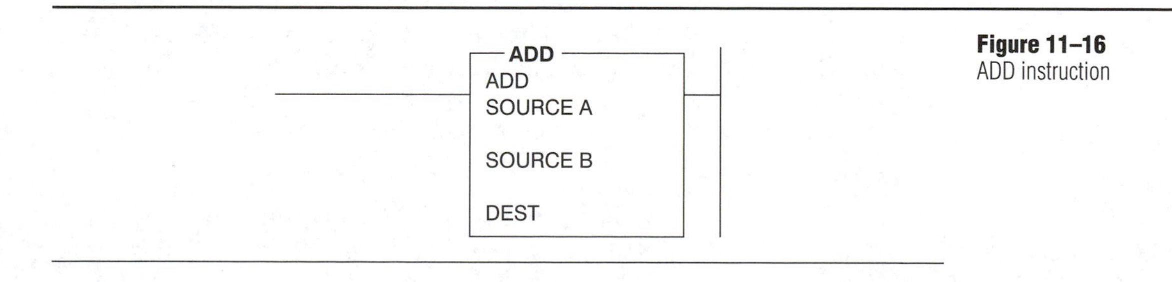

Figure 11–16
ADD instruction

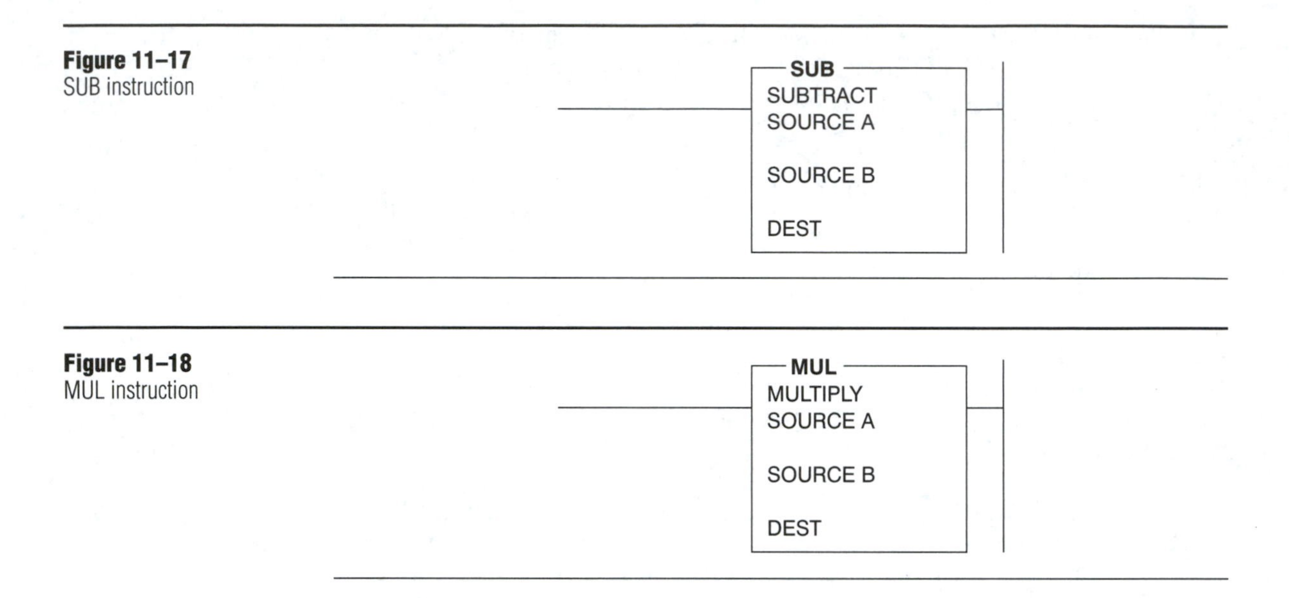

Figure 11–17
SUB instruction

Figure 11–18
MUL instruction

MUL Instruction

This instruction multiplies the value at source A by the value at source B and places the result in the destination address (Figure 11–18). The following arithmetic status bits are affected:

C: Always reset.
V: The overflow bit is set if the result overflows at the destination address.
Z: The zero bit is set if the result is zero.
S: The negative bit is set if the result is negative.
S:14, S:13: These bits contain the full 32-bit integer result of the multiply operation. This value can be used in the event of an overflow.

DIV Instruction

This instruction divides the value in source A by the value in source B with the rounded quotient being stored in the destination (Figure 11–19). If the remainder is 0.5 or greater, a round-up occurs in the destination. The unrounded quotient is stored in the most significant word of the math register. The remainder is placed in the least significant word of the math register.

The arithmetic bits are affected as follows:

C: Always reset.
V: The overflow bit is set if the result overflows at the destination address.

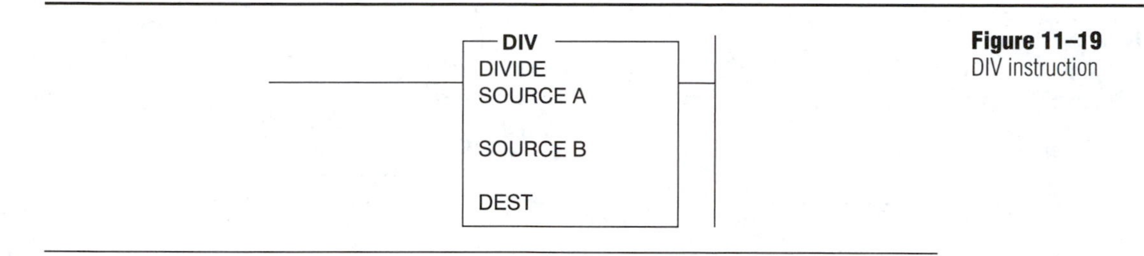

Figure 11–19
DIV instruction

Z:	The zero bit is set if the result is zero.
S:	The negative bit is set if the result is negative.
S:14, S:13:	The unrounded quotient is placed in the most significant word, the remainder is placed in the least significant word.

DDV Instruction

The DDV instruction divides the math register by the source value (Figure 11–20). The rounded quotient is placed in the destination address. If the remainder is 0.5 or greater, a round-up occurs at the destination address. The unrounded quotient is placed in the most significant word of the math register. The remainder is placed in the least significant word of the math register.

The arithmetic bits are affected as follows:

C:	Always reset.
V:	The overflow bit is set if division by zero or if the result is greater than 32,767 or less than –32,767; otherwise it is reset.
Z:	The zero bit is set if the result is zero.
S:	The negative bit is set if the result is negative.

NEG Instruction

The source value is subtracted from 0 and the result is stored in the destination location (Figure 11–21). The destination therefore contains

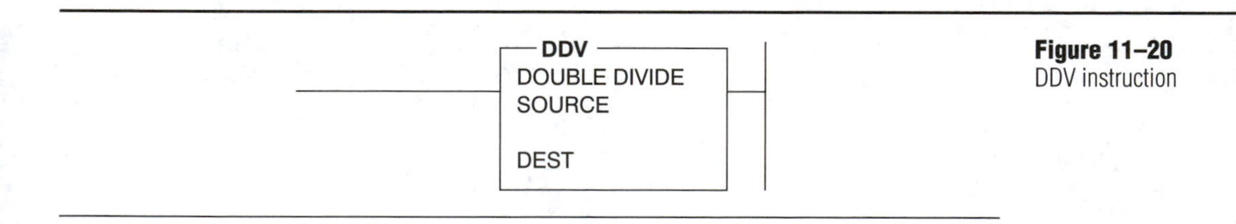

Figure 11–20
DDV instruction

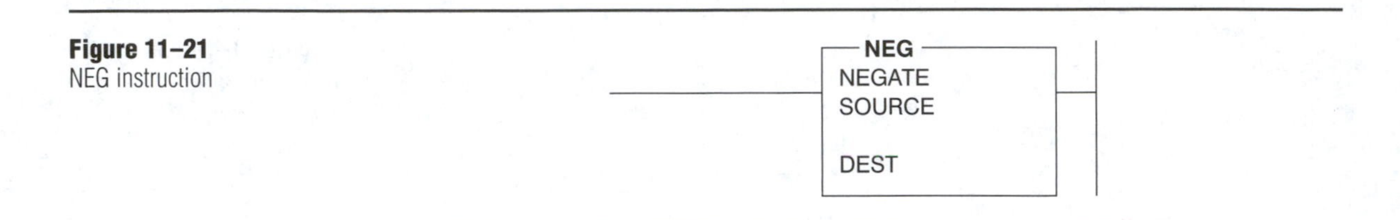

Figure 11–21
NEG instruction

the twos-complement of the source value and the arithmetic bits are affected as follows:

- **C:** Cleared if 0 or overflow; otherwise set.
- **V:** The overflow bit is set if an overflow occurred; otherwise reset.
- **Z:** The zero bit is set if the result is zero.
- **S:** The negative bit is set if the result is negative.

CLR Instruction

The destination address is cleared, set to all 0s (Figure 11–22). Bits C, V, and S are always reset; Z is always set.

TOD Instruction

The TOD instruction converts a decimal number into BCD format (Figure 11–23). The source must be a word address. The destination can be any word address or the math register S:13 and S:14 in the SLC-500/02 or higher controllers. In the SLC-500/01 controller, the destination can only be the math register.

As an example we will convert the decimal value of N7:6 to BCD in the math register.

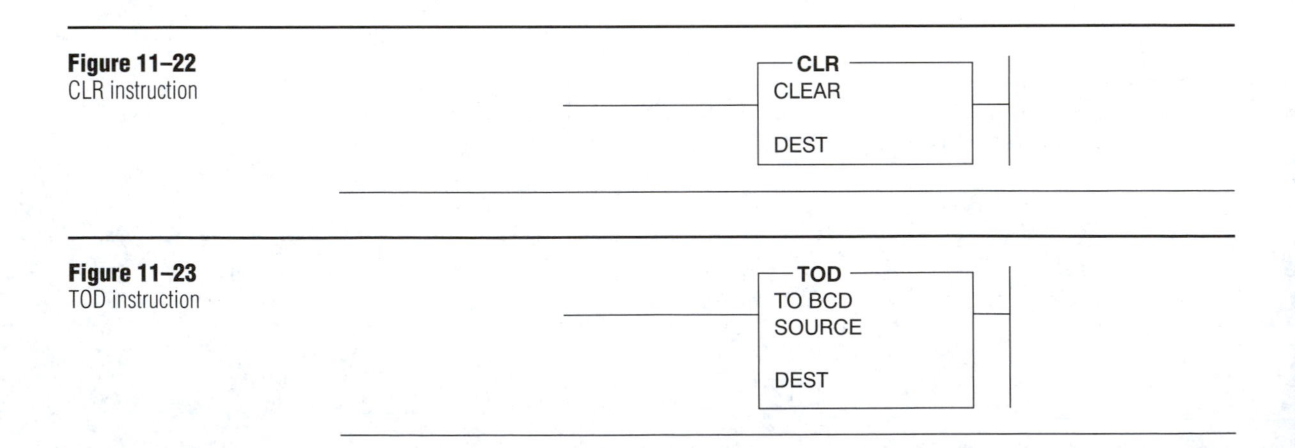

Figure 11–22
CLR instruction

Figure 11–23
TOD instruction

N7:6 = 12465

S:14 = 0 0 0 1   S:13 = 2 4 6 5

You can now move the math register to any location using the MOV instruction. Note how the arithmetic bits are affected:

C: Always reset.
V: The overflow bit is set if the BCD value is larger than 9999; otherwise reset.
Z: The zero bit is set if the destination value is zero.
S: The negative bit is set if the result is negative; otherwise reset.

FRD Instruction

This instruction is used to convert a BCD value to integer or decimal values (Figure 11–24). The source can be any word address or the math register S:13 and S:14 in the SLC-500/02 controller and higher. In the SLC-500/01 controller, the source can only be the math register.

As an example we will convert the contents of the math register to decimal and store it in address N7:7.

S:14 = 0 0 0 2   S:13 = 3 5 8 2

N7:7 = 23582

The arithmetic bits are affected as follows:

C: Always reset.
V: The overflow bit is set if a non-BCD value is contained at the source or the value to be converted is greater than 32,767; otherwise reset.
Z: The zero bit is set if the destination value is zero.
S: Always reset.

DCD Instruction

This instruction reads the low 4 bits of the source address and decodes the 4-bit binary value into 1 of 16 bits at the destination address (Figure 11–25). The source and destination variables must be word addresses. The contents

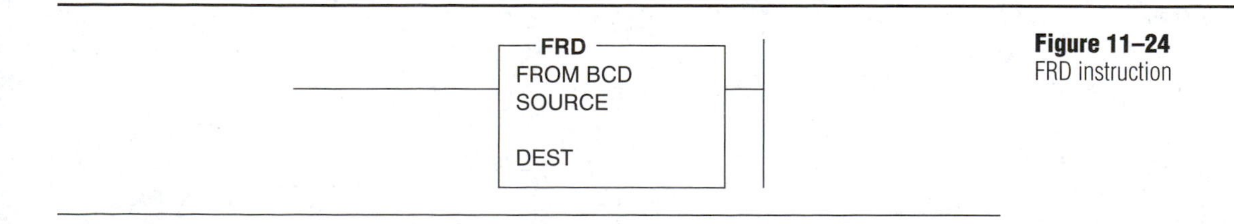

Figure 11–24
FRD instruction

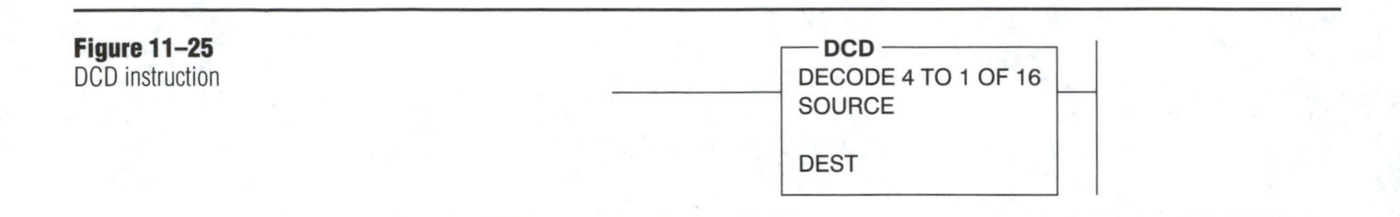

Figure 11–25
DCD instruction

of the high 12 bits of the source address are not changed or modified by this instruction.

Source The low 4 bits of data at this location are used as a binary value to be decoded.

Destination This location will have only one of its 16 bits set high depending on the value read from the source.

This instruction can be used when reading thumb wheel switches or other coded slide switch banks. When a programmer needs to strobe each switch in sequence and read the 4-bit return signals, this instruction is a good choice. It can also be used to sequence a set of lights to indicate what stage a mechanism is in for diagnostic purposes.

DATA FILE INSTRUCTIONS

Two instructions are available that allow you to operate on whole data files: the copy (COP) and the fill (FLL) instructions.

Copy (COP) Instruction

The copy instruction causes the entire data file, pointed to by the source address, to be copied to the destination addresses (Figure 11–26). The number of locations moved is determined by the length value.

The destination file type determines how many words are moved. For example, if the destination file type is timer and the source type is integer, then three words will be copied for each timer element. If both

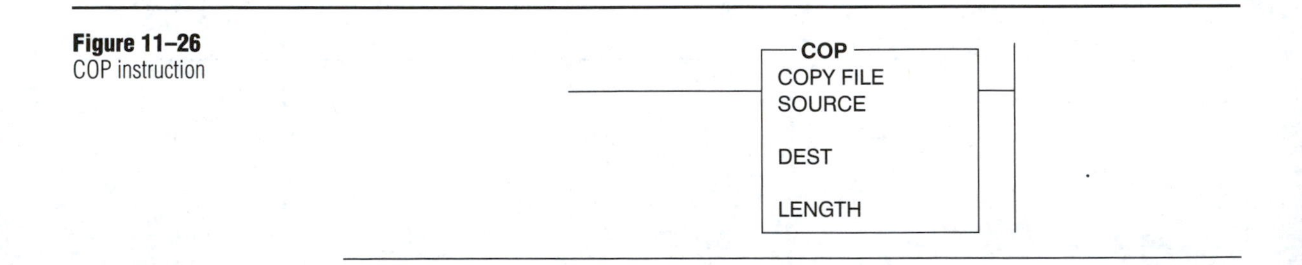

Figure 11–26
COP instruction

source and destination are integer types, then one word will be moved for each value set in the length variable.

Source	This is the beginning address of the file you want to copy. You must use the file indicator # in front of the address.
Destination	This is the beginning address of the place where you want the data copied. You must use the file indicator # in front of the address.
Length	This is the number of address locations that you want to copy. If the address destination is of a file type that requires three words per location the maximum number of locations you can move is 42. If the destination type is a single word, then the maximum file size is 128.

The entire file is copied during each program scan as long as the input condition is TRUE. Note that this can be a problem if you are using the copy command to make a shift register (see Chapter 13 program example). Data are copied in ascending order beginning at the source address location and ending when the length count reaches zero. No modification of the data is possible. The instruction will not write over a file boundary such as between N7:11 and N7:12.

Example

In control programs, different sets of variables are often required for different products being run down the same assembly line. Let's use an example of a paint mixing line. Containers of the base color are started down the line and pass under several color dispensers that meter color pigments into the base color.

We can control the amount of color pigment being dispensed by using a counter for each dispenser mechanism that counts each 0.1 ounce of color added. For each color run we will need to develop a recipe of counter preset values to be loaded into the dispenser counters. The easy way to do this is to have the recipes loaded into memory and use the copy command to download the proper recipe for the color we are going to run. When the new color recipe is downloaded, the entire line is changed over to run a new color in one scan time. This type of control problem is common to many processes in the medical, chemical, and food processing industries.

Fill (FLL) Instruction

The fill instruction is used to fill memory locations or registers with a common value (Figure 11–27). We often need to clear a block of memory to all "0000" or set it to all "FFFF" at the beginning of a program. The contents of

Figure 11–27
FLL instruction

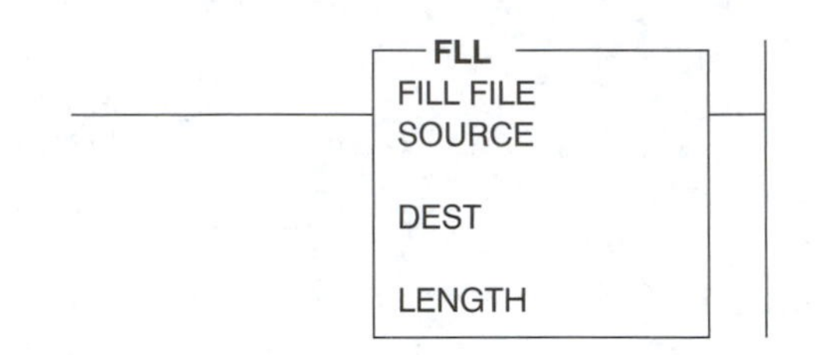

the source address or a constant value is copied to the destination addresses in an ascending order until the value entered in the length variable is zero.

Source This is the address where the fill value is stored, or you can enter a constant value to be used as the fill value. You do not use the file indicator # in front of the address.

Destination This is the beginning address of the place where you want the data copied. You must use the file indicator # in front of the address.

Length This is the number of address locations that you want to copy. If the address destination is of a file type that requires three words per location, the maximum number of locations you can move is 42. If the destination type is a single word, then the maximum file size is 128.

Addresses are filled in ascending order beginning at the destination address location and ending when the length count reaches zero. The instruction will not write over a file boundary such as between N7:11 and N7:12.

SPECIAL ADDRESSING MODES

Indexed Addressing Mode

This addressing mode is only available for use in SLC-500/03 and higher processors. The indexed addressing mode is signified by placing a # sign in front of the file indicator character. Indexed instructions all use the (S:24) word in the status file to get their pointer offset value. For example, if the S:24 register contained a value of 12 the instruction shown in Figure 11–28 would move the data at address N7:15 + 12 or address N7:27 and place the data in address N7:35. The value contained in the S:24 register is added to the instruction's source address and the data is read from the new address.

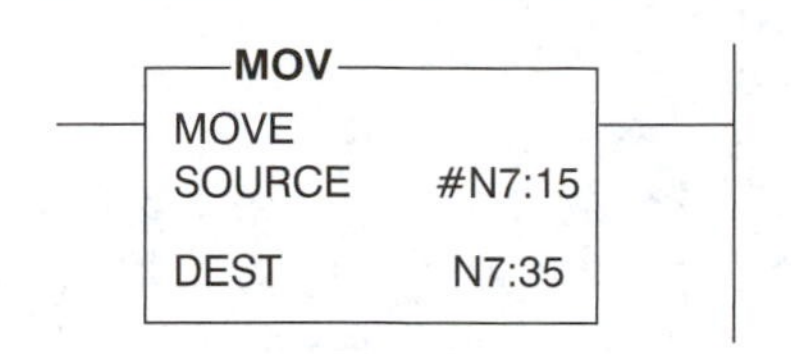

Figure 11–28
Move instruction using indexed addressing

You must be sure to set the value in the S:24 register just prior to using an indexed addressing instruction as many different instructions change the contents of the S:24 register. You must also be sure the combined file address is not larger than the file boundary. File boundaries are not automatically allocated when using an indexed instruction. You must set the upper file boundaries by opening the file you intend to use, in our example by double-clicking on the N7 file on the left edge of your programming screen, and then opening the properties window. In the example shown in Figure 11–29 we would need to change the *last* window value to a value greater than the highest address we will use. If N7:35 is the highest address you will use in the N7 file, then you should set the boundary to N7:40 to give yourself a little expansion room. Be aware that the value in the S:24 register can also be a negative number. You must be very careful not to offset the address below the lower file boundary. The address counter does not wrap around, as this

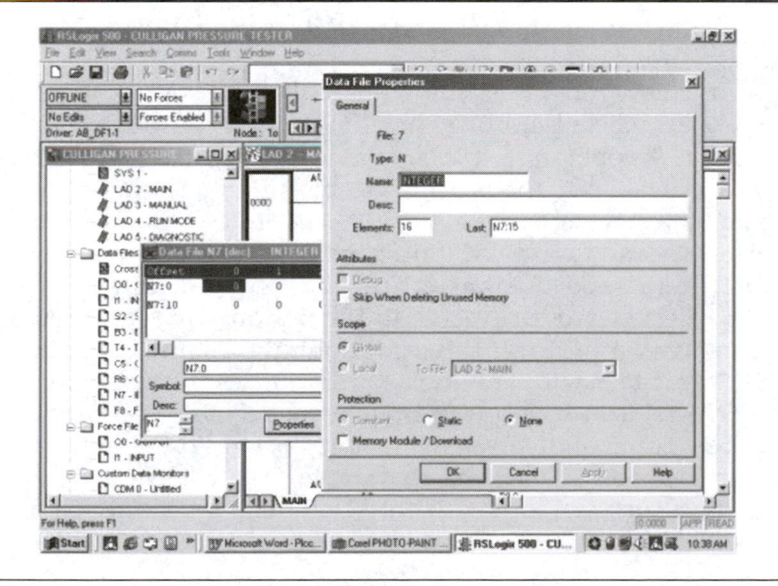

Figure 11–29
File boundary setup

condition could write data into the next lower file and could cause unpredictable problems.

One of the most common uses for indexed instructions is the loading of recipe data. It is common in process control projects to have a number of different products that are produced on the same set of equipment. We now look at an example of this type of design problem. Let's assume we have a batch mixing process at a personal care company. Let's say we are mixing up batches of hand cream to be dispensed into bottles and tubes. On this equipment we produce 10 different products that all start with the same base material but then get different additives and coloring mixed in. Each of these materials is pumped into the mixing vat through a separate metering valve that produces a pulse for each ounce of material that flows through it. The outputs of these metering valves are sent to nine counters, and the DN bits of these counters control the pumping motors for each material. See Figure 11–30 for the counter layout and recipe storage locations.

From a control standpoint all we have to do is enter the proper preset counts into the nine counters, reset the counters, then start the pumps and let each counter shut off its pump when the proper amount of material has been metered into the mixing vat. The problem here is a poor man-machine interface that requires the operator to enter nine different values each time they change products. This not only slows down the changeover process, it also opens the door to bad batches and wasted material if the operator enters the wrong data. A better method of operation would be to have all 10 sets of metering data stored into memory by the quality control manager and only require the operator to select which product is to be produced. Once the product is selected the control system should transfer the new counter values from the stored recipes to the counter presets.

Looking at Figure 11–30 you will see that the bottom 10 addresses of file N12 have been set up as the data transfer locations. The ladder program for this operation will have nine MOV instructions to transfer the nine data values to the proper counter preset locations. The 10 sets of recipe data start at address N12:10, and a new recipe starts every 10 address locations. Only the first three are shown in the diagram. We must find a way of transferring the required nine data values from the selected product recipe to the bottom nine locations in file N12. One way to perform this programming requirement is to use the COP command with an index source address. In this way we can set the S:24 register to point to the proper recipe selection and then let the COP command copy nine words of data to the bottom of the file. In order to keep this program sample simple let's give the operator a 10-position selector switch to select which product to run (see Figure 11–31).

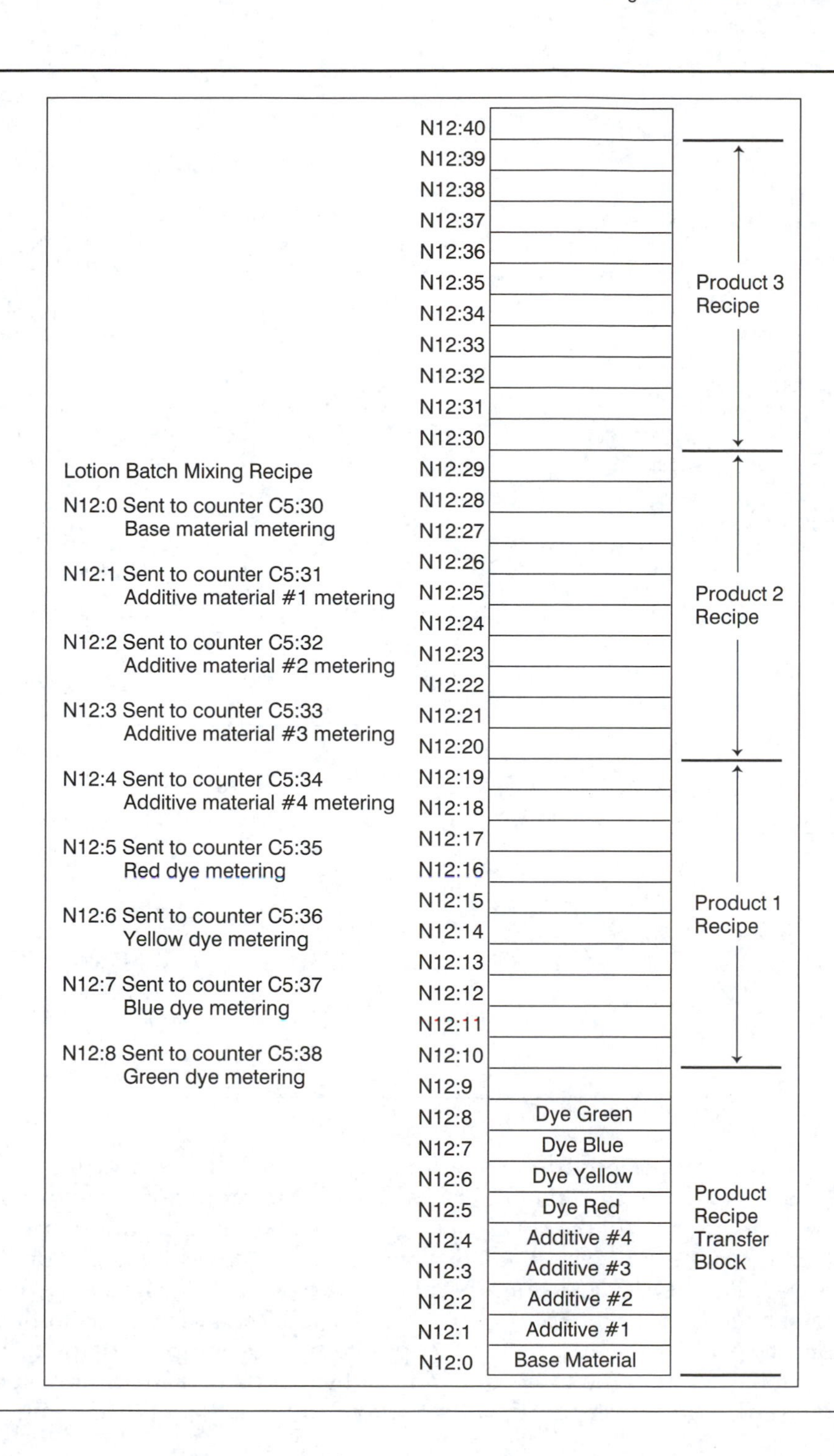

Figure 11–30
Address layout

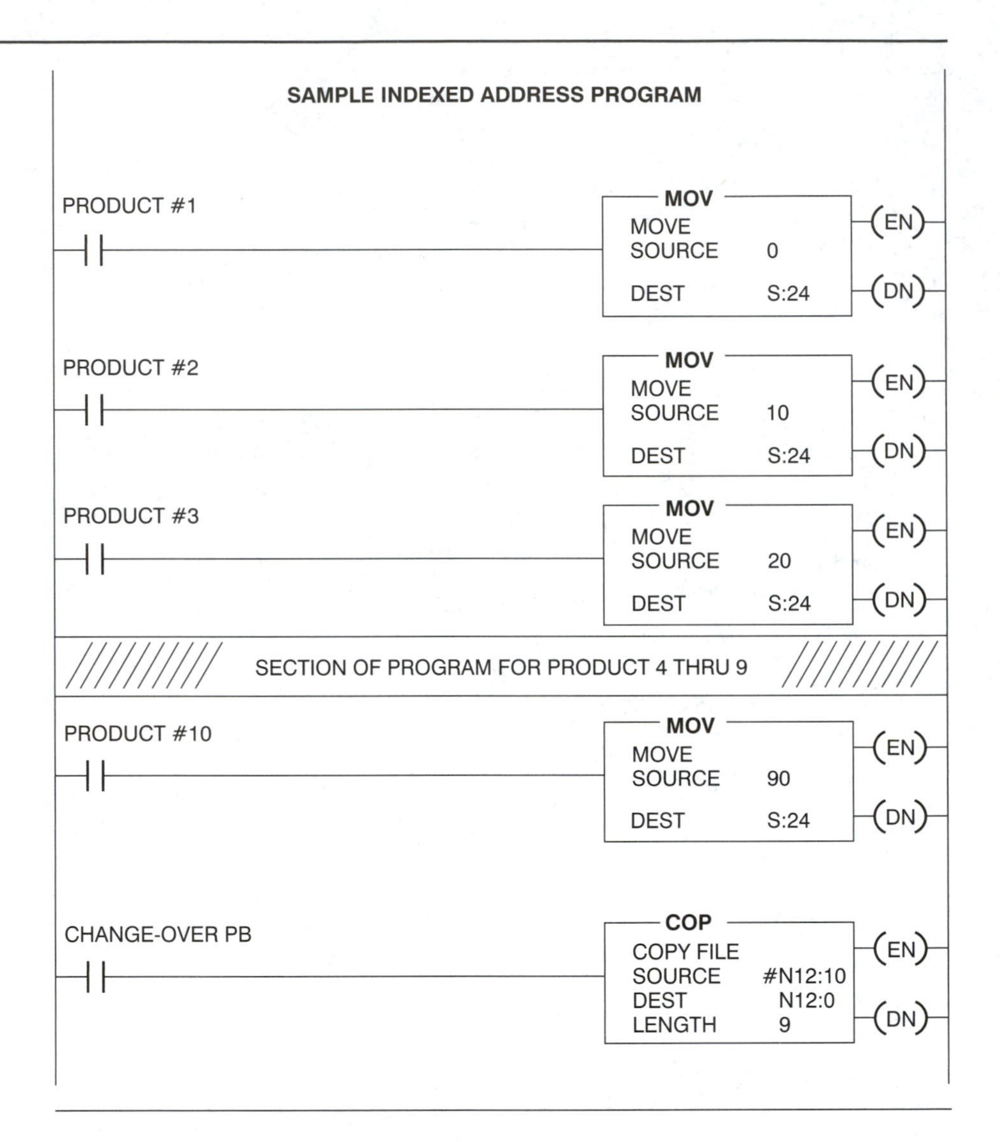

Figure 11–31
Sample program using indexed addressing

In the sample program shown in Figure 11–31 there are 10 MOV instructions that set up the S:24 register with the proper offset address. Note that only four of the ten are shown. At the bottom of the program the COP command will be activated when the changeover push button is pressed. The COP instruction uses an indexed address for its source address. Note that if the S:24 register contains a 0, then the data is pulled from the N12:10 address. As the S:24 pointer is moved up by 10 for each additional product the COP command pulls the data for the next recipe from the base address plus the offset amount. In this example one might ask, why not just use 10 different COP commands instead of the indexed addressing mode? In fact this would also work, but in a real application, the offset value will most likely come for a single address assigned to the HMI, Human Machine Interface, instead of a selector switch. In this case the indexed addressing mode makes a lot more sense, as only one move instruction would be necessary.

Indirect Addressing Mode

The indirect addressing mode can be used whenever you wish to use a file type and the contents of a word in memory to define the complete address within that file. You use the [] characters to bracket the substitute address element. This addressing mode is only available in SLC-500/03 and higher processors. You could use a counter's ACCUM value to define the word position in a file, for example, N7:[C5:12.ACC]. This would allow you to sequence up the N7 file one word at a time by incrementing the counter C5:12 value once for each operation. You could also let a word in memory contain the position count for a bit file, for example, B3/[N7:25]. In this case the value in N7:25 would be used to define the position of a bit in the B3 file. One can even use the value in a word to define a slot number for an I/O function, I:[N7:10].0/1. This instruction address would use the value in the N7:10 word to define which slot number to pull the data from. Another example of indirect addressing would be the sequential creation of a sequence of N files by using the format N[C5:15.ACC]. This would allow you to create sequential N files for saving batch processing data.

As was the case with indexed addressing, you must be careful not to cross file boundaries in indirect addressing mode. The same rules apply as previously defined in indexed addressing.

CONCLUSIONS

In conclusion, we have now looked at all the instructions in the SLC-500 that allow you to compare, change, modify, or move the data in any register or memory location. In most cases we have provided short examples as to how these instructions can be used to solve real-world programming problems. We have also introduced two new addressing modes that are available in the higher numbered processors.

In the following chapters we will show many examples of how this group of instructions is used in more complicated programs. If you have problems understanding how a particular instruction was used, you may want to refer to this chapter and review the particular instruction in more detail. You are now ready to complete the lab at the end of this chapter.

REVIEW QUESTIONS

1. Show an example of how you could use the LEQ and GEQ instructions together to turn on memory bit B3/7 if the counter C5:6, ACC value was between 5 and 12.
2. Show an example of how you would turn on memory bit B3/4 only when the timer T4:10, ACC value was 24 and input I:0.1/2 was on.
3. Show an example of how you would clear the most significant 8 bits of memory location N7:8.
4. Show how you would set the most significant bit of N7:19 to a value of 1.
5. Show how you would move the counter C8:5, ACCUM value to location N7:25 whenever B3/18 is on.
6. Show how you would add 15 seconds to the preset value of timer T4:12, PRE when bit B3/19 is on.
7. Show a rung that would move a mixing recipe beginning at address N7:10 to counters C5 through C14 when bit B3:15 is on.
8. Show a rung that would fill N:14,0 to N:14,15 with all zeros whenever input I:0.1/12 is on.
9. Show a rung that would change the preset value of timer T4:14 to "050" if input I:0.1/13 is on.
10. Show a rung that would turn on bit B3/19 if counter C5:3's value is under 15 and timer T4:13 is below 025.
11. At what address would data be read from if the S:24 register was set to 30 and the MOV instruction's source address was defined as #N7:25?
12. If counter C5:20 was at 15 and the move instruction's destination address was defined as N12:[C5:20.ACC] where would the data be stored?
13. Why must you load the S:24 register with the offset value immediately before using an indexed instruction?
14. What is meant by the phrase "crossing a file boundary"?
15. How can you set the file boundary to a large enough value to avoid boundary problems?

LAB ASSIGNMENT

Objective

Introduces the student to a continuously running program and methods for detecting states within the program cycle. We will use various compare instructions to detect program states.

1. Open a new file in RSLogix and save it as Chpt-11-lab. Although this lab was designed to run on the lab kit described in Appendix A, it is generic in form, and can be run on most lab kits by changing the I/O addresses. Be sure to enter the correct I/O configuration for the lab kit you are using.
2. Enter the rungs shown in Figure 11–32, then download and run the program. Turn on SEL SW 1. Verify that the program runs and increments or decrements counter C5:1 once every second. The counter should count up to 20, then down to zero, and then repeat the cycle as long as SEL SW 1 is on.
3. What function does B3/0 perform and why is it latched?
4. What would happen if we didn't latch B3/0?
5. What condition does lamp 1 display?

Figure 11–32

```
0   SEL SW 1 I:0.1 [ ] 4 ---- T4:1 [/] DN ---- FLASH TIMER 1
                                               TON
                                               TIMER DELAY ON
                                               TIMER        T4:0      (EN)
                                               TIME BASE    0.01
                                               PRESET       50        (DN)
                                               ACCUM        0

1   T4:0 [ ] DN ---------------------------- FLASH TIMER 2
                                               TON
                                               TIMER DELAY ON
                                               TIMER        T4:1      (EN)
                                               TIME BASE    0.01
                                               PRESET       50        (DN)
                                               ACCUM        0

2   T4:0 [ ] TT ---- B3 [/] 0 -------------- CTU
                                               COUNTER UP
                                               COUNTER      C5:1      (EN)
                                               PRESET       20
                                               ACCUM        0         (DN)

3   T4:0 [ ] TT ---- B3 [ ] 0 -------------- CTD
                                               COUNTER DOWN
                                               COUNTER      C5:1      (EN)
                                               PRESET       20
                                               ACCUM        0         (DN)

4   C5:1 [ ] DN ---- T4:1 [ ] TT ------------------------------------ B3 (L) 0

5   EQU
    EQUAL
    SOURCE A  C5:1.ACC  ---- T4:1 [ ] TT ---------------------------- B3 (U) 0
    SOURCE B  000

6   T4:1 [ ] TT ----------------------------------------------------- LAMP 1 O:0.0 ( ) 0

7   B3 [ ] 0 -------------------------------------------------------- LAMP 2 O:0.0 ( ) 1
```

6. What condition does lamp 2 display?

7. Add rungs to the program to perform the following functions:

   - Turn on lamp 3 when the count is equal to or above 10.
   - Turn on lamp 4 when the count is below 10.
   - Turn on lamp 5 when the count is between 5 and 15 inclusive.
   - When SEL SW 2 is ON, change the count rate to once every 2 seconds. Change it back to once a second if SEL SW 2 is OFF.

- When SEL SW 3 is ON, change the count preset value to 30. When SEL SW 3 is OFF, change it back to 20. Be sure not to reset the count to 20 if it is currently above 20 and counting up.
- Set address N7:0 equal to 40.
- Create a rung that takes the value in counter C5:1.ACC, adds it to the contents of N7:0, and places the answer in N7:1.
- Create a rung that subtracts the current value of C5:1.ACC from the contents of N7:0 and places the answer in N7:2.

8. Verify that all the conditions are working. Print out your program and demonstrate it to the instructor.

CHAPTER

12

Basic Machine Control Programming

LEARNING OBJECTIVES

After completing this chapter the reader should understand:

1. Program construction using subroutines for manual and automatic operation.
2. OSHA requirements for manually activated machines.
3. How to build safety into programs for machine operation.
4. Creating a watch dog timer to stop the run mode cycle if a jam-up occurs.
5. The use of a single run/manual selector switch to change operating modes.
6. The use of stack lights to display a machine's current operating mode.

It is difficult to generalize about machine control programming due to the almost infinite variety of machines now in common use. I can, however, list some basic rules that should be followed on most new equipment designs to meet OSHA and other regulatory organizations' laws and requirements with which the design engineer must be concerned.

One of the most important elements of all the laws and requirements is that the designer must consider the operator's safety in the original design. It is important to keep accurate records in the early design stages of all the effort that went into the safety considerations. In any litigation that may result from someone being harmed on a piece of equipment you designed, the court will be looking for proof that the designer considered the safety of the operator and implemented means to prevent any possible injury. Being able to provide documents that prove the operator's safety was considered, and all reasonable safety devices were installed, may mean the difference between a small award for real damages and an extremely large award to punish the organization for flagrant disregard for operator safety.

Most automation equipment can be divided into two major classifications: manually activated or continuously running equipment. Manually activated machines require the machine's operator to start each cycle, and they usually perform some function such as loading and unloading of parts. Regardless of the machine's function, when the cycle is complete, the machine stops and waits for the operator to start the next cycle. The typical example is a single-cycle punch press that must be activated by the operator each cycle.

Continuously running machines, once started, will cycle continuously until they are cycle stopped or shut down due to a diagnostic stop. Operators of machines of this class may only monitor the machine while supplying raw material and removing finished product. Continuously running machines will be covered in more detail in the next chapter.

We will investigate manually activated machines in this chapter, and finish by writing a program for this machine type. Manually activated machines are generally easier to program than continuously running machines. Manually activated machines are defined as equipment for which the cycle is physically started each time by an operator and then stops after the machine's cycle is complete. An example of this would be a punch press that has an operator who must load a part into the machine before the press cycles and then unload the part after the cycle is complete. The operator must repeat the process each time the machine completes the cycle. A second example might be an injection molding machine that molds plastic around some metal part that is loaded into the mold fixture. In each case the operator must load the machine, start the cycle, and then unload the finished product. To help illustrate these concepts, we will create an example machine and program it in detail.

OSHA requires that the designer comply with the following rules:

1. Both of the operator's hands must be in a known safe place to start the machine's cycle. One way to accomplish this is to have two separate start push-button switches spaced far enough apart that an operator *cannot* press both buttons with one hand. They should also be recessed so the operator cannot press one with a hand and the

other with the elbow of the same arm. Additional safety can be gained by requiring that both push buttons must be pressed within a specified time interval. Intervals of 0.25 to 0.5 seconds are common.

2. Both push buttons must be released at the end of the cycle before another cycle can be started. This is to prevent an operator from holding one button down with an object, and running the machine using only one hand to start each operation. You must also be sure the program will not allow the machine to recycle if both push buttons are still being held down at the end of the cycle.
3. After starting the cycle, both push buttons must be held down until the machine reaches a safe state. If either button is released before the safe state is reached, the machine must stop immediately.
4. You also must enclose any part of the machine where a pinch point is located. The enclosure must have some type of safety switches installed that sense if any part of the enclosure has been removed. In the event that one of these safety guard switches opens up during an assembly cycle the machine must stop operating immediately.

How you comply with the four main OSHA rules depends on the type of equipment you are building. To comply with rules 1 and 2 most people use two start push buttons spaced apart and usually recessed to prevent a long object from being placed across both buttons. In addition, an anti-tie-down circuit is required to complement the mechanical input devices. This circuit should require the operator to press both push buttons within a short time window to start the cycle and to release both buttons to start a new cycle. Due to carpal tunnel problems the two push buttons are often replaced with optical sensors that do not require the operator to apply any force to start each cycle.

Rule 3 requires that the control circuit sense when a safe state is reached. If no safe state exists, then the operator must hold the start buttons for the whole cycle. A safe state can be accomplished by having some type of wire screen or clear guard move in front of the operator. In the case of a molding operation, when the mold is closed, the operator can release the buttons.

Rule 4 is normally accomplished by placing a wire screen or Plexiglas enclosure around the dangerous area. These safety enclosures must include safety switches that will indicate if the enclosure is removed. The control circuit must shut down the cycle if any safety switch is opened. The machine can have a safety bypass switch that allows maintenance people to work on the equipment with the guards off, but it should be a keyed switch to prevent machine operators from bypassing the safety guards.

CONTROL PROGRAM LAYOUT

Up to this point, we have only written small pieces of programs to demonstrate particular instructions or methods of controlling certain mechanisms. In this chapter, we will soon be putting together our first really complete program, but there are a few basic program construction parameters that we need to present before we do that. How you construct your program has a great deal to do with how easy it will be to read, debug, and document. The proper layout will also enhance the safety systems needed in programming a manually activated machine. Most mechanical equipment must be controlled in two very different modes of operation. They are usually referred to as *setup* and *run* mode or something similar to those names. Setup mode is also sometimes called *jog* or *manual* mode.

In the setup mode, the machine is usually being fixed or adjusted by a maintenance man or maintenance person. He requires the ability to move each mechanism independently and as often as necessary. If he is adjusting a prox or limit switch on an air cylinder or some mechanism attached to it, he may need to cycle the cylinder numerous times until he has the adjustment right. We expect the maintenance person to have a very good understanding of the equipment and therefore we will not limit what he can do except in cases where damage would result.

To provide him with this manual control we must supply him with push button switches or some other type of input device to manually activate each moving element in the machine. He often will be working without machine guards in place and therefore we should provide him with a keyed switch to bypass the guard safety switches. We should also provide some type of visual device for telling the maintenance person when the machine is in the home or start position.

In run mode we assume the machine operator has little or no understanding of the equipment, and we therefore must control every step of the process in exacting detail from cycle start to cycle done. Because the production operator is normally not qualified to set up or repair the machine we must disable any manual mode switches used by the maintenance man and allow the operator merely to start the machine cycle with the safety switches always in place. We need to provide a diagnostic program to watch the operation and shut down the machine if it finds conditions that indicate something has failed. This will be the job of our watchdog timer in the run portion of the control program.

As you can see from the preceding descriptions, the two modes of operation are very different. Ladder programs written to control each mode of operation will be very different in structure and we must be sure that only one of them is ever active at a given time. Both of these programs will need to control the same sets of outputs and in some cases read the same inputs. We must provide a means for switching between the two modes easily and quickly. We usually accomplish this with a mode control switch mounted on the main control panel.

Ladder control programs can be written in a number of different ways to provide the different modes of operation. A basic approach to this programming problem will be explained next. A brute force way to achieve the dual program control is to write two parallel rungs for each output, that is, creating one rung for the setup mode and one rung for the run mode. While this method works, it has several drawbacks. The first is that all the rungs will be scanned all of the time. Because only one mode is ever active at a time, we will spend a good deal of time scanning rungs that are not active. A second drawback is that we will have a great deal of redundancy because we must separate each program function into setup and run mode. This means the first element in each rung must be the bit indicating which mode we are in. This will require a great deal of wasted memory just to separate the two mode functions.

A better way to separate the two programs is to use the subroutine functions supplied in the SLC-500. In Chapter 5 we stated that file 2 was the main ladder file and that files 3 through 255 were used for subroutines. In Figure 12–1 we show the basic block diagram for a control program. We will place all rungs that must be scanned all the time into the MAIN file 2 section. These will include any safety circuits, indicator lights, flash timers, and the rungs that determine what mode the machine is in. We will then place all the rungs that control the machine in setup mode in the file 3 section. All of the rungs that control the run mode will be written in the file 4 section. Four rungs in file 2 will now determine which of the other two subroutines will be scanned. Note that only the active mode subroutine is now scanned. This will save us a considerable amount of scan time.

In Figure 12–2 we can see that one selector switch controls which subroutine is active. It is important to understand that only one input address should be used for the run/setup mode selection. If two separate inputs were used, a possible failure mode could result in which both inputs could

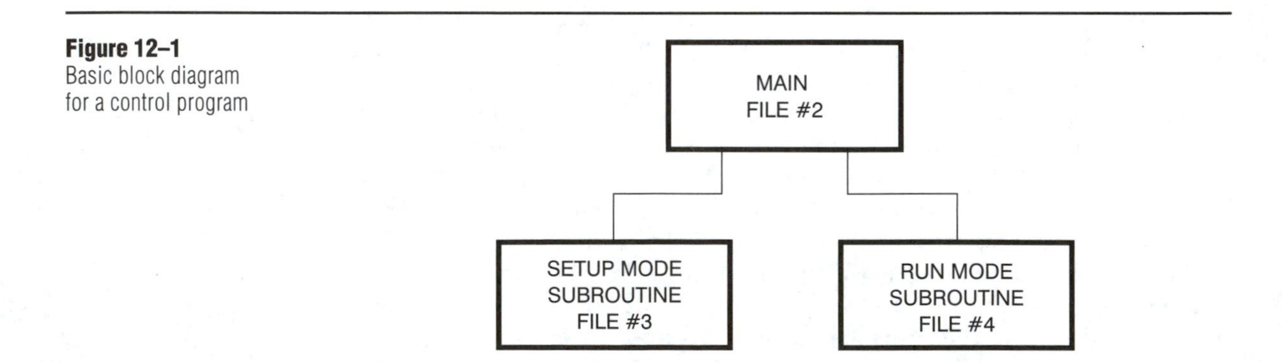

Figure 12–1 Basic block diagram for a control program

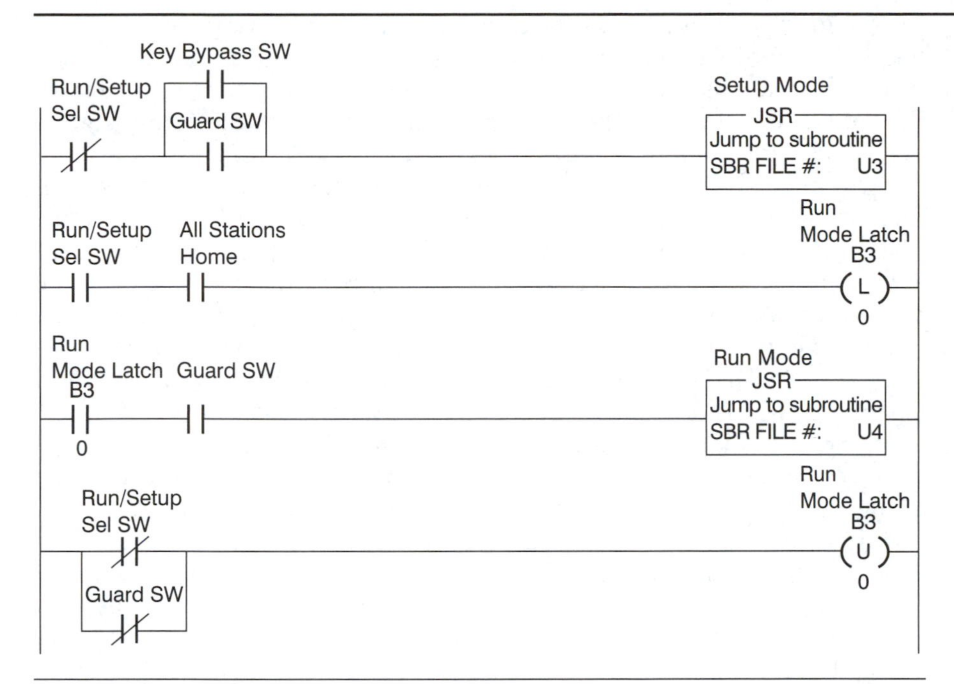

Figure 12–2
Section of ladder program showing subroutine selection

be on at the same time due to a shorted switch or wiring. This would cause dangerous control problems because both subroutines would be active at the same time. The machine would be very unstable and unpredictable. If we use only one input address for the run/setup mode selection this can never happen.

Note in Figure 12–2 that in the first rung the setup mode is active any time the selector switch is off, and the safety guard switches or the keyed bypass switch is on. It is important that no conditions be tied to activating the setup mode because the maintenance person must be able to activate it under any circumstances. The run mode should not be as easy to activate. We want to be sure the machine is in the proper startup condition before it begins operation. This means that all mechanisms must be in their proper positions. We normally have a program rung for each mechanism, located in file 2, to indicate when it is in its home position. These outputs are ANDed together to give a single output when all the mechanisms are in their home positions. In this example a single switch is used for the safety guard input. In a large machine, this would be an internal memory bit that was generated by ANDing all of the safety switches together in a single rung.

Note that we have the all home bit B3/5 address ANDed with the selector switch for run mode to ensure the run mode cannot be started if anything is out of place. Note also that this rung latches bit B3/0 and it is called the run mode latch bit. The bit must be latched because all home bit

B3/5 will shut off as soon as the cycle starts and the various mechanisms begin to move. The third rung activates the run mode subroutine if the run mode latch bit and the safety switch are on. The last rung unlatches the auto mode latch bit if the selector switch is moved to the manual mode position or we lose any of the safety guard switches.

Before we begin using subroutines, let's explore some of the requirements and conditions of using subroutines correctly. Early revisions of the SLC-500 software required a return instruction at the end of every subroutine. Later revisions automatically return at the end of the subroutine. The processor returns to the calling program file number when it encounters a return statement or the last rung of a subroutine. Subroutines can be nested up to four levels deep in the SLC-500/01 processors and up to eight levels deep in SLC-500/02 and higher processors. If you nest more routines than the limit, a run-time error will occur.

One problem most first-time ladder programmers run into when using subroutines is that when a subroutine is deactivated all the outputs that it controlled remain in their last state; also, all timers stop accumulating time but retain the ACCUM value. If the programmer reactivates the subroutine without clearing these functions, they will resume in the state they were in when it was deactivated. You can use unlatch instructions to turn off outputs or memory bits left on in a deactivated subroutine.

In addition you can use the reset or clear instruction to clear timer ACCUM values in deactivated subroutines. Failure to clear latched bits and timer ACCUM values can cause unstable operation when subroutines are reactivated. Some programmers like to use memory bits as outputs in both the setup and run subroutines and then combine them to drive the real outputs in a different file. This has the advantage of keeping all the real outputs in one file and therefore you never have to worry about two different rungs activating the same output address. An additional advantage is the ability to turn off one subroutine and deactivate all the real outputs. Although these are good advantages, it does make troubleshooting more difficult in that you must move between several files to trace the problem if an output does not perform correctly. It also uses considerably more memory because each real output will require several interim memory addresses. I prefer to use real outputs in each subroutine and take care that I never activate both subroutines at the same time.

SAMPLE MACHINE

Figure 12–3 shows a diagram of an insert molding machine that makes switch blade contacts. The operator must load two sets of blades and two spring strips into the mold fixture. When the machine cycle is complete and the operator has finished loading the blade parts, the operator

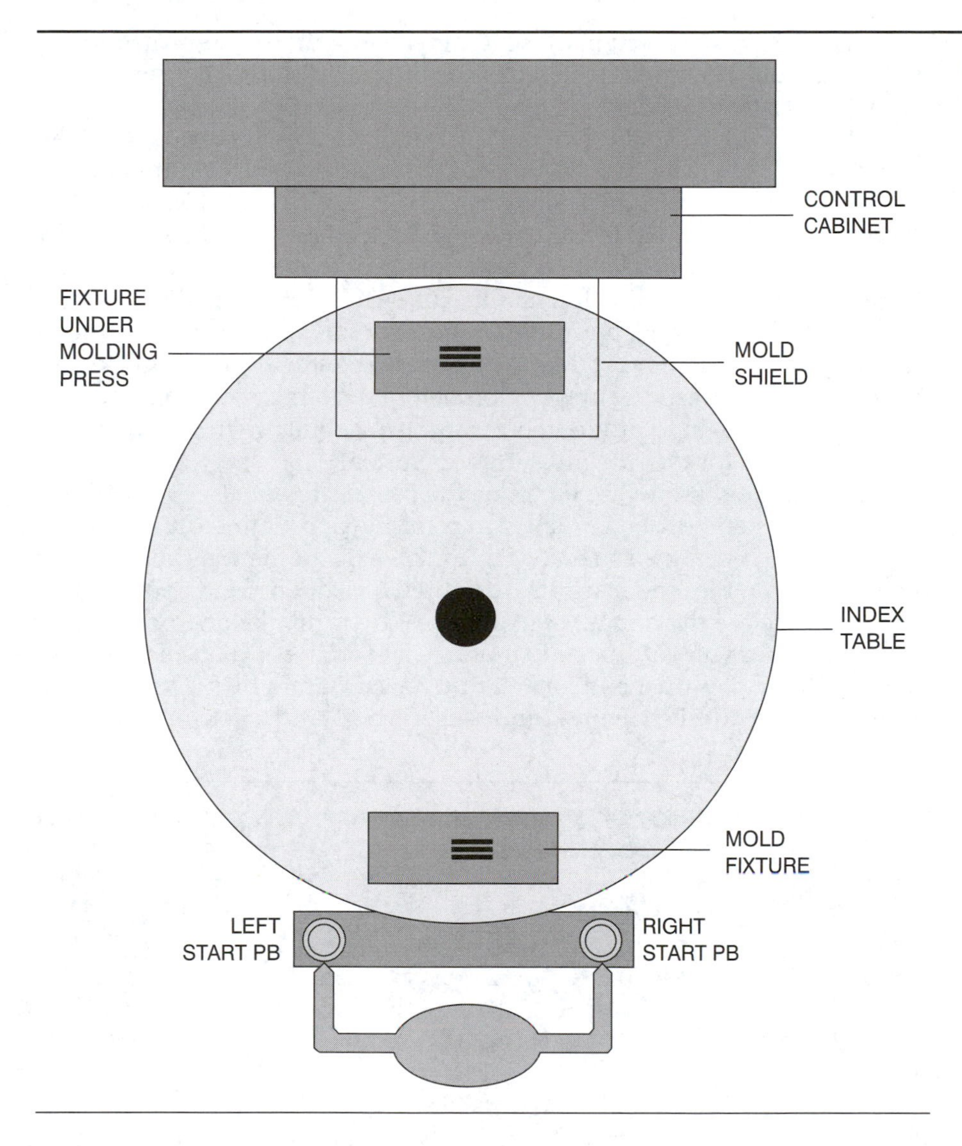

Figure 12–3
Insert molding machine

presses the two start buttons and holds them down while the table rotates 180 degrees. The mold shield then comes down to cover the mold operation. At this point the operator can let go of the start buttons. If the operator lets go of either button before the mold shield comes down, all movement stops. If the operator lets go of a button as the mold shield is closing, the shield must open again so as not to trap a hand under the shield.

Once the shield is in the down position, the rest of the molding cycle runs automatically. The mold halves are closed, then the injector ram is moved forward to push liquid plastic into the mold. The ram injector is held in the extended position for several seconds and then returned to the home position. The mold is held closed for an additional few seconds to allow the plastic to cool.

At the end of the cool-down cycle, the mold halves are opened and the shield is raised. At this point the machine is ready to start another cycle.

During the injection and cooling cycle the operator is unloading the finished parts from the mold in front of her and reloading the parts for the next cycle. This designer needs to be aware of several other requirements on this machine. This machine has an injector heater that melts the plastic to be injected and we must be sure the injector ram is never moved unless the temperature is at the set point. We must also be sure the mold shield and upper mold half are in the up position before the table can be rotated.

Before we begin to program our sample machine, we must create the subroutine files for the run and setup operations.

In the RSLogix software package, the process for creating new files is quite easy. When we open a new program file and define the processor type, we are shown the screen of Figure 12–4. Note that on the left edge below program files there are already three files open: SYS-0, SYS-1, and LAD-2. These are the default start-up files. By right clicking on the program file folder you get a pop-up menu. Click on the NEW selection. You now get a new pop-up menu for creating new program files (Figure 12–5). Enter the file number, name, and description and click on OK. The

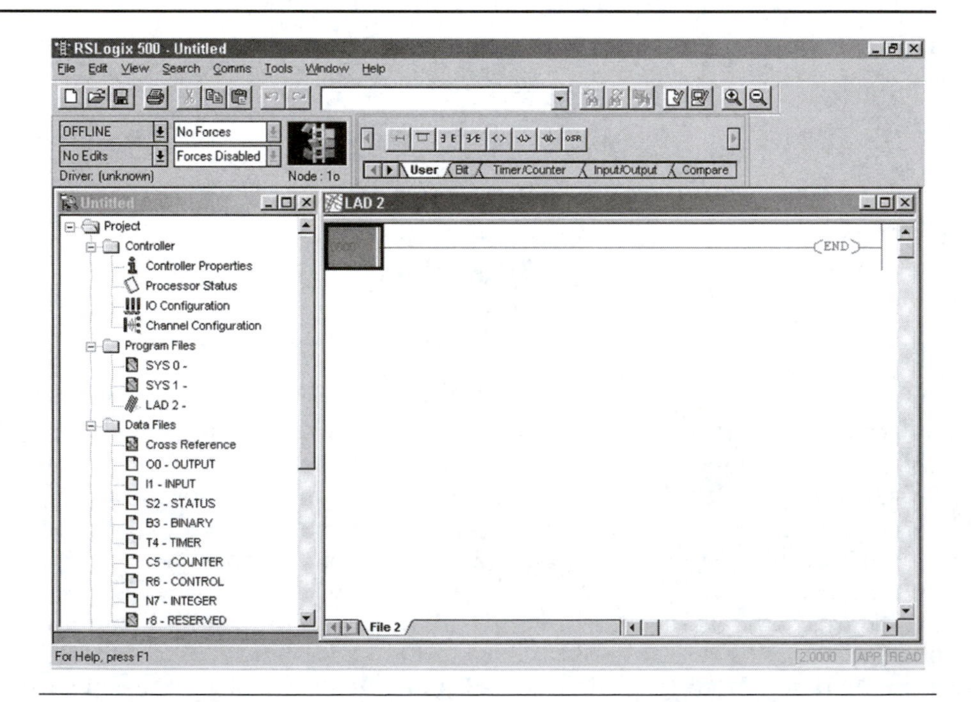

Figure 12–4
RSLogix screen for program editing

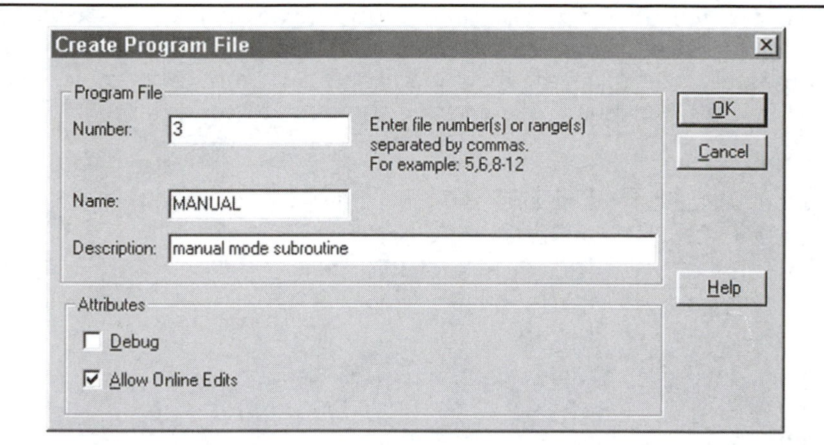

Figure 12–5
Creating a program file in RSLogix

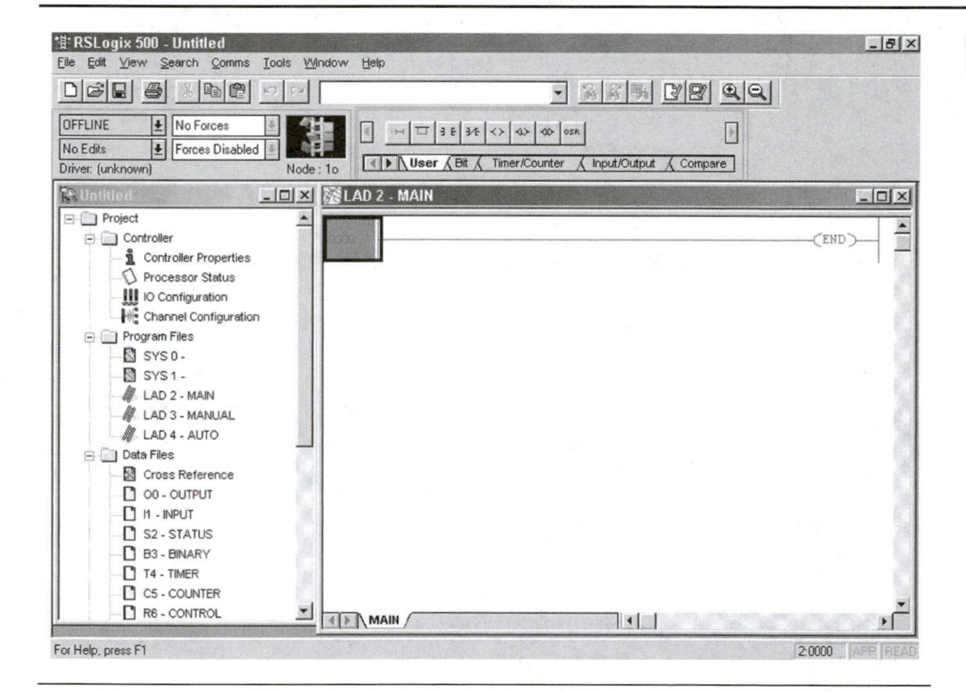

Figure 12–6
Renamed program files

programmer in this example called his subroutine manual in place of setup mode. You will need to create both files 3 and 4. Once you have created files 3 and 4, right click on the LAD-2 folder and click on Properties. Enter a new file name of Main. The screen should now appear like that shown in Figure 12–6. You can now move between the three files by double clicking on one of the file folder icons. This will put you into the selected file and allow you to begin editing it.

The next step in our project is to assign addresses to real I/O. We will add some indicator lights to display the molding machine's status. We do not require safety guards as the machine has its own moveable mold shield. Therefore, no safety guard switches are shown in the program.

Inputs:

I:0.0/0	LEFT START PB.
I:0.0/1	SHIELD DOWN L.S.
I:0.0/2	RIGHT START PB.
I:0.0/3	SHIELD UP L.S.
I:0.0/4	MANUAL INDEX PB.
I:0.0/5	MOLD CLOSED L.S.
I:0.0/6	MANUAL SHIELD DOWN PB.
I:0.0/7	MOLD OPEN L.S.
I:0.0/8	MANUAL SHIELD UP PB.
I:0.0/9	RAM EXTENDED L.S.
I:0.0/10	MANUAL CLOSE MOLD PB.
I:0.0/11	RAM RETURNED L.S.
I:0.0/12	MANUAL OPEN MOLD PB.
I:0.0/13	NOT USED
I:0.0/14	MANUAL RAM FORWARD PB.
I:0.0/15	NOT USED
I:0.1/0	MANUAL RAM RETURN PB.
I:0.1/2	NOT USED
I:0.1/3	INDEX CAM L.S.
I:0.1/4	RUN / MANUAL SEL. SW.
I:0.1/5	HEATER AT SET POINT
I:0.1/6	HEATER OVER TEMP
I:0.1/7	NOT USED

Outputs:

O:0.0/0	MANUAL MODE LAMP
O:0.0/1	RUN MODE LAMP
O:0.0/2	EXTRUDER TEMP AT SET POINT LAMP
O:0.0/3	EXTRUDER OVER TEMP LAMP
O:0.0/4	ALARM LAMP
O:0.0/5	TABLE IN POSITION LAMP
O:0.0/6–15	NOT USED
O:1.0/0	INDEX MOTOR ON RELAY
O:1.0/1	SHIELD DOWN SOL.
O:1.0/2	NOT USED
O:1.0/3	SHIELD UP SOL.

O:1.0/4	NOT USED
O:1.0/5	MOLD CLOSE SOL.
O:1.0/6	NOT USED
O:1.0/7	MOLD OPEN SOL.
O:1.0/8	NOT USED
O:1.0/9	RAM FORWARD SOL.
O:1.0/10	NOT USED
O:1.0/11	RAM RETURN SOL.
O:1.0/12–15	NOT USED

With our I/O list complete, we should now lay out a flowchart to define the program in detail. We will create three main sections of the flowchart to match the main, run, and setup modes.

Main Program Flowchart

Note in Figure 12–7 how the main program flowchart controls only a few lights and the decision as to which subroutine to run. We have made a decision to run the temperature controller as a separate function and not include it in the PLC program. At this point in our understanding of PLC controllers we have not covered analog input and output modules and it was felt that introducing them at this time would be too confusing to the reader. For this reason the temperature controller is simulated as two input addresses in this exercise.

Note also that the run mode cannot be activated unless the injector ram temperature is up to set point and not above. The run mode, setup mode, and alarm lights are usually all combined into a stack light, with run mode being green, setup mode blue, and the alarm light red (Figure 12–8). In this way anyone can tell the status of the machine by looking at the stack light.

We can add some additional information to the stack lights by making them flash for certain conditions. In the setup mode, the blue light will flash if one of the main moving elements is not in its home position. This will help the setup and maintenance people by indicating when they have returned everything to home position.

In run mode, the green light will flash anytime the machine is running a cycle. The operator can tell when the cycle is over by looking for a solid green light. As a safety device we have added a timer to check for problems by timing each cycle. We have determined that the normal cycle time is about 13 seconds. In the program we will set the cycle "too long" timer for 20 seconds. In normal operation, this timer should never time out. If it does time out, we need to immediately deactivate the run mode and light the alarm lamp. This alarm can be cleared by momentarily switching to the setup mode with the mode selector switch. A timer used in this fashion is often called a watch dog timer.

Figure 12–7
Flowchart for the main mode

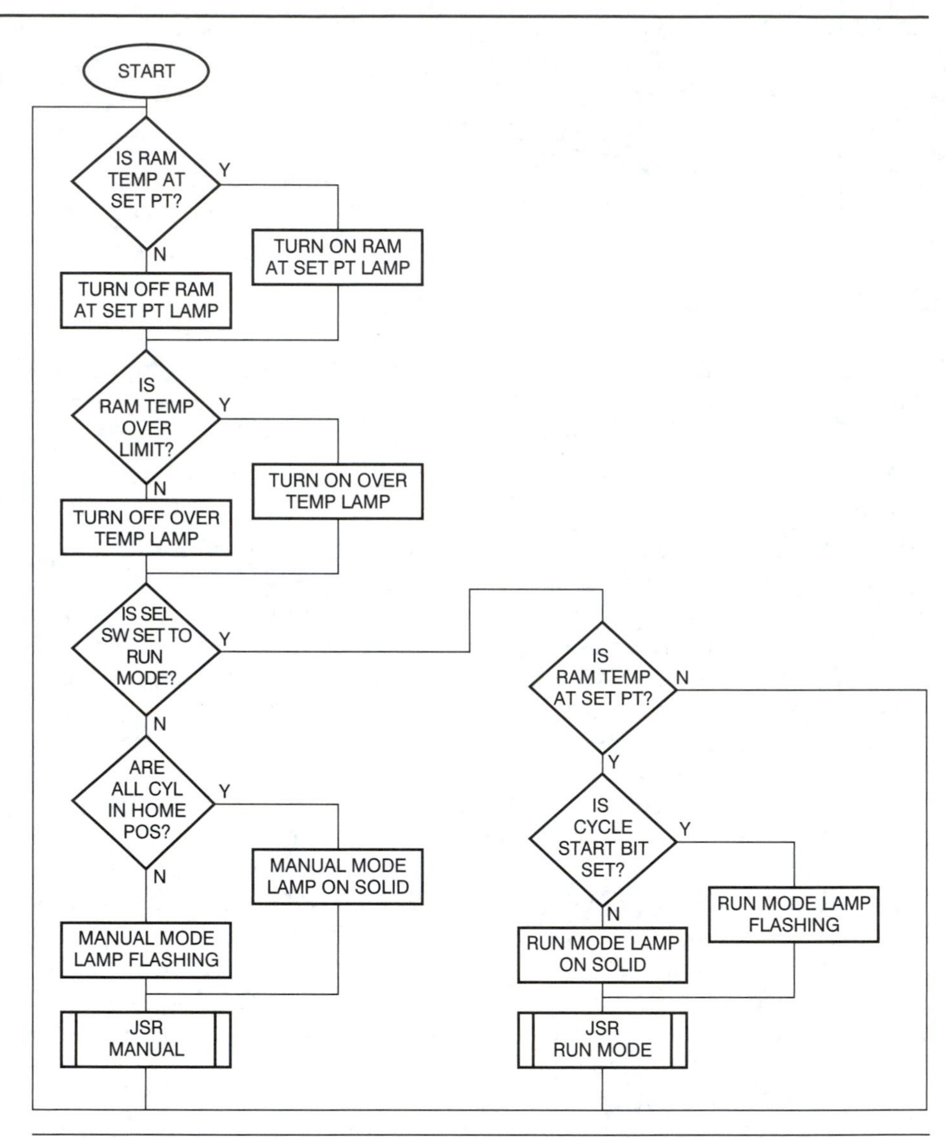

Manual Mode Flowchart

Looking at the manual mode flowchart (Figures 12–9 and 12–10), we see seven parallel paths. The first is the manual index function. The table can be turned anytime the manual index PB is pressed with both the shield being up and the mold halves open. We have provided a table in position lamp to help the setup and maintenance personnel tell when the table is in one of the two mold positions.

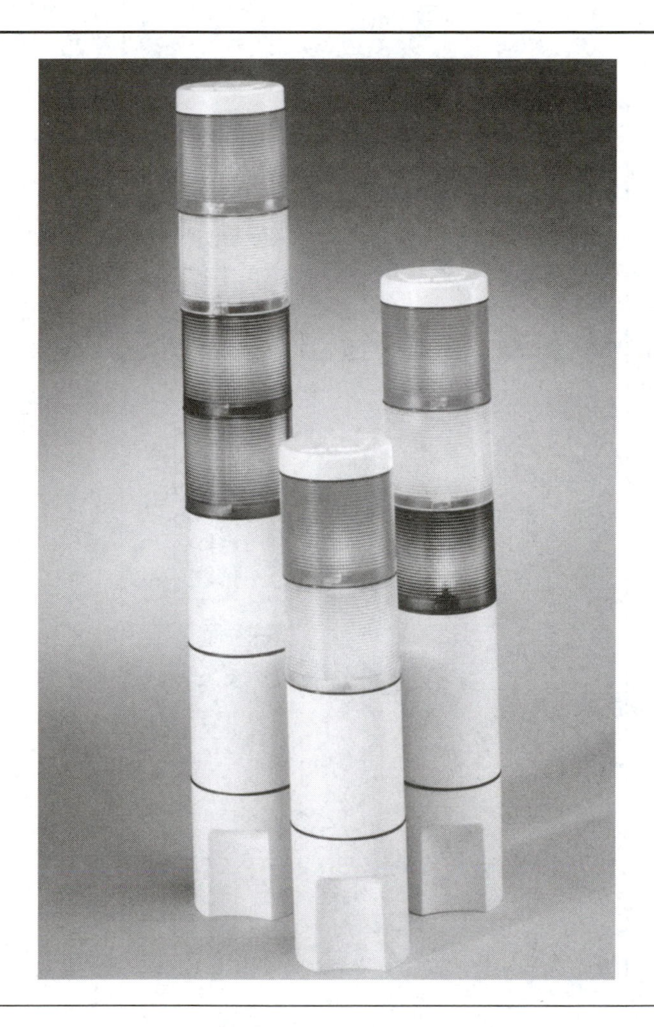

Figure 12–8
Stack lights.
Courtesy: Federal Signal Corporation, University Park, Illinois.

The next two paths control the opening and closing of the mold halves. Conditions for closing the mold are the close PB being pressed and the index table in position. To open the mold, only the open mold PB must be pressed. The shield open and close functions are next and they are identical to the mold close and open functions. Our last two paths control the injector ram extend and return functions. Note that the table does not have to be in position for these functions to operate. The only condition is that the ram temperature be up to the set point.

We have not included the table in the position switch because the setup person must purge the injector barrel when changing plastic color or starting up from a cold condition. To accomplish this, she turns the table halfway, places a container under the upper mold half, and cycles the injector ram several times. This allows the purged plastic to be collected and recycled.

Figure 12–9
Manual mode flowchart

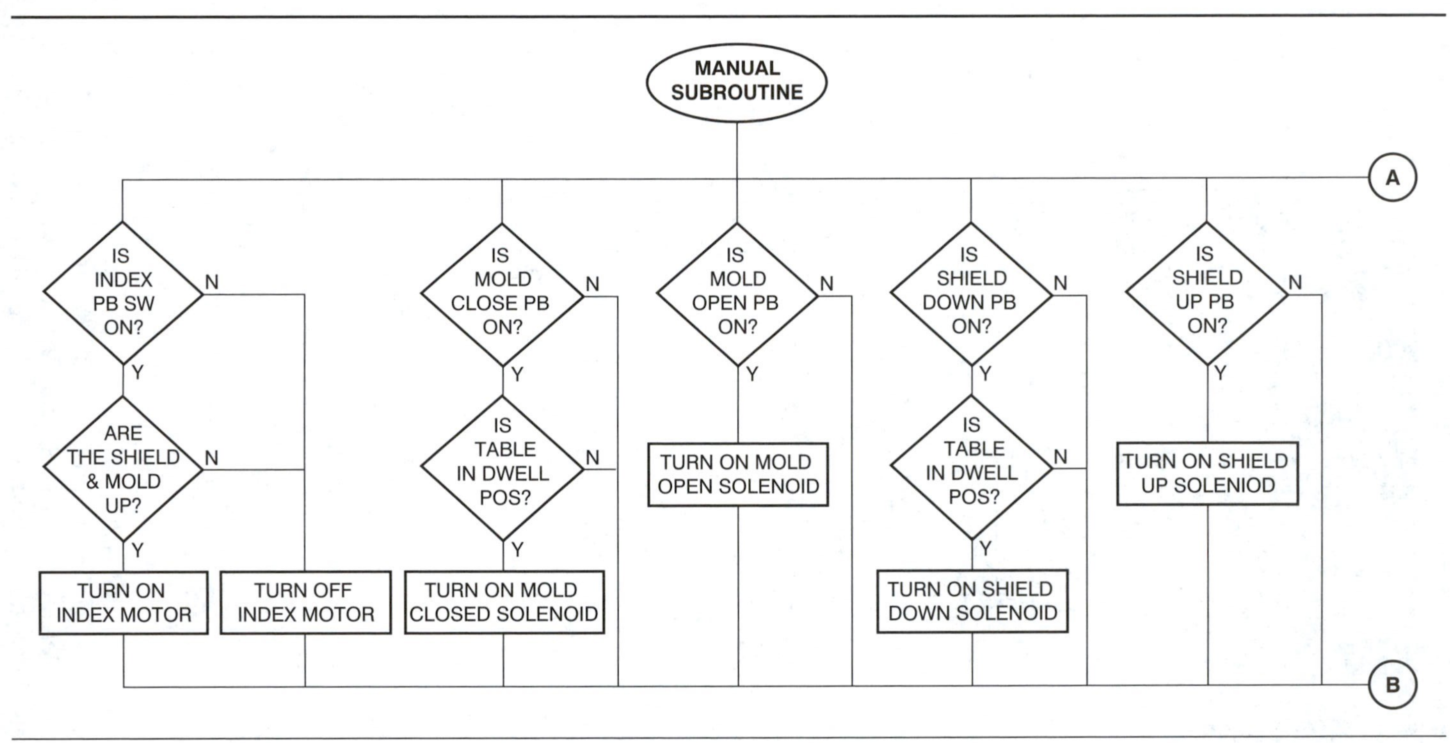

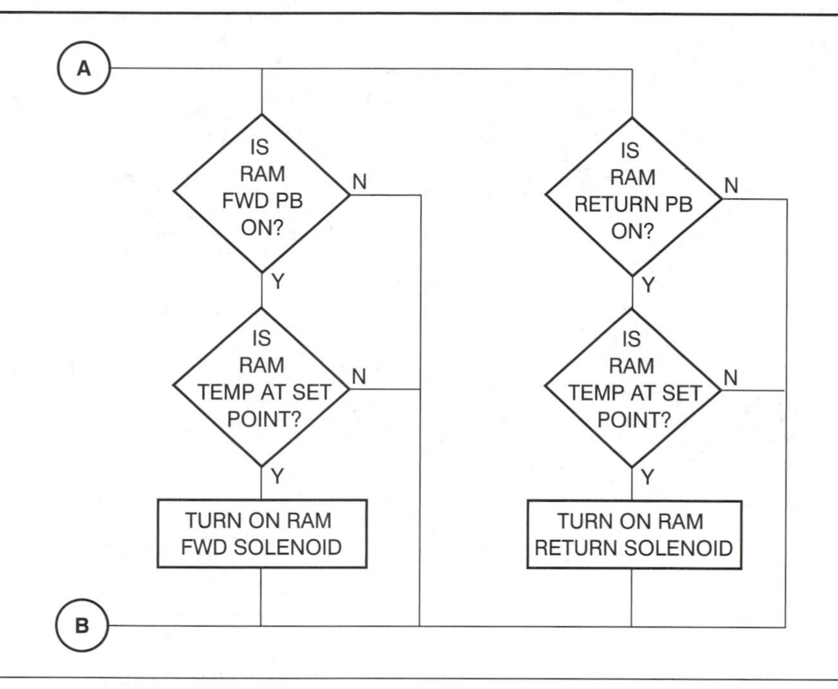

Figure 12–10
Manual mode flowchart, *continued*

Run Mode Flowchart

In run mode (Figure 12–11) we have an immediate problem. We want to rotate the table and then close the mold shield. The problem is that if we use the cam switch being off to start the shield closing, we have to block the function as soon as the start buttons are pressed because the cam will be off for a short time until the table starts to move. To get around this problem, we created a cycle start latch bit that is turned on when the cam switch activates as the table turns. By using this bit and with the cam switch returning to OFF again, we can tell if a full turn of the cam has been completed. This is a typical method of handling the problem when your start conditions are the same as your done conditions.

Note that if either PB is released before the table is in position, the motor is turned off. Once the index is complete, we can start the mold shield down. Again we must protect against either PB being released before the shield down L.S. is made. If either button is released, we have to open the mold shield again. This is to prevent an operator from being trapped by reaching under the shield as it closes.

Once the shield is closed a second latch bit disables the start PB until the cycle is completed. Once the mold cycle start latch bit is set, the cycle is automatic. We close the mold halves, move the ram forward, wait for 3 seconds, return the ram, wait for an additional 3 seconds. We now test to see if both start PBs are open. If not, the cycle waits for them both to be

Figure 12–11
Run mode flowchart

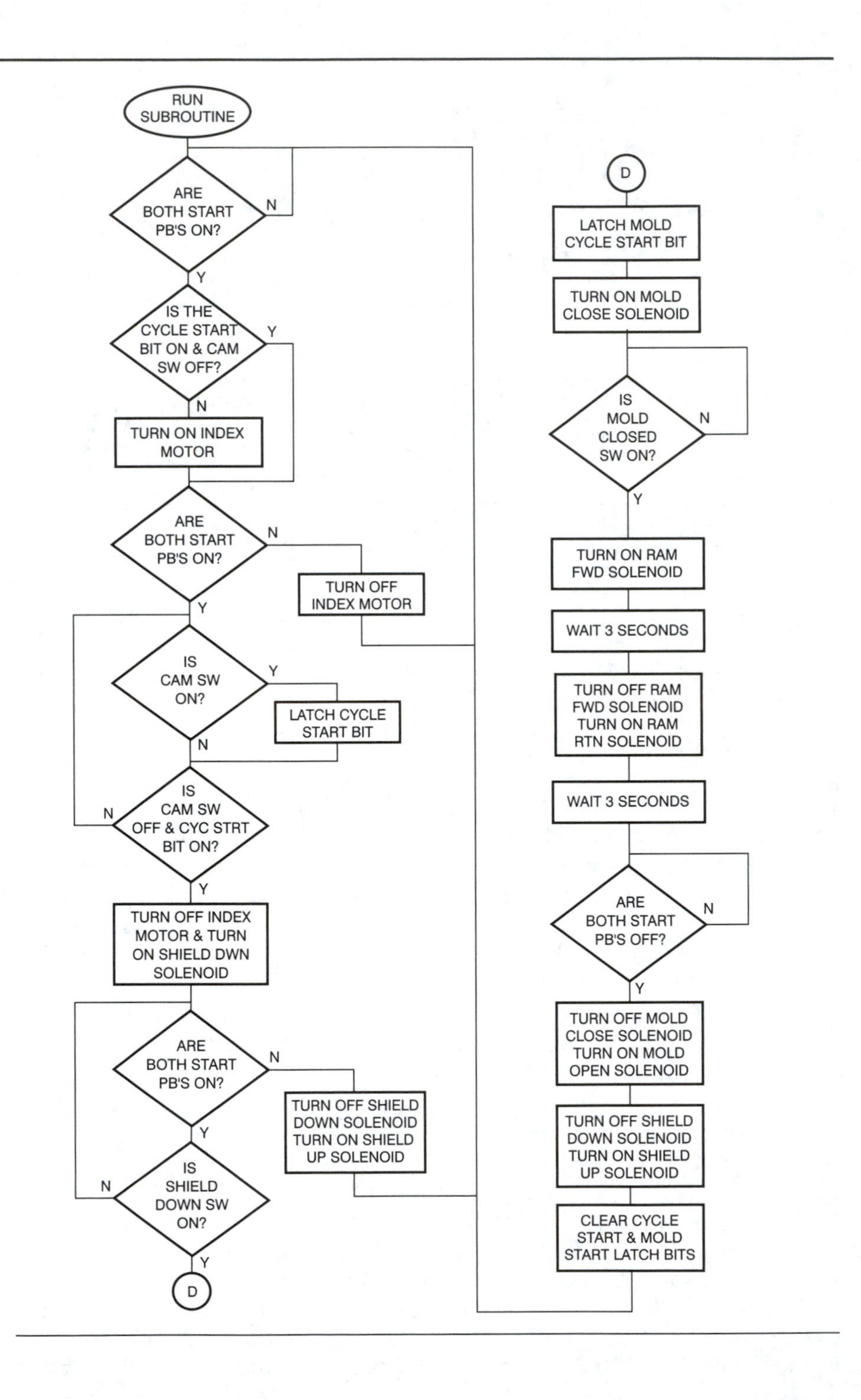

released. This prevents a one-hand operation if the operator tries to tie down one PB and run with only one start PB active. If both start PBs are open, the mold halves are opened and the shield is raised. At this point both latch bits are reset and a new cycle can be started.

We have now completed our three flowcharts, so now we're ready to start examining the ladder code.

Main Program Ladder File 2

We start with a rung-by-rung description of the ladder program. Refer to Figures 12–12 and 12–13 as you read the following descriptions:

Rung 0 Activates the ram temperature at the set point lamp.

Rung 1 Contains a "too long" cycle timer that looks at the cycle start latch bit and times out if it is on longer than 20 seconds. Because a normal cycle is 13 seconds, the latch bit should turn off and reset the timer long before it times out.

Rung 2 Turns on the alarm lamp if the "too long" cycle timer times out.

Rung 3 Lights the table in position lamp if the cam switch is off, indicating it is in the dwell position.

Rung 4 Controls the ram over temp high limit lamp.

Rung 5 Looks at all the moving elements and sets B3/1 on if they are all in home position.

Rung 6 Looks at the selector switch, ram temp inputs, and the alarm timer. If all are OK, it sets B3/2 on to activate the run mode.

Rung 7 The first of two timers that interact to generate a 0.4-second on/off flash bit.

Rung 8 The second half of the flash timer pair.

Rung 9 Controls the run mode lamp. If the index cycle start bit is on, the lamp is flashed by T4:1/TT. If it is off, the lamp is on steady.

Rung 10 Causes the program to jump to the run mode subroutine if the run mode bit B3/2 is on.

Rung 11 Causes the program to jump to the manual mode subroutine if the mode selector switch is not set to run mode.

Rung 12 Controls the manual mode lamp. The lamp is flashed by T4:1/TT if the B3/1 all home bit is not set. If B3/1 is set, the lamp is on steady.

Rung 13 Unlatches the alarm lamp if we switch to manual mode.

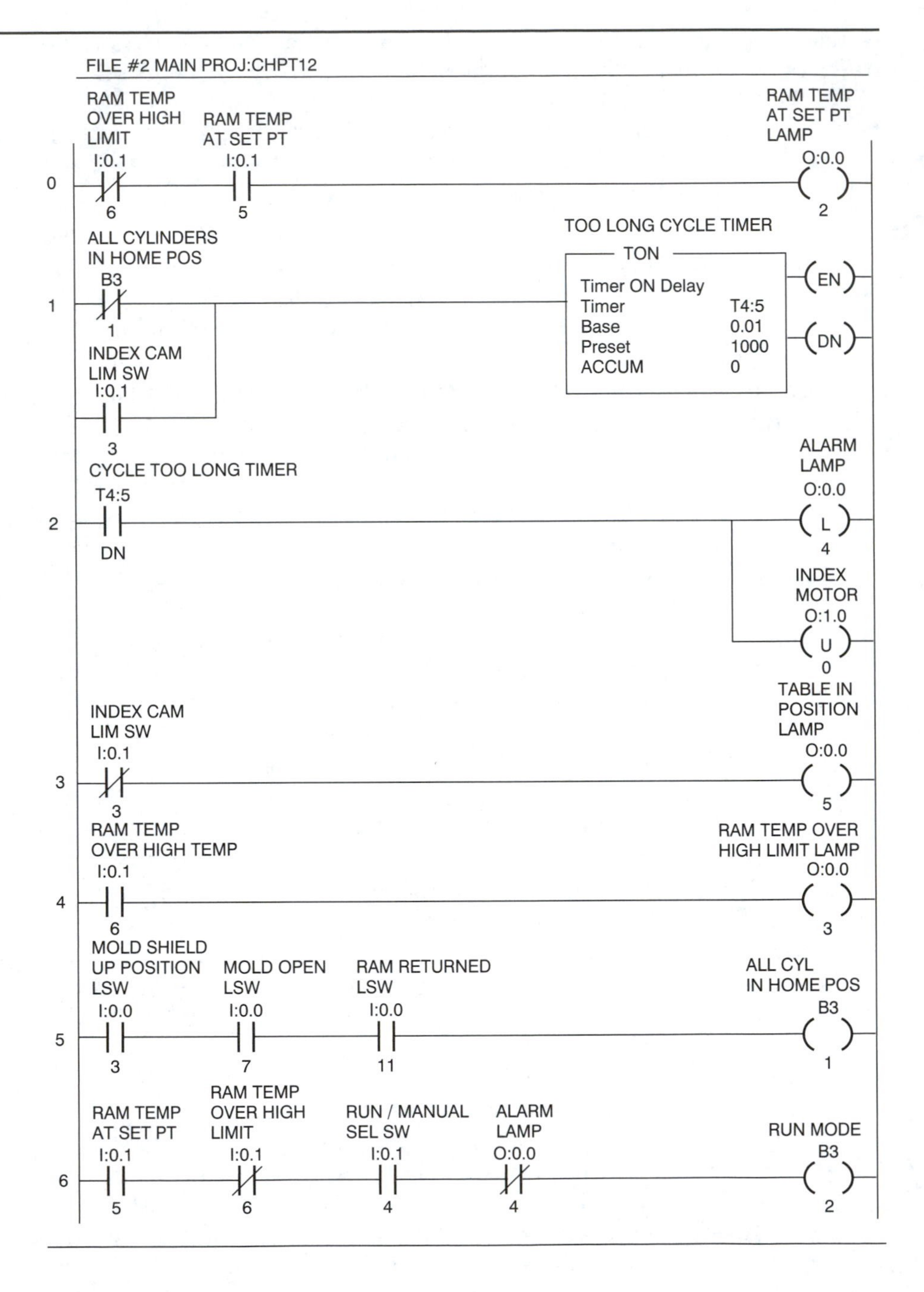

Figure 12–12
Main program ladder file 2

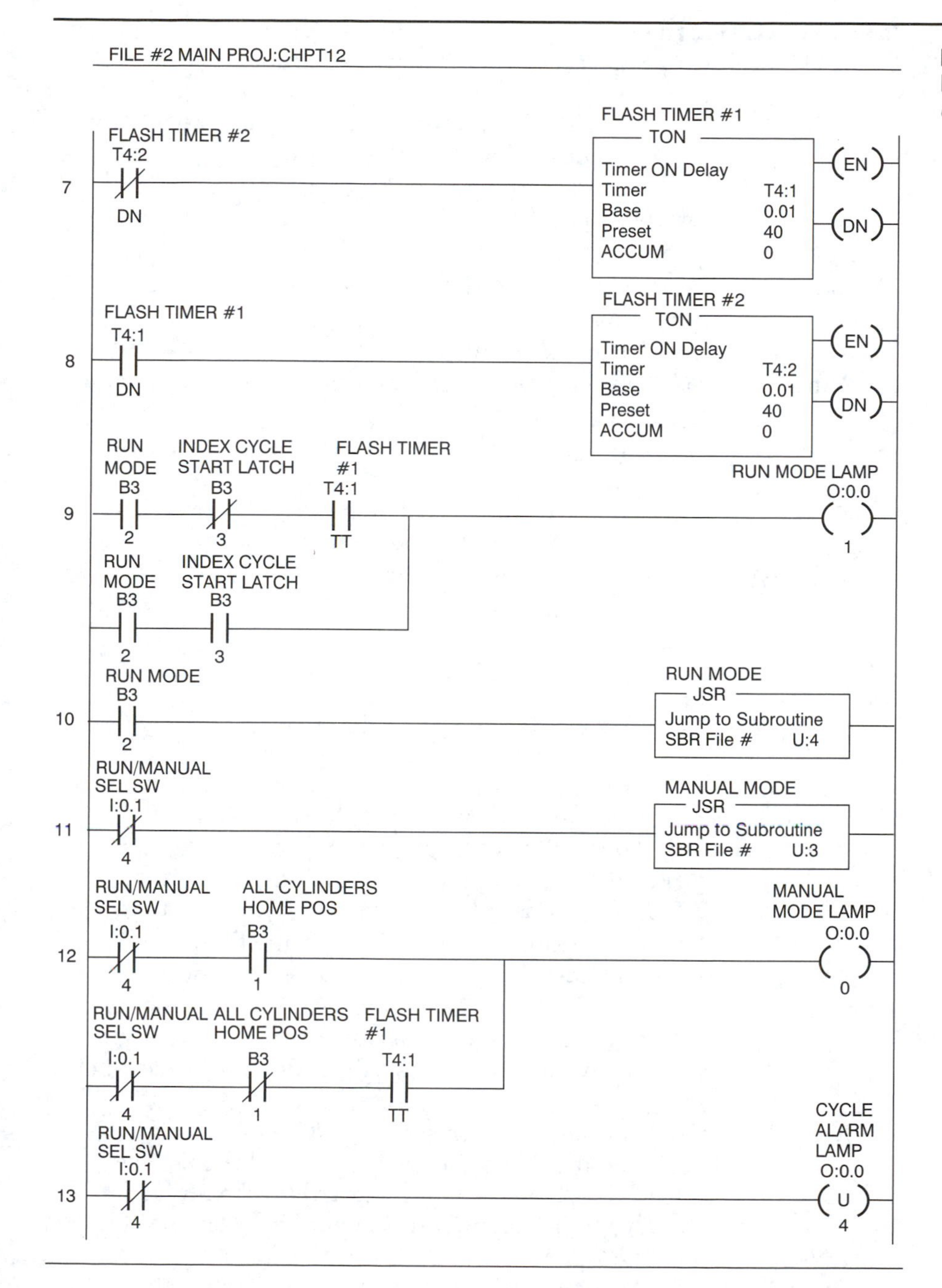

Figure 12–13
Main program ladder file 2, *continued*

Manual Mode Ladder File 3

Refer to Figure 12–14 as you read these rung descriptions:

Rung 0	Controls the index motor. The shield must be up and the mold open to activate the motor.
Rung 1	Controls the close mold halves solenoid. If the table is in position, the mold close solenoid is activated.
Rung 2	Controls the mold open solenoid with no conditions.
Rung 3	Controls the shield down solenoid if the table is in position.
Rung 4	Controls the shield up solenoid with no conditions.
Rung 5	Controls the injector ram forward solenoid if the ram temp is at set point.
Rung 6	Controls the injector ram return solenoid if the ram temp is at set point.
Rung 7	Clears the two-run mode latch bits B3/3 and B3/4, which in turn clears the alarm timer.
Rung 8	Returns to the main program file.

Run Mode Ladder File 4

Refer to Figure 12–15 as you read these rung descriptions:

Rung 0	Controls the index motor. We must have both start buttons on and the mold cycle latch bit B3/4 off. The top branch allows the motor to run before the cam switch activates. The lower branch shuts the motor off after the cam falls back into the dwell position.
Rung 1	Latches the index-cycle-started latch bit B3/3 on when the cam comes out of the dwell position.
Rung 2	Controls the shield down solenoid. The solenoid is activated if both start switches are on, the cam switch is in the dwell position, and the index start cycle latch bit B3/3 is set.
Rung 3	Controls the shield up solenoid. The solenoid is activated if either start SW is open and the mold cycle latch bit is not set. We have added the up limit switch to prevent the solenoid from being held on once the shield is open.
Rung 4	Latches mold cycle start latch bit B3/4 when the shield is in the down position.
Rung 5	Controls the mold close solenoid anytime the mold cycle latch bit is set.

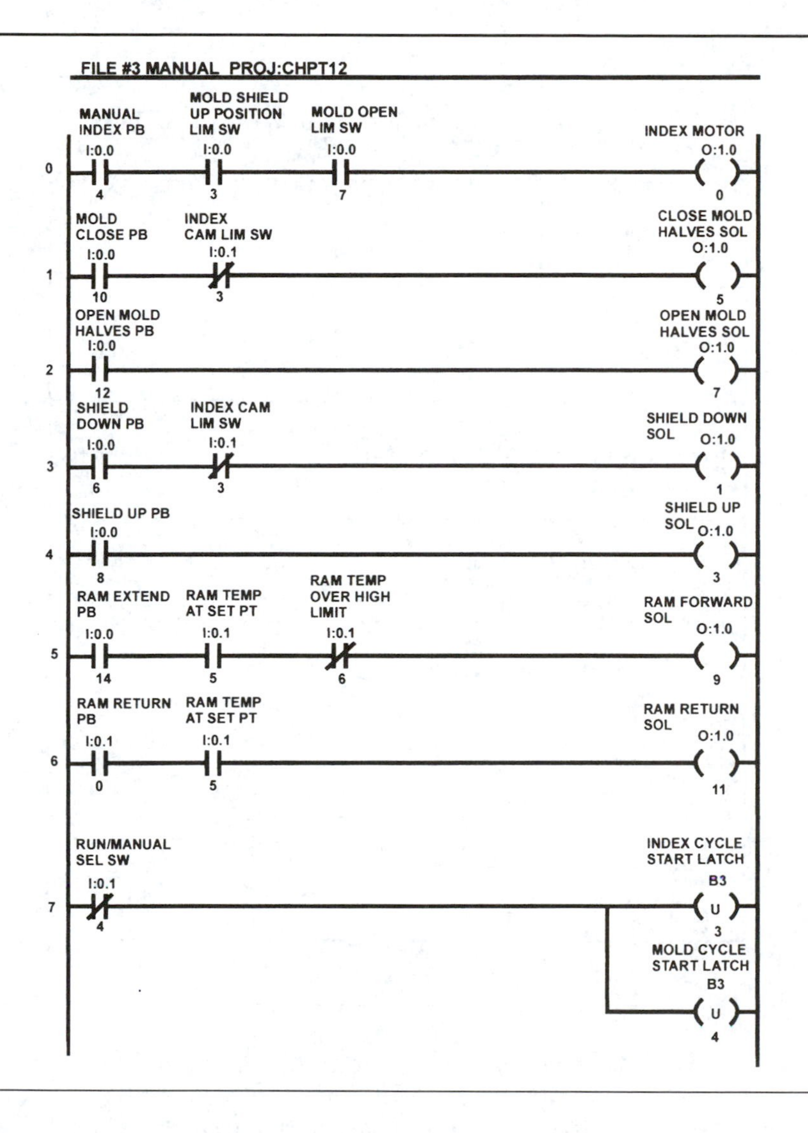

Figure 12–14
Manual mode ladder file

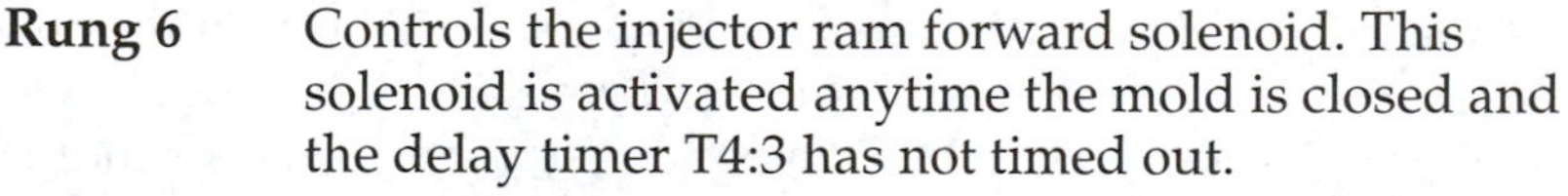

Rung 6 Controls the injector ram forward solenoid. This solenoid is activated anytime the mold is closed and the delay timer T4:3 has not timed out.

Rung 7 Controls the injector forward delay timer T4:3.

Rung 8 Controls the injector ram return solenoid. It is activated when the injector delay timer T4:3 times out.

Rung 9 Controls the mold cooling time. This timer T4:4 is started when the injector timer times out and the ram returns home.

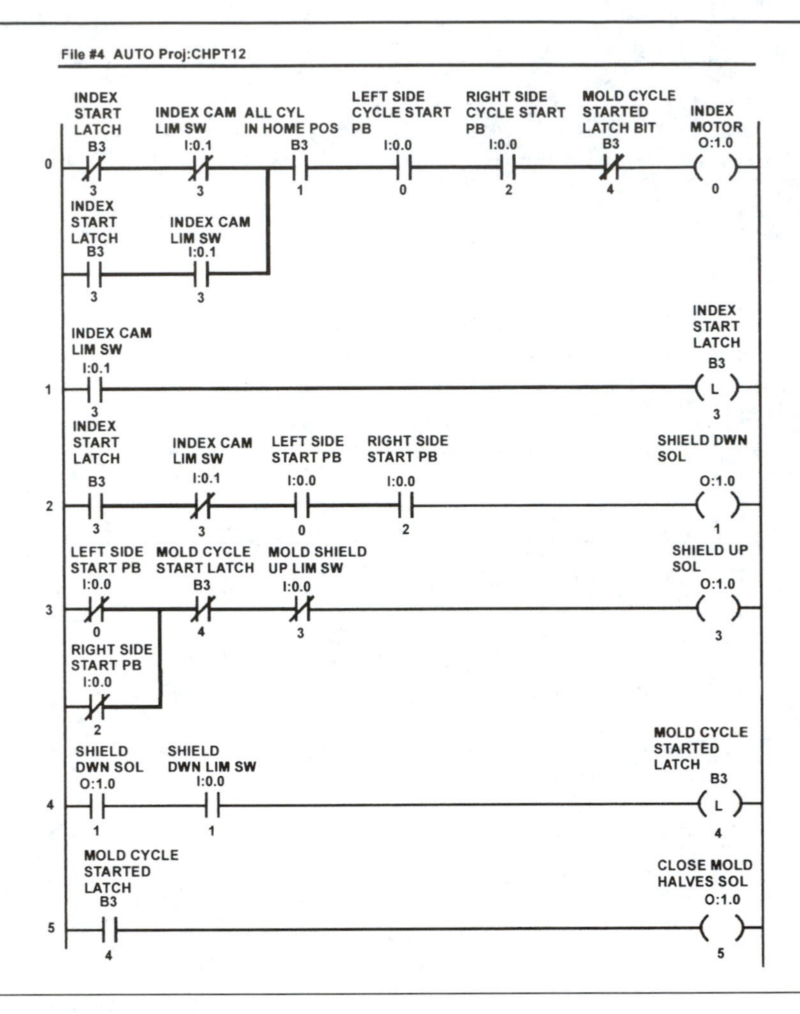

Figure 12–15
Run mode ladder program

Rung 10 Controls the end of cycle. When the mold cooling delay timer times out and if both start PBs are released, the rung resets both run mode latch bits B3/3 and B3/4.

Rung 11 Controls the open mold solenoid. The mold is opened when cycle start bit B3/3 is reset. We have added the mold open limit switch to prevent the solenoid from being held on when the mold is open.

This concludes our sample manually activated machine design project. Students should look over the documentation included with the program listing to familiarize themselves with the cross-reference listing and I/O map list.

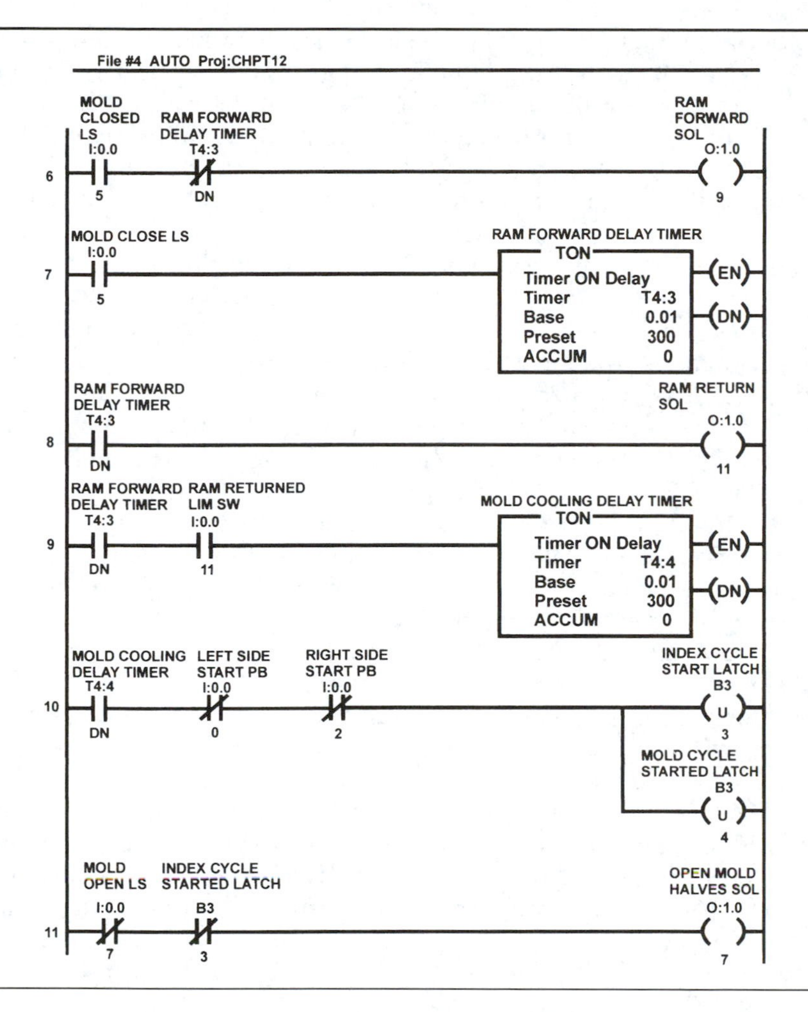

Figure 12–16
Run mode ladder program, *continued*

CONCLUSIONS

The concepts of subroutines were introduced in this chapter and were then used to separate the manual and automatic functions of the sample program. Several other programming functions are aided by the use of subroutine files. For instance, we often need to selectively turn on or off sections of a program that are controlling a machine.

As an example, if we have an assembly machine built around a rotary table with eight positions, and four of those positions are product dependent, we need a way to selectively activate each of those four stations depending on the current product type being assembled. If we use the traditional ladder programming technique, we need to place the product type selector SW in the first rung of each station's program. Even if the station is not selected, we will still scan all the rungs for the deselected station.

A more efficient method is to place each of the product-dependent station's ladder rungs in a separate subroutine. This subroutine will only be activated if the selector SW is set to enable the subroutine. In this way we can cut down on the scan time because only the stations selected are scanned.

In a later chapter we will learn how to build a shift register that can be used to track product data as the product moves down an assembly line. In many cases some of the stations on the assembly line will be product dependent as to their function. We would like to have the ability to produce any mix of product in any order on the line. To do this, we need to set up several bits in the shift register to determine the product type being run. We will load the product-type bit pattern into the first position in our shift register, which is set up to correspond to the load station on our assembly line. As the product moves down the assembly line, this product type data will move with it. Each station will then look at its shift register position with its corresponding product-type bits to decide if it should activate the control subroutine this cycle.

We now have the ability to run a lot size of one down the line and each station will decide to function or not function depending on the product-type setting. Because each station's control rungs are in separate subroutines, we will only be scanning the minimum number of ladder rungs to get the job done each cycle.

The reader is now ready to run Lab 12, which requires you to program your own manually operated machine. The sample program used in this chapter was laid out so that it can be run on the class kit. It is a good idea to load this sample program into the kit and verify that all functions work.

REVIEW QUESTIONS

1. What is the definition of a manually activated machine?
2. Why should we always use two start push buttons for this machine type?
3. What other precautions should be incorporated into the start-up controls?
4. What are the three main sections into which a program should be divided?
5. How does using subroutines save cycle time and memory locations?
6. Why should only a single input address be used to select between the setup and run modes?
7. Why is it usually necessary to clear latch bits and sometimes timers in disabled subroutines?
8. On the sample machine, why was it necessary to stop the table rotation if either start PB was released before the index was complete?
9. Why must we not allow the ram to move if the temperature is not up to the set point?

10. Why do we not place any conditions on entering the manual or setup mode?

11. If you were a production manager, how would the use of stack lights on all your machines help you know what was happening on the shop floor?

12. What is the function of a watch dog timer in a machine control program?

13. What does the term "anti-tie-down" mean?

14. Why should the start push buttons be recessed on the machine's control panel?

15. How would you program a set of start push buttons so that the operator has to press both push buttons within a 0.7-second window to start the cycle?

LAB ASSIGNMENT

Objective

Introduces program subroutines and methods for controlling a machine in manual mode and run mode. We will begin by demonstrating a program that has the typical two-subroutine setup for machine control. The student will then be asked to write a complete program to control a manually activated machine.

1. Open a new file in RSLogix and save it as Chpt-12-lab. While this lab was designed to run on the lab kit described in Appendix A, it is somewhat generic in form, and can be run on most lab kits by changing the I/O addresses. Be sure to enter the correct I/O configuration for the lab kit you are using.
2. To begin writing this program we must create two additional program files, #3 and #4. To do this, right click on the *Programs* folder at the left edge of the screen and then the *NEW* file selection. Enter files #3 and #4 along with the file names and descriptions. Use Figures 12–5 and 12–6 for reference in creating the new files.
3. The main file, #2, should contain all the ladder rungs that will be scanned all the time regardless of the mode we are in. This includes the lights that indicate which mode we are running in, manual or auto, and all the control rungs used to determine which mode the program will be operating in.
4. The I/O list for this program is shown below. Enter the Main program as shown in Figure 12–17.

I:0.1/4	SW1	AUTO / MANUAL MODE SELECTOR SW
I:0.0/0	PB1	CYCLE START AUTO MODE PB
I:0.0/2	PB2	CYCLE STOP AUTO MODE PB
I:0.0/4	PB3	MANUAL LAMP 1 ON PB
I:0.0/6	PB4	MANUAL LAMP 2 ON PB
I:0.0/8	PB5	MANUAL LAMP 3 ON PB
I:0.0/10	PB6	MANUAL RESET
0:0.0/0	LAMP 1	STEP 1
0:0.0/1	LAMP 2	STEP 2
0:0.0/2	LAMP 3	STEP 3
0:0.0/8	LAMP 9	AUTO MODE ON
0:0.0/9	LAMP 10	MANUAL MODE ON
0:0.0/10	LAMP 11	ALL HOME POSITION

Figure 12–17

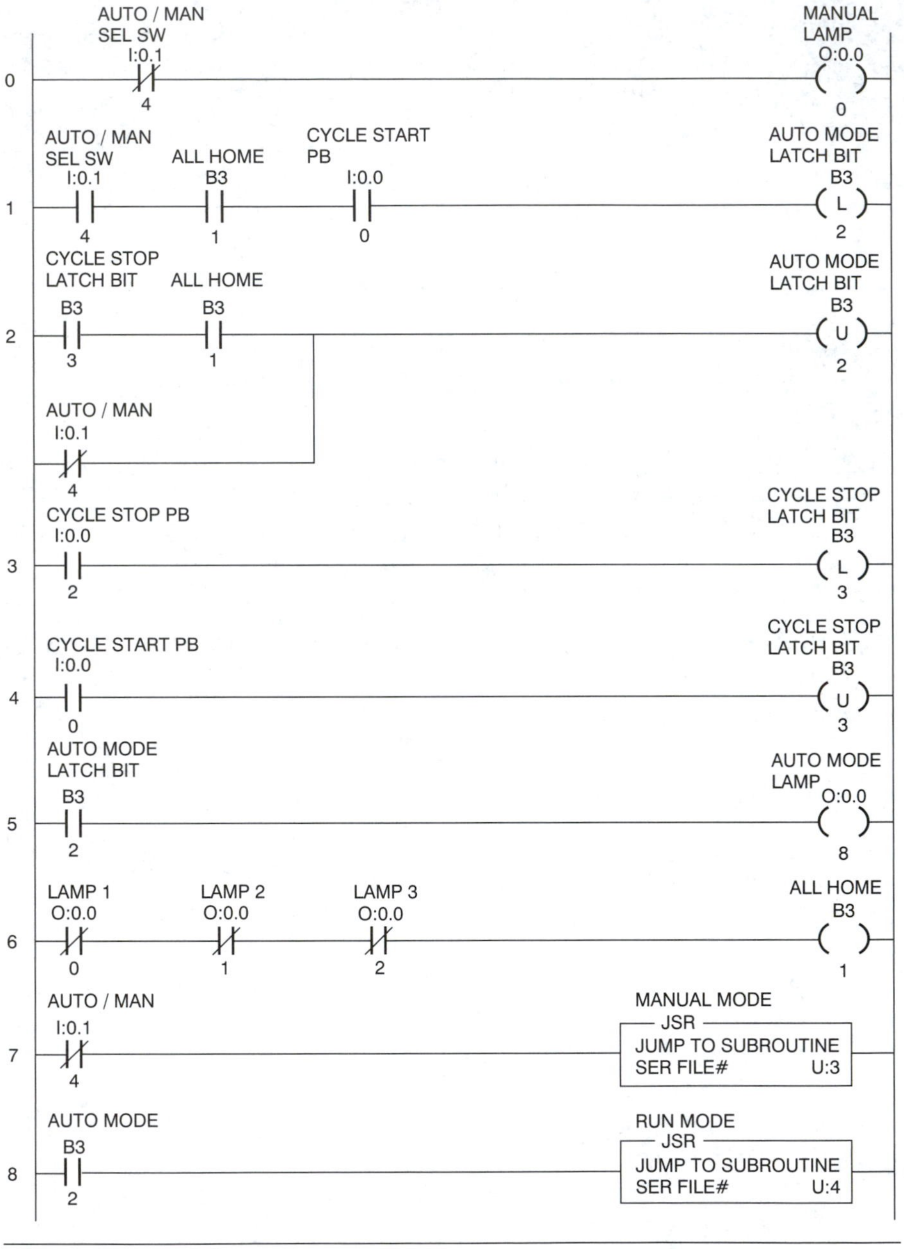

Rung 0: AUTO / MAN SEL SW I:0.1 4 — MANUAL LAMP O:0.0 0

Rung 1: AUTO / MAN SEL SW I:0.1 4 — ALL HOME B3 1 — CYCLE START PB I:0.0 0 — AUTO MODE LATCH BIT B3 (L) 2

Rung 2: CYCLE STOP LATCH BIT B3 3 — ALL HOME B3 1 — AUTO MODE LATCH BIT B3 (U) 2
AUTO / MAN I:0.1 4

Rung 3: CYCLE STOP PB I:0.0 2 — CYCLE STOP LATCH BIT B3 (L) 3

Rung 4: CYCLE START PB I:0.0 0 — CYCLE STOP LATCH BIT B3 (U) 3

Rung 5: AUTO MODE LATCH BIT B3 2 — AUTO MODE LAMP O:0.0 8

Rung 6: LAMP 1 O:0.0 0 — LAMP 2 O:0.0 1 — LAMP 3 O:0.0 2 — ALL HOME B3 1

Rung 7: AUTO / MAN I:0.1 4 — MANUAL MODE JSR JUMP TO SUBROUTINE SER FILE# U:3

Rung 8: AUTO MODE B3 2 — RUN MODE JSR JUMP TO SUBROUTINE SER FILE# U:4

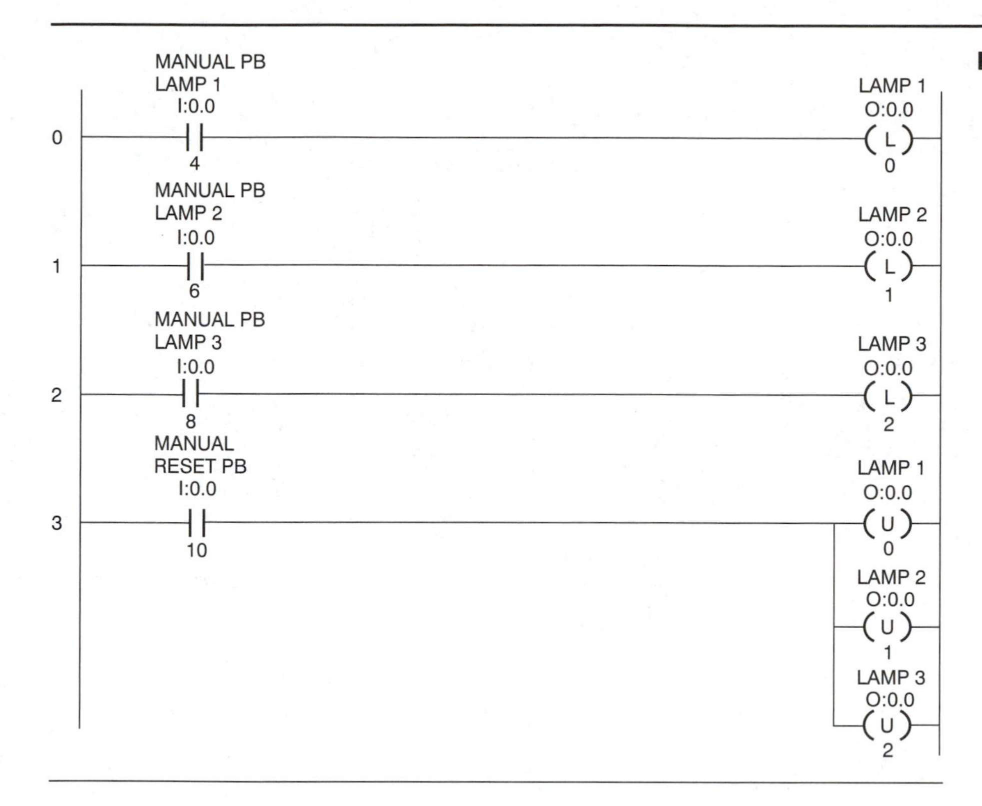

Figure 12–18

5. Enter the program shown in Figure 12–18 into file #3, the manual file.
6. Enter the program shown in Figure 12–19 into file #4, the run mode file. Once the program is entered, download the program and run it. Note, you can monitor only one file at a time while the program is running. The program will light the three lamps in sequence in the auto mode and then clear them. The sequence repeats indefinitely until you press the cycle stop button. Although this program does not perform any useful function, it does demonstrate how to construct a program where the two control programs are in different subroutines.
7. What happens if you change to manual mode while one or more of the three lights are on?
8. What happens if you switch back to auto mode after clearing the three lamps? Does the auto cycle start at the beginning?
9. What rungs could you add to the manual program to ensure the auto mode always starts at count 000?
10. What happens if you light one of the lamps in manual mode and then switch to auto mode? What causes this action?
11. Change the program to sequence four lights. Print out your final copy and demonstrate it to the instructor.
12. Close the program and create a new program called Chpt-12-labA. This will be the first program you will write from a specification. The project specification and I/O list are given below.

Figure 12–19

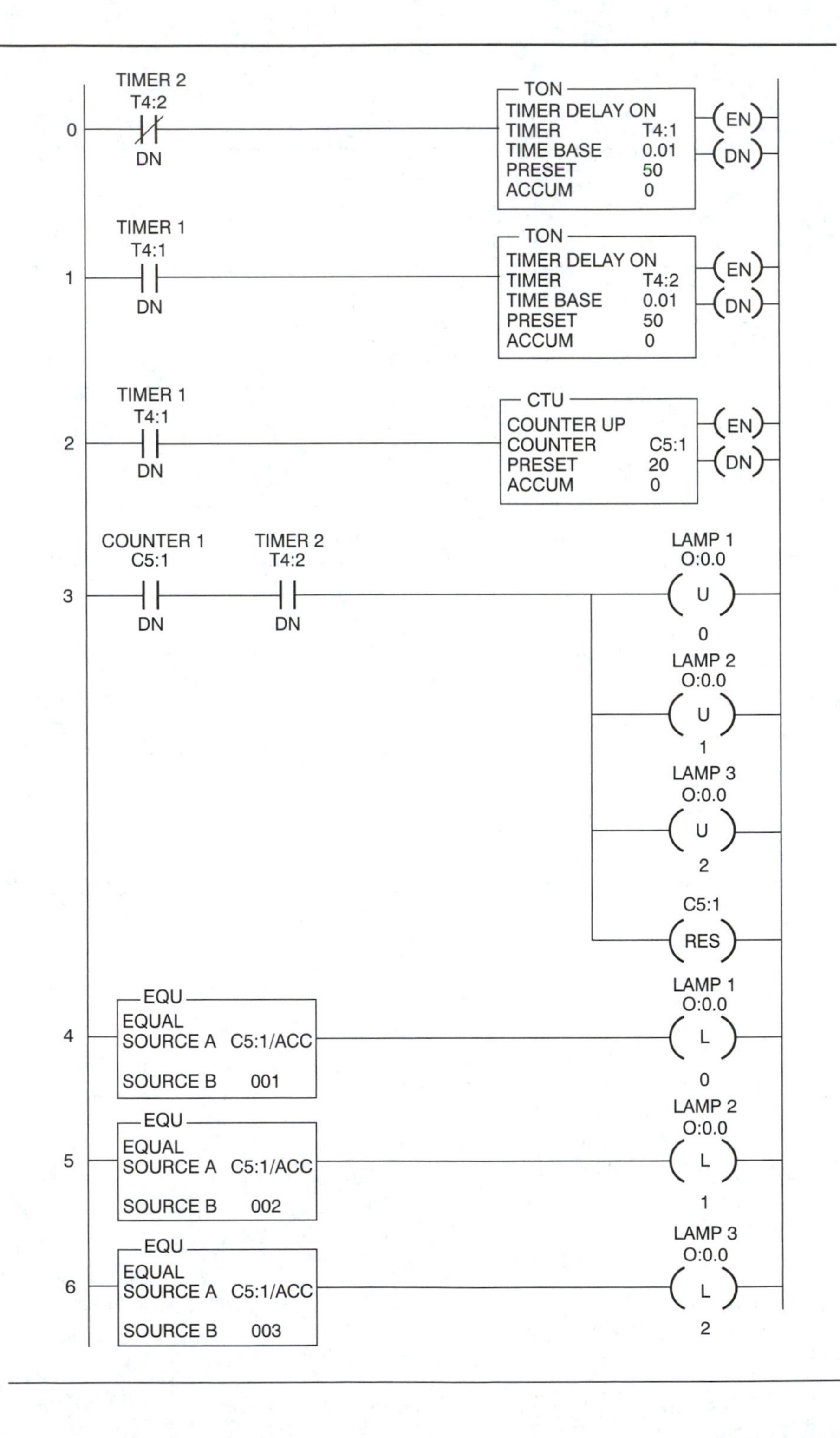
TIMER 2
T4:2
DN
0
TON
TIMER DELAY ON
TIMER T4:1
TIME BASE 0.01
PRESET 50
ACCUM 0
EN
DN
TIMER 1
T4:1
DN
1
TON
TIMER DELAY ON
TIMER T4:2
TIME BASE 0.01
PRESET 50
ACCUM 0
EN
DN
TIMER 1
T4:1
DN
2
CTU
COUNTER UP
COUNTER C5:1
PRESET 20
ACCUM 0
EN
DN
COUNTER 1
C5:1
DN
TIMER 2
T4:2
DN
3
LAMP 1
O:0.0
U
0
LAMP 2
O:0.0
U
1
LAMP 3
O:0.0
U
2
C5:1
RES
4
EQU
EQUAL
SOURCE A C5:1/ACC
SOURCE B 001
LAMP 1
O:0.0
L
0
5
EQU
EQUAL
SOURCE A C5:1/ACC
SOURCE B 002
LAMP 2
O:0.0
L
1
6
EQU
EQUAL
SOURCE A C5:1/ACC
SOURCE B 003
LAMP 3
O:0.0
L
2

Figure 12–20
Simulated machining center

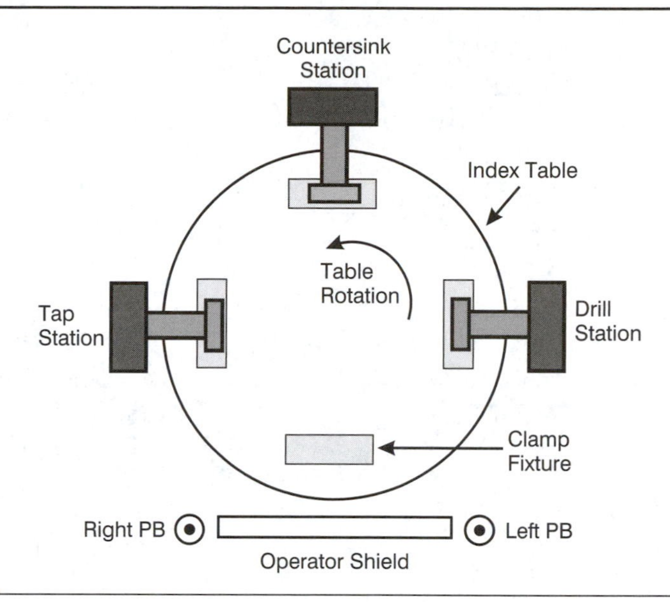

Project Specification

For this lab we will create a simulated machine and program it according to the written specs. The machine is a dial-indexed machine that drills, countersinks, and taps a hole in the object placed in the fixture. See Figure 12–20. There are four positions on the dial plate, and the index motor rotates the plate 90° each index cycle. The machine requires four cycles to make a complete part.

You must have run mode and manual mode subroutines.

In the manual mode, all the cylinders must be capable of being moved in both directions using push buttons or switches. You must be able to manually index the table only if all the stations except the shield are in the up position.

The auto mode cannot become active unless all the stations are in home position. The cycle starts when the operator presses both start push buttons. The shield moves down. If either push button is released before the shield hits the down prox switch, the shield must return to the up position. Once the shield down prox switch is made, the operator can release the start push buttons because the cycle will complete automatically. At this point the ready light will go off and the table will index one time. On completion of the index cycle, all three operations will begin and the shield can return to the up position. When all the operations are complete and both start push buttons are released, the cycle is complete and the ready light should come back on.

The manual and auto mode lamps must operate in accordance with the directions given in this chapter. A "too long" cycle timer must be incorporated, the alarm light must be lit, and the cycle stopped if the timer ever times out.

Be sure to follow all the rules for a two-hand manual operation.

Start this lab by designing either a flowchart or a sequence of state diagrams for each major section of the program. Show your diagrams to the instructor before you begin to write your ladder code.

I/O List

Inputs:

I:0.0/00	START PB LEFT
I:0.0/01	DRILL STATION DOWN L.S.
I:0.0/02	START PB RIGHT
I:0.0/03	DRILL STATION UP L.S.
I:0.0/04	DRILL UP MANUAL PB
I:0.0/05	C-SINK STATION DOWN L.S.
I:0.0/06	DRILL DOWN MANUAL PB
I:0.0/07	C-SINK STATION UP L.S.
I:0.0/08	C-SINK UP MANUAL PB
I:0.0/09	TAP STATION DOWN L.S.
I:0.0/10	C-SINK DOWN MANUAL PB

I:0.0/11	TAP STATION UP L.S.
I:0.0/12	TAP UP MANUAL PB
I:0.0/13	SHIELD DOWN L.S.
I:0.0/14	TAP DOWN MANUAL PB
I:0.0/15	SHIELD UP L.S.
I:0.1/00	MANUAL TABLE INDEX PB
I:0.1/01	NOT USED
I:0.1/02	NOT USED
I:0.1/03	INDEX CAM L.S.
I:0.1/04	AUTO / MANUAL SELECTOR SW
I:0.1/05	SHIELD DOWN MANUAL SW
I:0.1/06	NOT USED
I:0.0/07	NOT USED

Outputs:

0:0.0/00	AUTO MODE LAMP
0:0.0/01	MANUAL MODE LAMP
0:0.0/02	READY LAMP
0:0.0/03	ALARM LAMP
0:0.0/04	TABLE IN DWELL POSITION LAMP
0:0.0/05	NOT USED
0:0.0/06	NOT USED
0:0.0/07	NOT USED
0:0.0/08	NOT USED
0:0.0/09	NOT USED
0:0.0/10	NOT USED
0:0.0/11	NOT USED
0:0.0/12	NOT USED
0:0.0/13	NOT USED
0:0.0/14	NOT USED
0:0.0/15	NOT USED
0:1.0/00	INDEX MOTOR
0:1.0/01	DRILL STATION DOWN SOL.
0:1.0/02	NOT USED
0:1.0/03	DRILL STATION UP SOL.
0:1.0/04	NOT USED
0:1.0/05	C-SINK DOWN SOL.
0:1.0/06	NOT USED
0:1.0/07	C-SINK UP SOL.
0:1.0/08	NOT USED
0:1.0/09	TAP DOWN SOL.
0:1.0/10	NOT USED
0:1.0/11	TAP UP SOL.
0:1.0/12	NOT USED
0:1.0/13	SHIELD DOWN SOL.
0:1.0/14	NOT USED
0:1.0/15	SHIELD UP SOL.

Program Checkout

Once you have the program working, check out the following operations.

1. Switch to setup mode and verify that you can move all the major elements.
2. Verify that you cannot index the table when one of the cylinders is extended.
3. Verify that extending any of the cylinders during the index will stop the table.
4. Switch to run mode and verify that you cannot start a cycle by pressing only one button.
5. Verify that you cannot start a cycle by pressing one start button and then pressing the second start button 2 or 3 seconds later.
6. Verify that if you release one of the start buttons before the screen reaches the down position it returns to the up position.
7. Run a cycle and hold one of the cylinder rods so it cannot return to home. Verify that the watch dog timer shuts down the auto mode after the time-out period and lights the alarm lamp.
8. Switch to setup mode and reset all the cylinders. Verify that the alarm lamp is cleared.
9. Switch to auto mode and verify that you can start a new cycle without any erroneous movements.

Once you have the program working correctly, print out a copy of the ladder and demonstrate the running program to the instructor. Be sure to block one of the cylinders to show that the watch dog timer works and the program shuts down the auto mode.

CHAPTER

13

Continuously Running Machines

LEARNING OBJECTIVES

After completing this chapter the reader should understand:

1. Program construction using subroutines for automatic operation.
2. OSHA requirements for continuously running machines.
3. How to build safety into programs for continuous machine operation.
4. Creating a watch dog timer circuit to stop the run mode cycle if a jam-up occurs.
5. Setting up the timing loop for continuously running machines.
6. The use of word wide shift registers to move data along an indexed assembly line.
7. How to capture and move data on an asynchronous pallet line.

Continuously running machines are defined as machines that, once started, will continue to cycle until halted by a cycle stop push button or error condition. Examples of this type of automation equipment are assembly lines, continuous cycle molding machines, bobbin pinning machines, and most types of container-filling equipment to name just a few. Control system programs for these machines should be designed around the three-state control system that is explained in detail later in this chapter.

We also have some additional safety problems to deal with in this type of machine. A major safety problem with continuously running machines can occur when a jam-up condition arises. The machine can become a trap if the auto cycle run mode is not shut down. For example, if a machine has a loose prox switch that has moved a small amount, the next cycle may not complete because the switch is out of position and will not activate. The machine is now stalled in the auto cycle run mode and will wait indefinitely for the switch to activate. When the operator reports the problem and a maintenance person arrives and moves the switch, the machine will take off running again and possibly cause injury to the maintenance person. To prevent this kind of trap condition, all programs must incorporate a "too long" cycle timer or watch dog timer. This timer will disable the auto cycle run mode if the machine does not complete its cycle in the normal time frame.

A second safety concern with large equipment (equipment for which the operator cannot see the whole machine from the control panel) can occur when someone is in a dangerous position and the operator starts up the machine. To provide for a safe start-up, we need a warning alarm and a time delay before the machine begins running in auto cycle run mode.

THE THREE-STATE SYSTEM

A block diagram of the desired three-state system is shown in Figure 13–1. The operator interface most often used for mode control and to start and stop continuously running machines is shown in Figure 13–2. It consists of a mode selector switch and two push-button switches.

As you can see from Figure 13–1, our proposed operating system has three operating states. File 2 is the main program file and is active all the

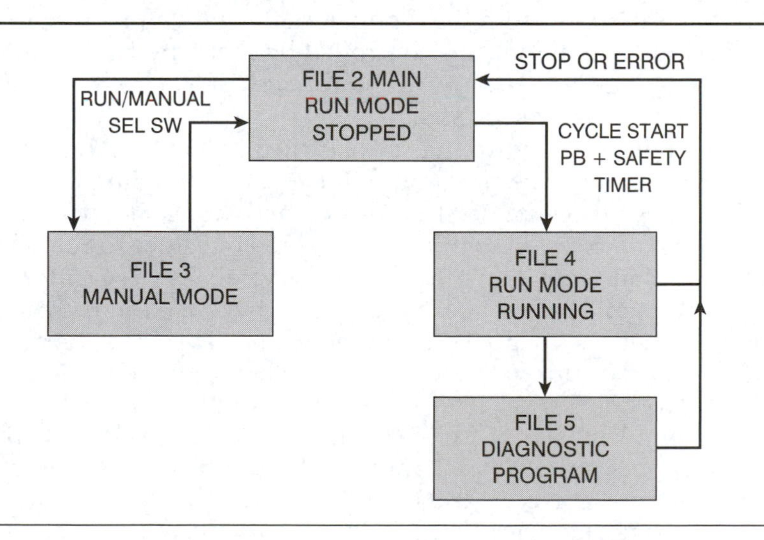

Figure 13–1
Block diagram of a three-state system

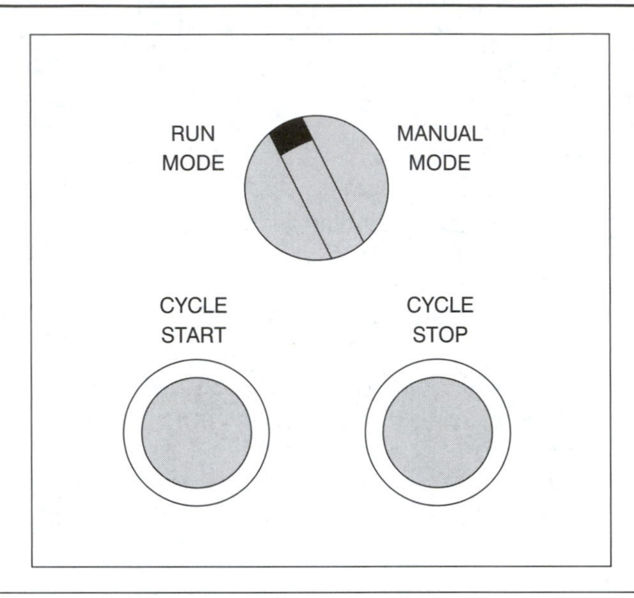

Figure 13–2
Operator interface for three-state system

time. File 3 is our manual or setup file and is called by file 2 if the run/manual switch is set for manual. File 4 is our run mode file and is called from file 2 if the run/manual switch is set to run and the cycle start push button has been held in for the required safety time. Note that file 5, our diagnostic file, is called from file 4 and therefore runs only when file 4 is active.

The safe machine state is defined as having only file 2 active. You are in this state if the machine's mode selector switch is set to run mode but the cycle start push button has not been held in for the required start-up time. In this safe state, both the manual mode and run mode programs are deactivated and all outputs controlled by them are held in their last state. The safe state is also the state you return to if a cycle stop, E stop, or diagnostic error condition occurs.

If the mode selector switch is now turned to MANUAL MODE, the manual mode subroutine program will become active immediately. We normally do not place any restrictions on activating the manual mode subroutine because the operator will need to activate it to recover from jam-up conditions. Just as in the previous chapter, the programmer must be sure to reset all latches, alarm bits, and timers used in the run mode subroutine when the manual mode is activated. This must be done due to the fact that the auto run mode may have been exited by the diagnostic program or an E stop. If this was the case, the last cycle may not have been completed and therefore latch bits may have been left on. If the programmer does not clear all the auto cycle latch bits, then the run mode may not reactivate correctly and may behave erratically. When the selector switch is switched back to the RUN MODE, the machine should not immediately start up. The cycle start PB must again be pushed and held in for the safety period.

As previously stated, on the main control panel there should be two push-button switches marked CYCLE START and CYCLE STOP located next to or below the mode selector switch. To satisfy our start-up safety concerns, the operator must press the cycle start PB and *hold it in* for the required warning time. During this start-up time if someone makes their presence known in a dangerous position, the operator can release the cycle start PB and the machine will not activate the automatic cycle. Only if the cycle start PB is held in for the entire delay time will the auto cycle run mode latch in and the machine begin running. If the button is released and pressed again, it must be held in for the entire start-up time again. The length of this warning time is usually between 3 and 5 seconds. Some type of loud warning alarm should sound during this period.

Once the machine is latched into auto cycle run mode, there are three ways to deactivate it:

1. If someone presses an E stop push button, the machine must immediately return to the safe mode state, in which only file 2 is scanned, and all motors and outputs must be killed.
2. If someone presses the cycle stop PB, our control program must finish the current machine cycle before we stop the machine. If we stopped immediately, we could have half-assembled parts on the line every time the machine was stopped for breaks. We then return to the safe mode stopped state and are ready to restart auto run mode if the cycle start PB is held in again for the required safety delay time. This cycle stop PB input must activate a latched logic function because the machine's cycle time may be a long period of time and the operator should be able to quickly press and release the cycle stop button at any time during a machine cycle.
3. If a diagnostic program is running on our machine, it will stop the machine if it detects an error condition. All continuously running machines must have at least a "too long" cycle timer to shut the auto cycle run mode down if a problem or jam-up condition occurs. This will deactivate the run mode subroutine and return us to the auto mode stopped state, which is our safe condition.

Our control system design should also provide a stack light with at least three colors for displaying the machine's system status to the operator. This stack light arrangement was described in detail in the previous chapter.

We now discuss a few concerns on placing and wiring E stops and how to program their inputs on this class of machines. On all automation equipment, the system designer must provide some type of E stop device within easy reach of each operator working on the machine. Every E stop device should have two sets of contacts, one normally open and one normally closed. The normally closed contacts must be part of the

MCR relay hold in the circuit. It is this circuit that holds the main power distribution circuit on. Each normally open contact set should be used as an input to the PLC. This will make it easy to identify which E stop device was activated. Remember that the E stop function cannot be a logic function in the PLC. E stops must be hard wired to shut down the output power circuit.

In the case of an assembly line it is also a good idea to provide an additional paddle switch or light beam device in front of each station on the line to detect an out-of-place product that might jam a station on the next index. Some type of sensor device should also be located at the end of the assembly line to detect unremoved product or misplaced tools before they can jam the index motor and gearing.

ASSEMBLY LINE PROGRAMS

At this time we should discuss some important considerations that only pertain to assembly line control programs. Most assembly lines contain a combination of automation stations and manual operator workstations. The two most common assembly line arrangements are *synchronous indexed* and *asynchronous pallet* lines. Which type you use will depend on the number and type of automation/manual workstations being used on the line.

Indexed lines are most often used when the assembly line can be balanced so that every station's operation time is closely matched to that of all other stations. From a programmer's standpoint this is the simpler control system to work with. Note that an indexed assembly line can be built using carriers attached to a chain or pallets that ride on a moving belt. The condition that defines a line as an indexed type line is that all the positions move together during a common index.

Pallet pull style assembly lines are most often used when each station's time to complete an assigned task varies by a substantial amount on each product or when certain assembly stations need two to three times more operational time than others need. In this type of line, each operator has a release button and the pallet being worked on is released only when the operator is finished. This type of line is usually paced by the slowest operator or workstation unless parallel branches are used. The main advantage of this type of assembly line is flexibility, because workstations can be reconfigured for different products easily without the requirement of close line balancing. Also, the assembly line may have several side branch lines that split from the main assembly line and loop back to it. These parallel branches may be needed for rework operations or parallel functions that require more time to complete their functions. Although this flexibility is good for production planning, it makes tracking product data difficult for the programmer because any pallet can be released from any station at any time. Also, products may take a branch route several times before completing the assembly operation.

As we will soon discover, writing programs to control these two very different types of assembly equipment requires us to use completely different methods of data tracking and storage.

SYSTEM TIMING

When writing programs for synchronous continuously running machines one of the first design tasks is to develop the timing sequence that will enable the equipment to cycle in a continuous manner. The first decision to make is where the assembly cycle will be in the timing sequence. We can index the line first and then operate the assembly stations, or we can operate the assembly stations first and then index the line. There are certain advantages to indexing the line first and then following up with the assembly cycle. In most automation designs it is usually better to index first and then perform the assembly cycle. In the event we have an E stop or diagnostic stop in the middle of the assembly cycle, we will not have a problem restarting the line if we index first. If we were to perform the assembly function first and then index the assembly line, we could have a major problem if the assembly line was E stopped in the middle of the assembly cycle. On the next start-up we would again perform the assembly function. If our assembly process requires some part to be placed onto the pallet or a crimped connector to be placed on a wire, and the previously aborted assembly cycle already has placed a unit or crimp there, then the assembly line will try to place a second unit or crimped lug into the same place. This may cause major jam-up problems and damage mechanisms. By indexing first we may have a few partially assembled units to reject, but we will never cause the assembly function to run twice on the same position.

For an index-first control program, we need to create a sequence of timing pulses to accomplish the following train of events. This is a very common requirement in any continuously running type of program. We will explain in detail how we generated this timing sequence a little later in the chapter. For now we only need to understand the sequence of events that cause the program to enter run mode and once there, continue to cycle until a cycle stop input is detected. Our program timing sequence is as follows:

1. Wait for start button to be pressed.
2. Wait for 3-second warning delay.
3. Latch run mode.
4. Index the line.
5. Start the assembly cycle.
6. Check for stop bit set; if not set go back to step 4.
7. If stop bit is set, unlatch run mode and go to step 1.

Notice the sequence starts in run mode with an index function. Also notice the stop cycle function can happen at any time but we only look for it at the end of the assembly cycle.

At this point we will design a short program to demonstrate how to set up a timing sequence for a continuously running type of machine. We must have the following conditions satisfied by our program.

1. Cycle start PB must be held on for 3 seconds before the cycle can start. During this time a flashing light will warn anyone that the machine is about to start.
2. The cycle stop PB can be pressed and released at any time during the cycle and the cycle will finish its current cycle and then stop.
3. A "too long" cycle timer will time each cycle and if the cycle does not complete within this time it will shut down the auto run mode.
4. A 1-second pulse will be generated after the index is complete and this pulse will be used to start all assembly stations.
5. Often in assembly equipment we have people working at stations on the assembly equipment along with the automation stations. In order to provide time for the human workers on the equipment to finish their task, we will have a cycle timer set for 10 seconds. Even though the automated assembly equipment is complete at 5 seconds we will hold off the next index for this additional time.

We will write this sample program in the same manner as we wrote the manually activated program. File 2 (Figures 13–4 through 13–6) will contain all the rungs that must be scanned all the time. File 3 (Figure 13–7) will contain the rungs controlling the continuous run mode. For this sample program our automatic cycle will consist of extending cylinders 1, 2, and 3 and then returning them. The main purpose of this sample program is to demonstrate how to set up the timing sequence for a synchronous, continuously running assembly line or assembly equipment.

Figure 13–3 shows the timing chart for this program. Note that after the index cycle is complete we generate an assembly cycle start pulse of about 1 second. This pulse will be used to trigger the assembly operation on every station on the assembly line or assembly equipment.

The assembly cycle start pulse also starts the cycle timer. This timer's job is to time the assembly cycle. If we have human operators working on the assembly line or equipment this timer will be set for a time duration that is long enough for the human operators to finish their work. Note that in our timing diagram the automated assembly stations return to their home position after about 5 seconds. In many cases this timer's value is adjustable from some kind of an operator interface panel. As an example, if we have an assembly line that is normally manned by 15 people, but due to weather problems only 7 show up, we would like to be able to slow down the cycle time. Conversely, if we have a rush order and put 5 extra

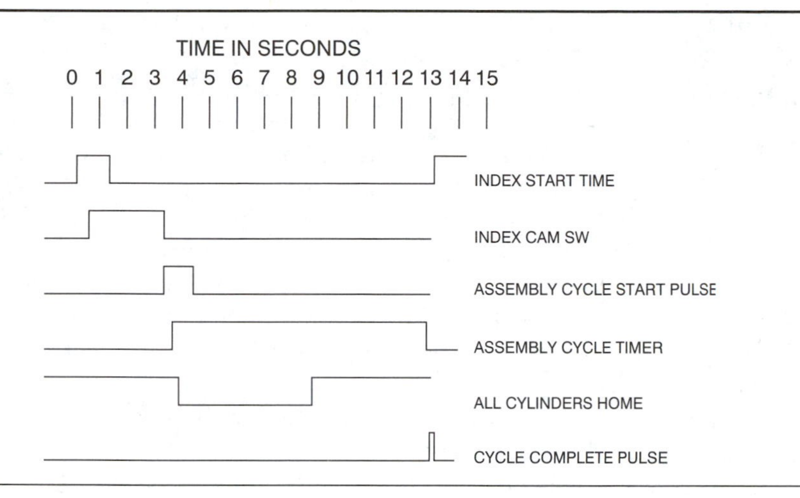

Figure 13–3 Timing diagram for continuous cycle

people on the line, we might want to be able to speed up the cycle time. The assembly cycle is finished when the cycle timer times out and all stations are home.

When the assembly cycle timer times out and all stations are home we next generate a 50 ms cycle done pulse. This pulse will be used to check the cycle stop condition and shut the line down if the cycle stop PB was pressed during the last cycle. If the cycle stop PB was not pressed, the cycle done pulse will trip the index start timer to begin another index and assembly cycle.

The following pages contain the sample program and a rung-by-rung description of how it works.

Rung 0 First flash timer in the two-timer flash circuit to provide ½-second pulses for flashing lamps.

Rung 1 Second flash timer in the two-timer flash circuit to provide ½-second pulses for flashing lamps.

Rung 2 Turns on the manual light if we are in manual mode and all the cylinders are home. If any of the cylinders is not in the home position the light is flashed.

Rung 3 Flashes the auto mode light if we are ready to run. When we start the cycle running, the lamp is held on solid.

Rung 4 This is the safety delay circuit. When the cycle start PB is pressed, all stations are home, and there is no alarm, we start the safety delay timer. The cycle start PB must be held in for the entire 3 seconds.

Rung 5 Latches in the auto cycle latch bit when the safety timer times out. This will activate the auto mode.

Rung 6 Jumps to file 4, run mode, if the auto cycle latch bit is set on.

Rung 7 Jumps to file 3, manual mode, if the auto/manual selector switch is turned off.

Rung 8 Contains the conditions that shut off the auto mode. Changing the auto/manual selector switch to off unlatches the auto mode. If the cycle stop latch bit is on, and the cycle done pulse comes on, we also unlatch the auto run mode. The "too long" cycle timer and the alarm latch also unlatch the auto run mode.

Rung 9 Latches the cycle stop latch bit if the cycle stop PB is pressed.

Rung 10 Unlatches the cycle stop latch when the cycle start PB is pressed to start a new cycle.

Rung 11 Turns on the "all home" bit if all three cylinders are in their home position. In a large program you would have one rung to indicate that each station on the equipment is home and then sum all the station home bits together in this "all home" rung.

Rung 12 Contains our "too long" cycle timer or watch dog timer. It starts timing when the cycle start latch bit is set. In a normal cycle it should be reset long before it times out.

Rung 13 Latches the alarm latch if the "too long" cycle timer ever does time out.

Rung 14 Unlatches the alarm condition when you turn the auto/manual selector switch to the manual position.

Rung 15 Flashes the alarm lamp.

Rung 16 Flashes the safety warning lamp during the 3-second start-up delay.

Let us now look at the rungs of the manual mode shown in Figure 13–7.

Rung 0 Unlatches the cycle start latch bit that was set in the auto mode file 4. It also resets timer T4:3, the index start timer. This rung cleans up bits that may have been left on if we had an alarm condition in the run mode and didn't complete the cycle. By switching to the manual mode, you reset all the latch bits and the timer. This ensures that when you start auto mode again, everything has been reset.

Rung 1 Allows the setup person to manually activate the index motor.

Rung 2 Returns cylinder #1 to its home position.

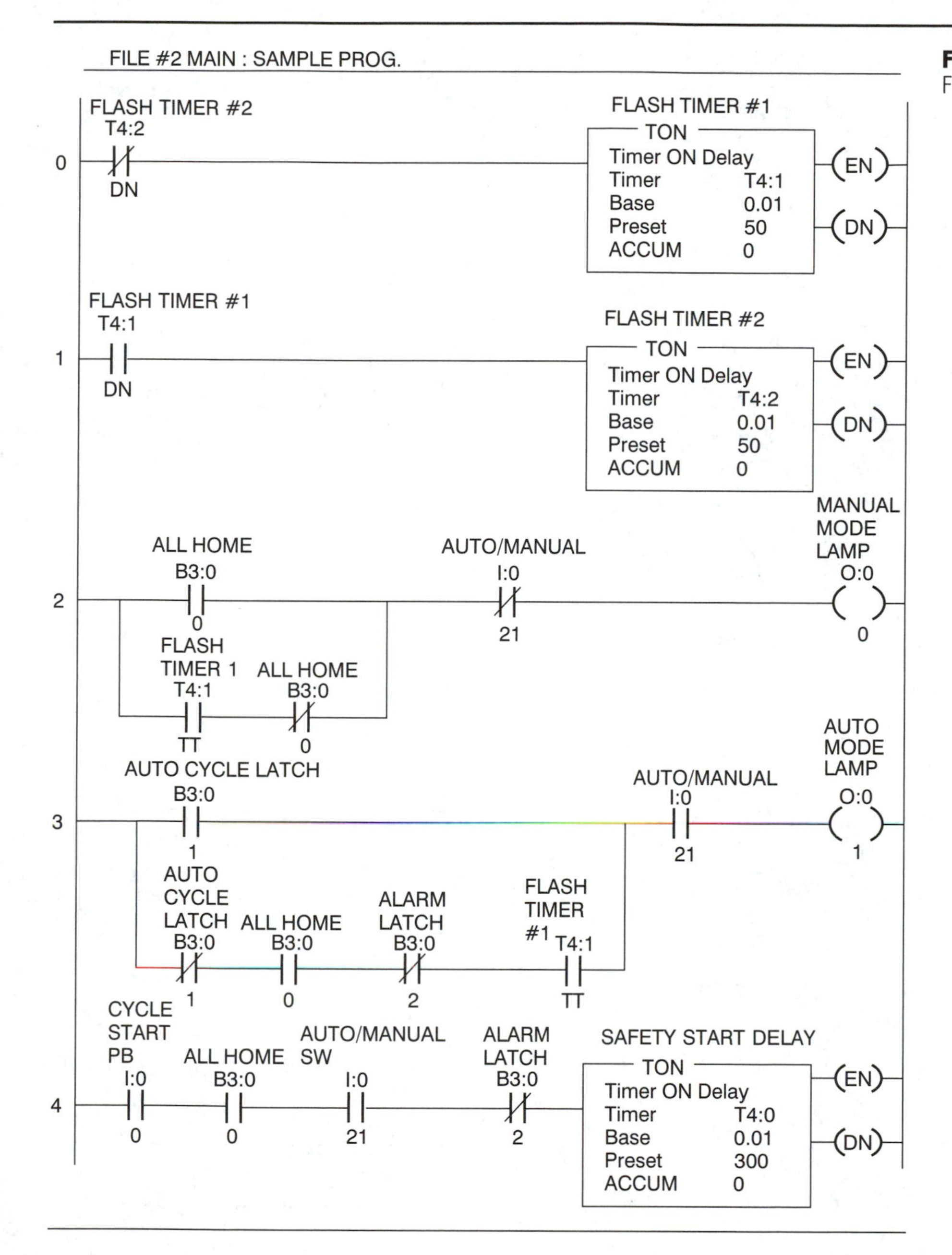

Figure 13–4
File 2 main program

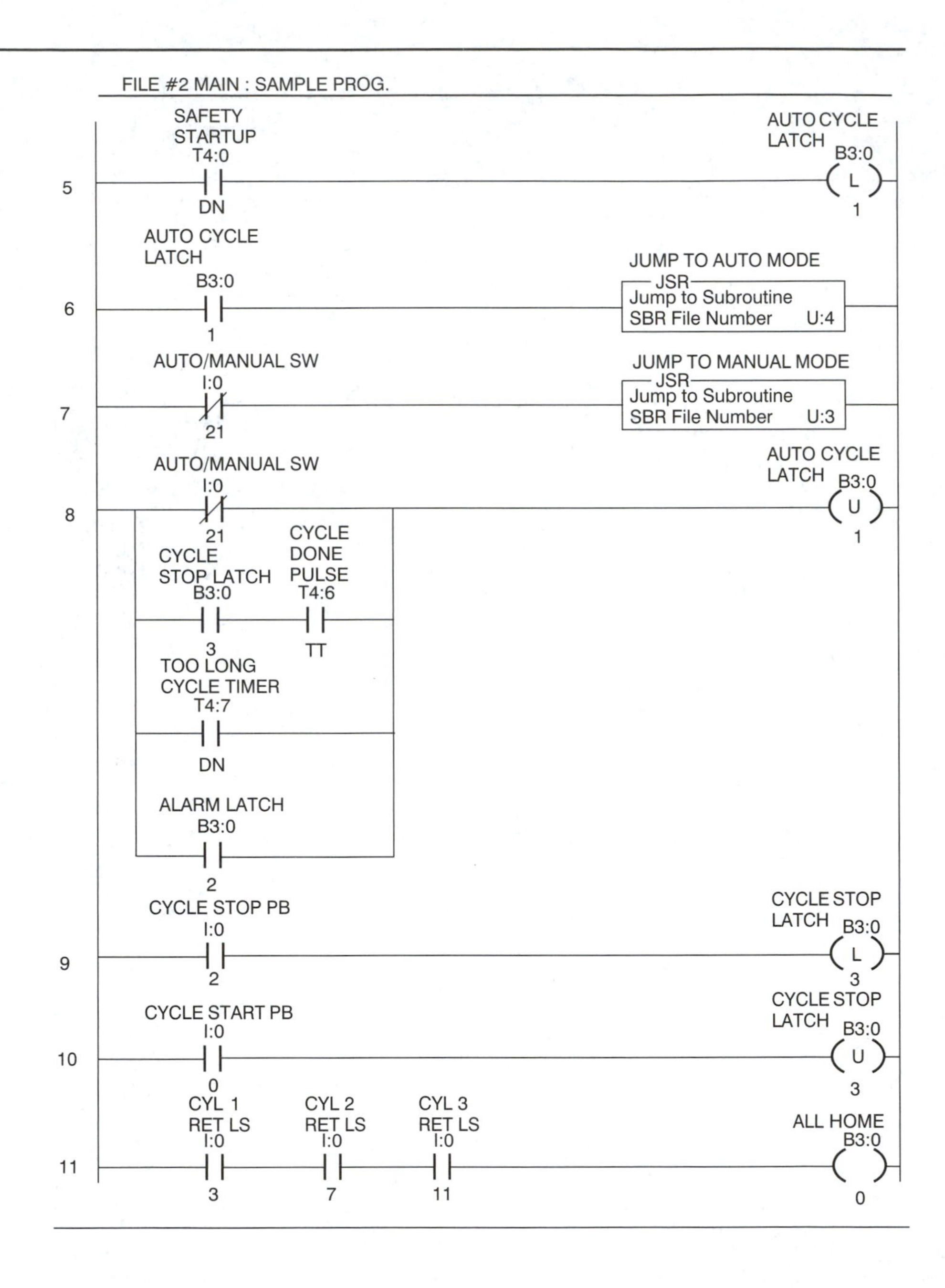

Figure 13–5
File 2 main program, *continued*

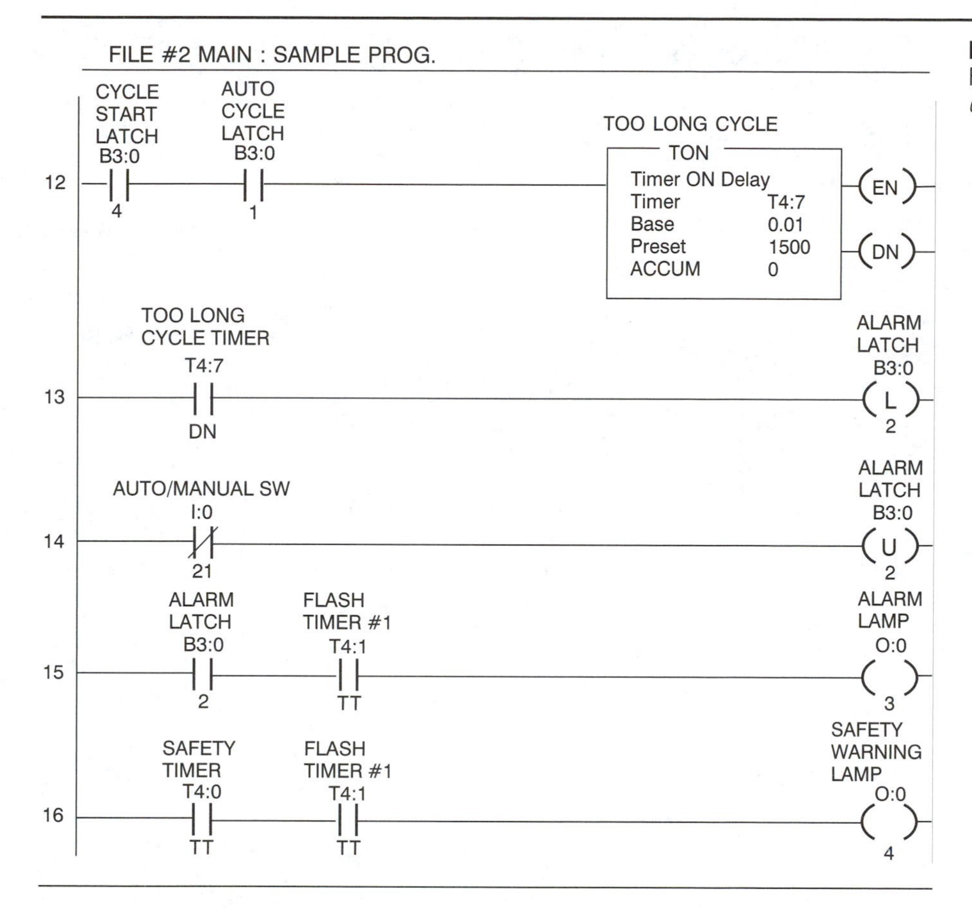

Figure 13–6
File 2 main program, *continued*

Rung 3 Returns cylinder #2 to its home position.

Rung 4 Returns cylinder #3 to its home position.

Now let us look over File 4 (Figures 13–8 and 13–9), the automatic run section of this program.

Rung 0 Controls the index start timer. This timer is the first step in our timing sequence. Note how the cycle done pulse is used to reset this timer for the next cycle.

Rung 1 This is the standard index motor control rung we have used throughout this text beginning in Chapter 9.

Rung 2 Generates the cycle start pulse when the index function is complete. This pulse would be used to start the assembly cycle on all stations along the line if this were a large project.

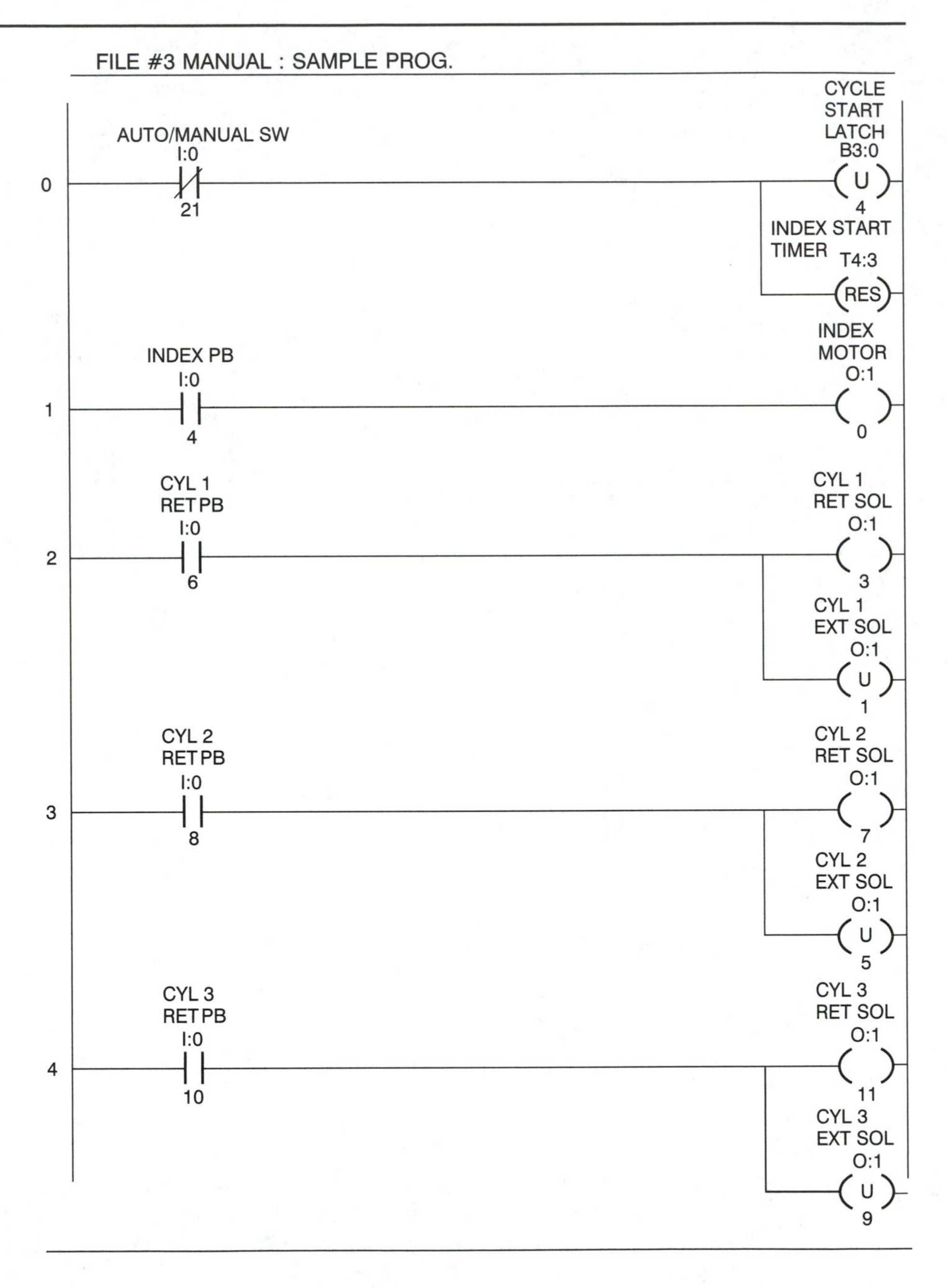

Figure 13–7
File 3, manual mode rungs

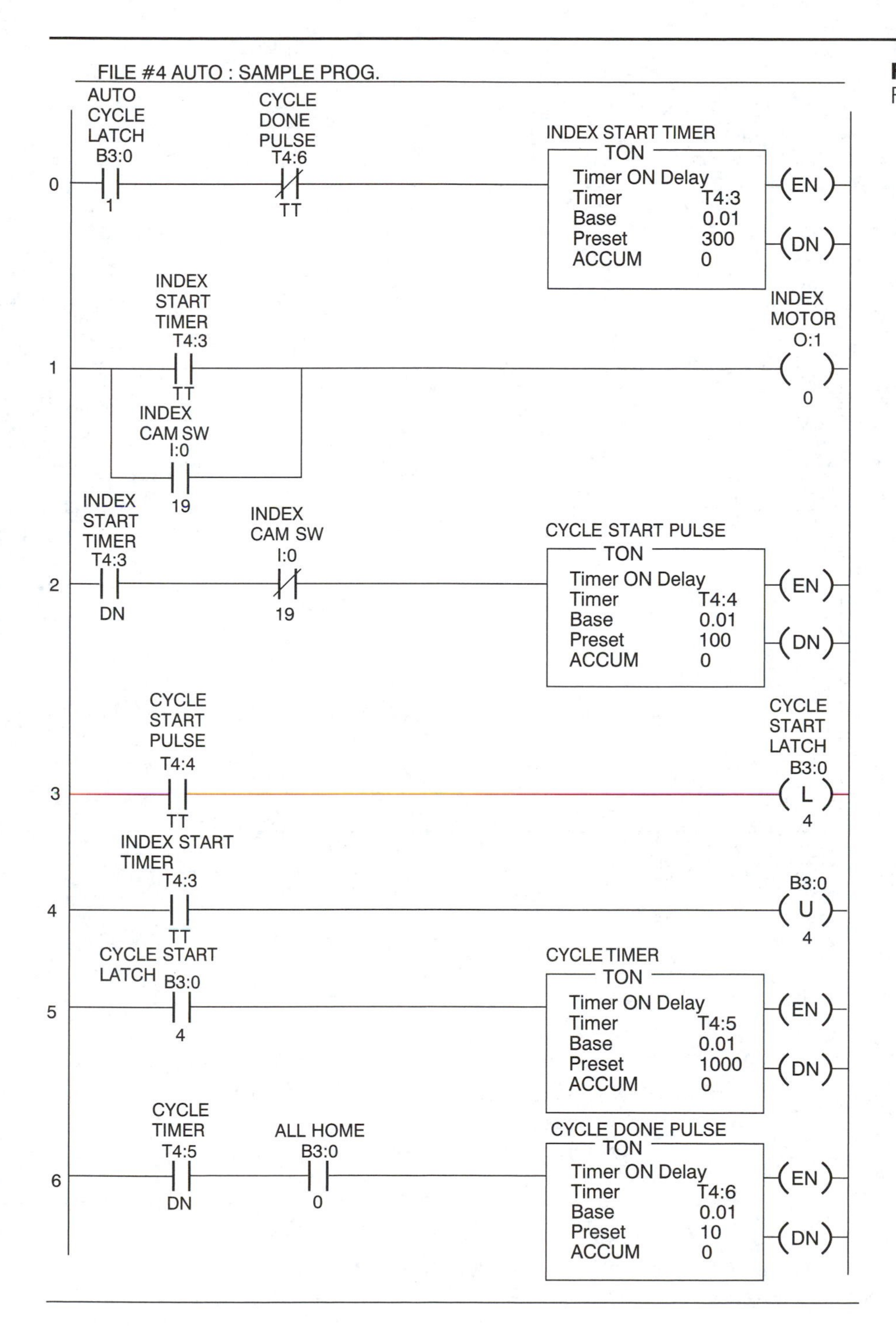

Figure 13–8
File 4, auto run mode rungs

Figure 13–9
File 4, auto mode rungs, *continued*

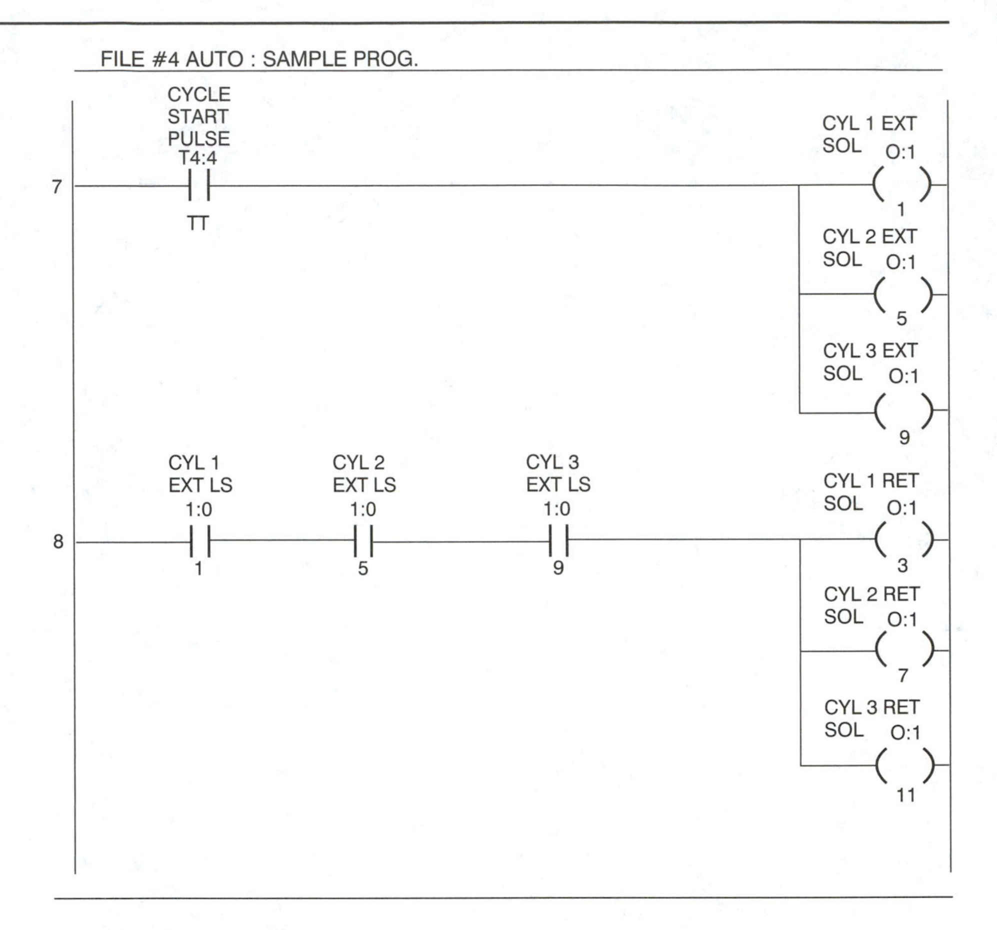

Rung 3 Latches the cycle start latch bit to keep track of the fact that we have started the assembly cycle.

Rung 4 Unlatches the cycle start bit when the next index is performed.

Rung 5 Contains the cycle timer function. In our case it is set for 10 seconds.

Rung 6 Contains the cycle done pulse function. Note that it is activated when the cycle timer times out and all stations are home. The "all home" bit is placed here to prevent another cycle from starting if the current cycle does not complete. If one station does not return home, the program will wait until the "too long" timer shuts down the auto run mode.

Rung 7 Extends all three cylinders.

Rung 8 Returns all three cylinders when all three extend switches have turned on.

This program is a good tool to experiment with to see how changes in the process cycle affect the program's timing.

Start the auto cycle and allow several cycles to complete. On the next cycle hold one of the cylinders to prevent it from extending. After 15 seconds the alarm should come on and the auto run mode drop out. Try to restart the auto cycle. What prevents you from restarting?

Switch to manual mode and reset the cylinders to home. Switch back to run mode and restart the auto run cycle. The cycle should restart correctly now.

Try removing the reset timer T4:3 instruction from rung 0 of the manual mode file 3. Repeat the above alarm condition. What happens when you switch to manual, clear the cylinders, and then try to restart the auto run mode?

TRACKING PRODUCT DATA ON INDEXED LINES

It is not uncommon on assembly lines, assembly machines, and other types of continuously running equipment that the programmer will need to track information about the product as it moves down the line. If only one bit of information is needed you can use a bit shift instruction to keep track of this data. If, as is usually the case, you need to collect a number of bits of information about the product, you will need to use a 16-bit word wide shift register. This will allow you to keep track of 16 bits of information about the product as it moves down the assembly line. Higher-level PLCs like PLC-5 or Control Logix processors will allow 32-bit word wide shift registers.

Data collection and tracking programs for indexed type assembly lines require an accurately documented workstation map of the assembly line. The programmer must know the exact number of carriers on the line and the position of each automation station in relation to the first carrier position. We must follow all the rules given at the beginning of this chapter for worker safety and operator interface functions. In Figure 13–10 we have diagrammed a small assembly line with 13 index positions and 7 automated workstations.

Because our lab kit is limited to only five air cylinders, some of the functions on our sample assembly line have been greatly simplified. No air cylinders were assigned to the test functions as would be normal on a real line. In our example I have assigned two selector switches to simulate the testing functions. To simulate the test function, two selector switches were assigned to provide pass or fail test functions. We have assigned only one air cylinder to each of the other working stations on our example line. In an actual automation design project, most stations would require several air cylinders to accomplish their machine functions. While greatly simplified in order to fit into the

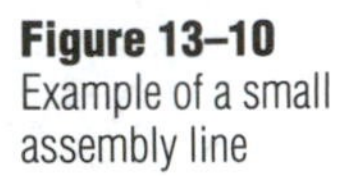

Figure 13–10
Example of a small assembly line

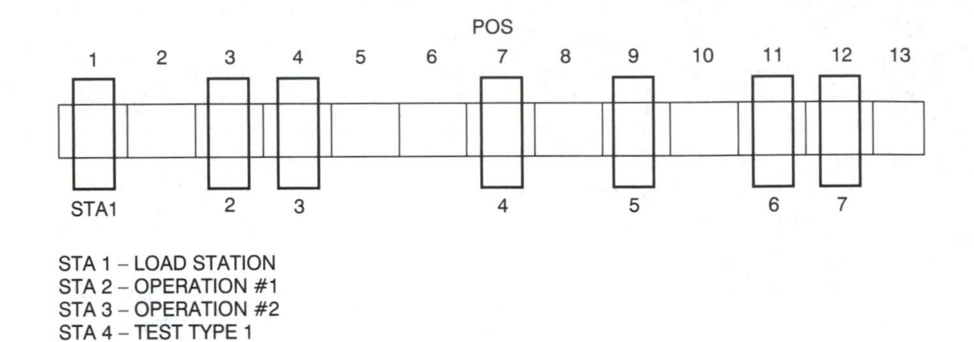

STA 1 – LOAD STATION
STA 2 – OPERATION #1
STA 3 – OPERATION #2
STA 4 – TEST TYPE 1
STA 5 – TEST TYPE 2
STA 6 – REJECT
STA 7 – UNLOAD GOOD PARTS

limited I/O functions of the lab kit, our sample assembly line can still demonstrate all the data handling functions that would be used on a real indexed style assembly line.

ASSEMBLY LINE PROGRAM SPECIFICATIONS

Our simulated indexed assembly line runs three different types of products. The product type being run should normally be selected at the main control panel by a selector switch that has three positions: product A, B, or C. Because we only have toggle switches available on the lab kit we assigned selector SW inputs I:0.1/6 for product A and I:0.1/7 for product B to simulate this function. Turning on just one switch will define the product type being processed. Turning both on will simulate selection of product C.

Product A gets operation #1 but not #2 and test #1 only. Product B gets operation #2 but not #1 and has only test #2 performed. Product C gets both operations and both tests performed. We must write a control program that will check the control panel switch settings and activate the proper stations for the product being run. We must also be able to store the test results of each product and decide if the product should be rejected or unloaded at the end of the assembly line. Doing so will require the use of a word wide shift register. If you have looked for an instruction that shifts words in the SLC-500 you will not find one. There are bit shift instructions, which will be explained in the next chapter, but no word shift instructions.

To create a word shift register, we use the copy file instruction. We start by assigning one word address to each assembly line position (see Figure 13–11). Because the copy file instruction begins the copy data function from the lowest word address first, we must assign the start of the

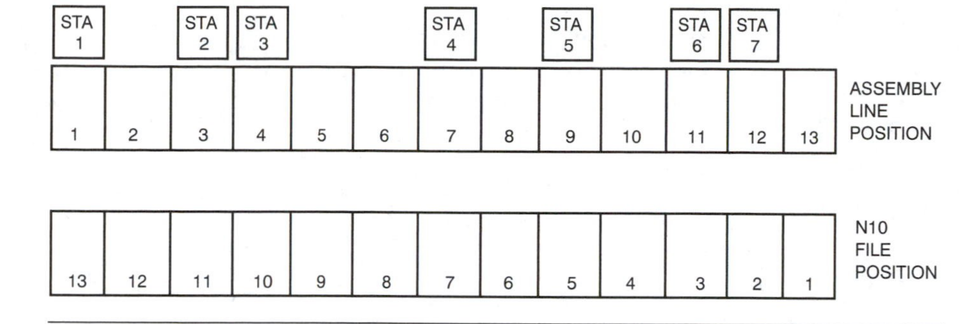

Figure 13–11
Map of memory addresses and assigned physical line positions

assembly line to the highest address. We have arbitrarily picked file #N10 as the word shift register file. Any integer file number above #9 could have been used. Unfortunately, the copy file instruction copies all of the data from the file address list during each program scan for as long as the rung is true. Because we only want to copy the data once each time the index cam turns on, we will need to use a one-shot instruction in this program rung. This will cause the copy command to only copy data (shift in our case) once each time the rung goes from FALSE to TRUE.

To get a better understanding of how we will use the copy file instruction to make a word shift register, let us first look in detail at how the instruction operates. Figure 13–12 shows a sample rung containing a copy instruction and a view of a section of file N10. As long as the "A" input remains high (1), the instruction will copy the data located at N10:20 through N10:24 to address locations N10:30 through N10:34 during each program scan. Note that the first address to be transferred is word N10:20 to N10:30, then N10:21 to N10:31, and so on until all five addresses are transferred. It is important to understand the order of the actual copy data transfer sequence because it will be important in the upcoming discussion about the data movement in a word wide shift register.

Figure 13–13 shows our shift register memory block and how the data will be moved from address to address in our actual shift register. Note that the input elements to the copy instruction are the cam switch and a one-shot instruction. Because we only want to move the data once for each cycle of the index cam, we need to add this one-shot instruction. Without it, the data would be moved once during each program scan and that could be many thousands of times during the time the cam switch is on. The one-shot instruction causes the copy instruction to be executed only once each time the rung goes true.

Note that the data source is set for N10:2 and the destination is N10:1. This means the first data move will be from N10:2 to N10:1 and then each additional move will be advanced up one address location. This is why it is so important to always have the end of your physical line be the lowest

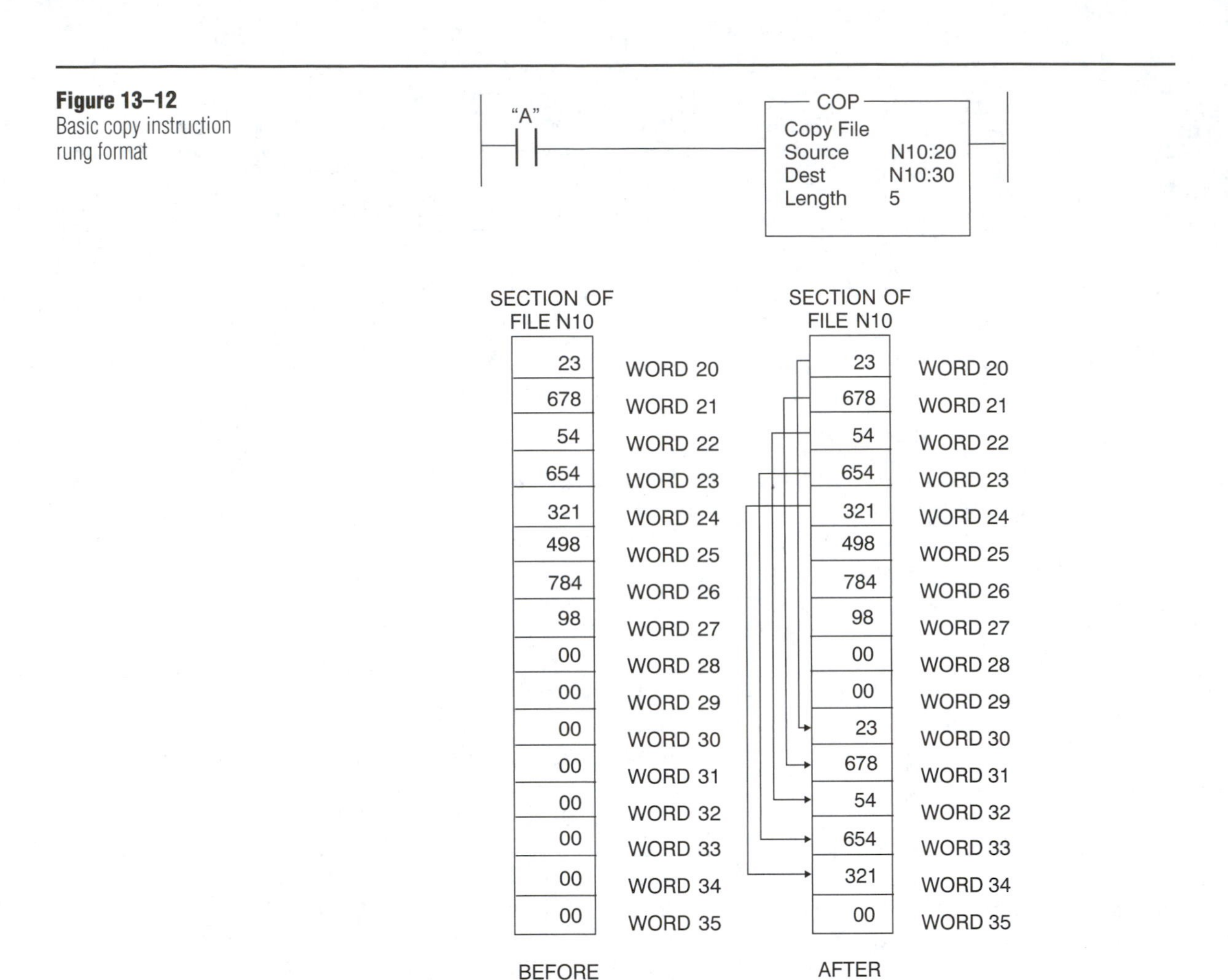

Figure 13–12
Basic copy instruction rung format

memory location. Data must be moved in the proper sequence so as not to write over data. In our example new data are started at address N10:13 and exit at location N10:1. Note that Length is set to 12 because there are only 12 moves even though we have 13 locations. It could be set for 13 if we always keep N10:14 set to all zeros; in this way we would always clear N10:13 for each shift. If you do not do this, however, you must be sure to always clear the old data in N10:13 before latching new data there. Now let's look at how each bit is assigned a function.

Note in Figure 13–11 that a word address was assigned to each assembly line position even if that position did not have a workstation assigned to it. The reason for this is that we need to have a place to hold the data for each product even though nothing is happening to the product at this location. The next procedure will be to assign bit functions to our

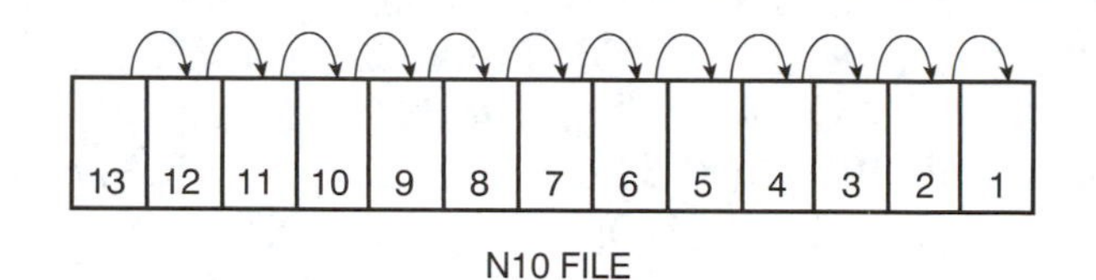

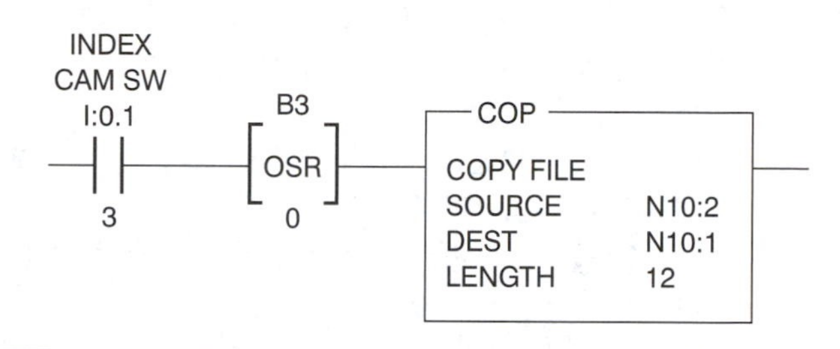

Figure 13–13
Shift register instruction and data movement map

shift register words. We have 16 bits to work with and we begin assigning functions at bit 00. We assign each assembly line data function that we wish to keep track of to a bit location.

N10:X/00	base product loaded
N10:X/01	selector switch set for product A
N10:X/02	selector switch set for product B
N10:X/03	test #1 passed
N10:X/04	test #1 failed
N10:X/05	test #2 passed
N10:X/06	test #2 failed

Remember that product C is run if both product A and B inputs are on. Bit 00 would be set by a prox switch located in the load station on a real assembly line. For our simulation we will use switch input I:0.1/4. This input is required on a real assembly line to prevent an assembly operation from being performed on an empty carrier location. This could occur if the base supply ran out or the loader missed picking up a base for one cycle.

We now assign the rest of the I/O for this simulated program:

Inputs:

I:0.0/0	CYCLE START PB
I:0.0/1	LOAD STATION CYLINDER EXTEND SW
I:0.0/2	CYCLE STOP PB
I:0.0/3	LOAD STATION CYLINDER RETURN SW
I:0.0/4	LOAD STATION MANUAL EXTEND PB
I:0.0/5	OPERATION #1 CYLINDER EXTEND SW

I:0.0/6	OPERATION #1 MANUAL EXTEND PB
I:0.0/7	OPERATION #1 CYLINDER RETURN SW
I:0.0/8	OPERATION #2 MANUAL EXTEND PB
I:0.0/9	OPERATION #2 CYLINDER EXTEND SW
I:0.0/10	REJECT STATION MANUAL EXTEND PB
I:0.0/11	OPERATION #2 CYLINDER RETURN SW
I:0.0/12	UNLOAD STATION MANUAL EXTEND PB
I:0.0/13	REJECT STATION CYLINDER EXTEND SW
I:0.0/14	TEST STATION #1 FAIL PB
I:0.0/15	REJECT STATION CYLINDER RETURN SW
I:0.1/0	TEST STATION #2 FAIL PB
I:0.1/1	UNLOAD STATION CYLINDER RETURNED SW
I:0.1/2	NOT USED
I:0.1/3	INDEX CAM SWITCH
I:0.1/4	LOAD STATION PART IN PLACE SW
I:0.1/5	AUTO/MANUAL SELECTOR SW
I:0.1/6	PRODUCT A SELECTOR SW
I:0.1/7	PRODUCT B SELECTOR SW

Outputs:

O:0.0/0	AUTO RUN MODE LAMP
O:0.0/1	MANUAL MODE LAMP
O:0.0/2	START-UP WARNING LAMP
O:0.0/3	ALARM LAMP
O:0.0/4	PRODUCT A LAMP
O:0.0/5	PRODUCT B LAMP
O:0.0/6	PRODUCT C LAMP
O:0.0/7	NOT USED
O:1.0/0	INDEX MOTOR
O:1.0/1	LOAD STATION EXTEND SOL
O:1.0/2	NOT USED
O:1.0/3	LOAD STATION RETURN SOL
O:1.0/4	UNLOAD STATION EXTEND SOL
O:1.0/5	OPERATION #1 EXTEND SOL
O:1.0/6	NOT USED
O:1.0/7	OPERATION #1 RETURN SOL
O:1.0/8	NOT USED
O:1.0/9	OPERATION #2 EXTEND SOL
O:1.0/10	NOT USED
O:1.0/11	OPERATION #2 RETURN SOL
O:1.0/12	NOT USED

O:1.0/13	REJECT STATION EXTEND SOL
O:1.0/14	NOT USED
O:1.0/15	REJECT STATION RETURNED SOL

In auto run mode each station on the assembly line, with the exception of the loader station, will only function if the shift register data contained in that station's address matches its start-up requirements. Here is a list of each station:

Operation #1 station will only cycle if the shift register word N10:11 has a part present bit set and the product A bit set.

Operation #2 station will only cycle if the N10:10 word has a part present bit set and the product B bit set.

Test station #1 will only function if the N10:7 word has a part present bit set and the product A bit set. If it functions, it will in turn set N10:7 bit 3 or 4 on, depending on the test results.

Test station #2 will only function if the N10:5 word has a part present bit set and the product B bit set. If it functions, it will in turn set N10:5 bit 5 or 6 on, depending on the test results.

The reject station will cycle only if word N10:3 has a part present bit set and either fail bits 4 or 6 are set.

The unload station will cycle only if word N10:2 has a part present bit set and no failure bits 4 or 6 set.

We now generate a flowchart for this project even though each station has only one cylinder and the functions are very simple. The main purpose of this example is to demonstrate programming concepts that incorporate the safety requirements required on a continuously running machine and the data movements of an assembly line into one program. As in the previous example we will write this program in three files. File 2 is the main program containing all the rungs that are common to both modes. File 3 is the manual mode program. File 4 is the run mode program.

In a larger program we may have written each workstation's ladder control rungs in a separate subroutine nested to the run mode file, file 4.

SAMPLE PROGRAM

Main Program Flowchart

Looking at the MAIN program flowchart (Figure 13–14), we can see that our first main decision is which position the mode switch is in. If the switch is set to manual, we clear all the latch bits that might have been left on in run mode. We then clear all alarm bits and change the stack light setup. We will

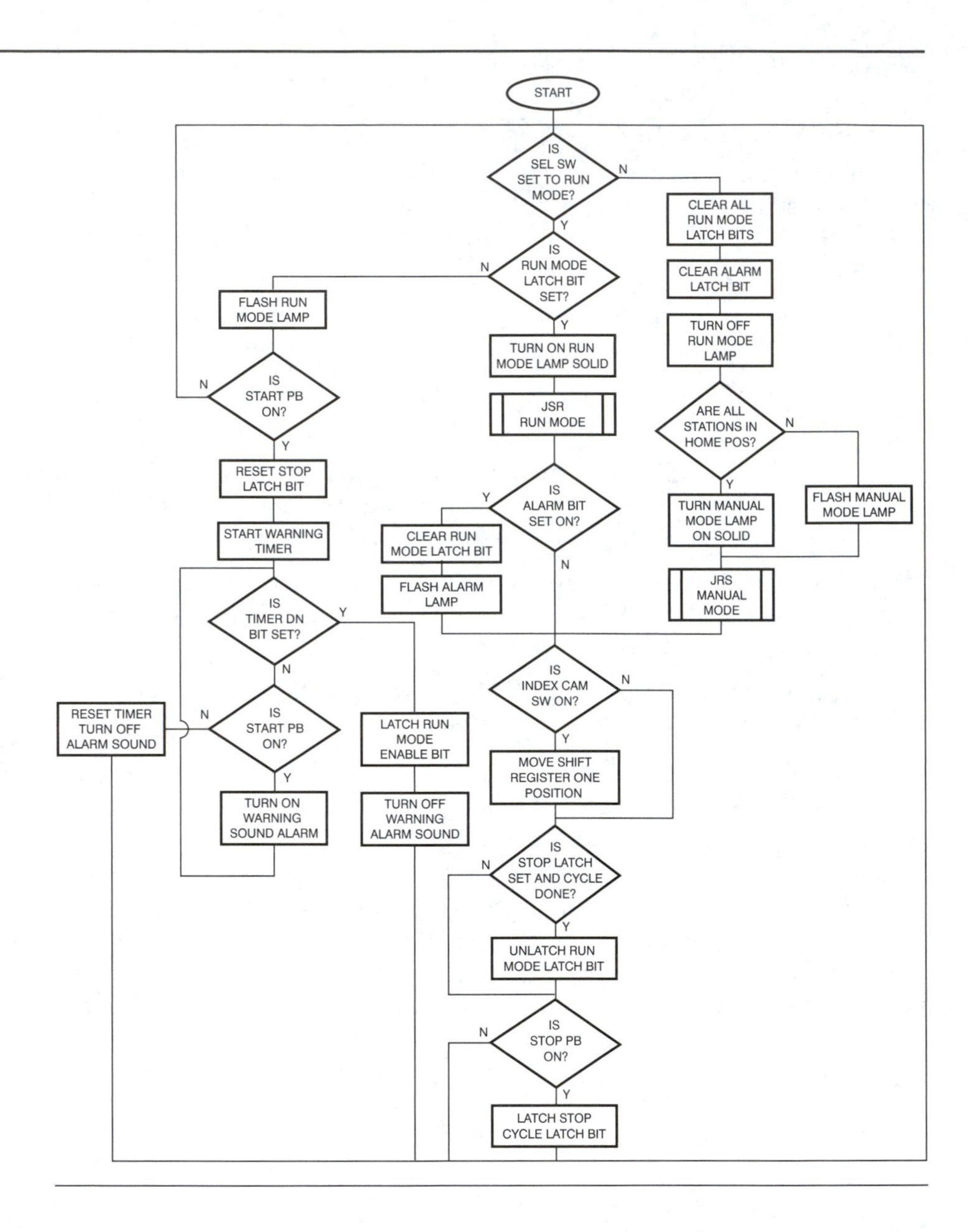

Figure 13–14
MAIN file flowchart for sample program

continue to scan the manual mode subroutine until the mode switch is changed. If the switch is set to run, then the next decision is concerned with the status of the run mode latch bit. If it is not set on, we loop, looking for the cycle start PB. Note that if we find the cycle start PB on, we must still wait for the warning timer to time out before we latch the run mode enable bit on. If the cycle start PB is released any time before the warning timer times out, the timer is reset and the run mode enable bit is not latched on.

Once we get the run mode enable bit latched on we begin to scan the run mode subroutine. Once set on, only an alarm bit, stop latch bit, or changing the mode selector switch will drop out the run mode.

Manual Mode Subroutine Flowchart

In the manual mode we have a parallel path for each station on the assembly line (Figure 13–15). Note that the shift register is moved in the main file. This was done so we could index the line from manual or run mode and always move the shift register to follow the part movement. The manual push buttons that move the stations are only enabled if the index cam is in dwell position. The manual index can only be run if all the stations are in home position.

Run Mode Subroutine Flowchart

We start run mode by indexing the line #1 position (Figures 13–16 and 13–17). It is a good idea to always index the line first because the program does not know if an index has taken place since the last assembly cycle was performed. This is very important if the assembly process places a part on the line or has a wire lug crimping operation. Performing a second assembly operation on the same parts may cause extensive damage to your tooling.

Once the index is complete, we start seven parallel operations depending on the shift register data present at each station. Once the assembly cycle has started, we wait for all stations to return home and the cycle timer to time out. If the stop bit is set, we reset the run mode enable bit. If it is not set, we reset the cycle timer and index start timer and begin a new cycle.

Main File

At this time I will describe each rung of the ladder program beginning with the main file, file 2. Figures 13–18 through 13–22 show the main file; Figures 13–23 and 13–24 show file 3; and Figures 13–25 through 13–29 show file 4.

Rung 0	Since the cycle stop function is latched, we need to unlatch it when we restart the line. This rung uses the cycle start PB input to unlatch the cycle stop bit.
Rung 1	Controls the run mode lamp. If the line is switched to run mode and all the cylinders are home, this lamp will flash. When the cycle start PB is pressed and held in until run mode latches, the light will become solid.
Rung 2	Controls the manual mode lamp. If all cylinders are home and the index cam switch is off, the light will be on solid. If any cylinder is out of home position or the cam switch is on, the light will flash.
Rung 3	The first flash timer function.

Figure 13–15
Manual mode subroutine flowchart for sample program

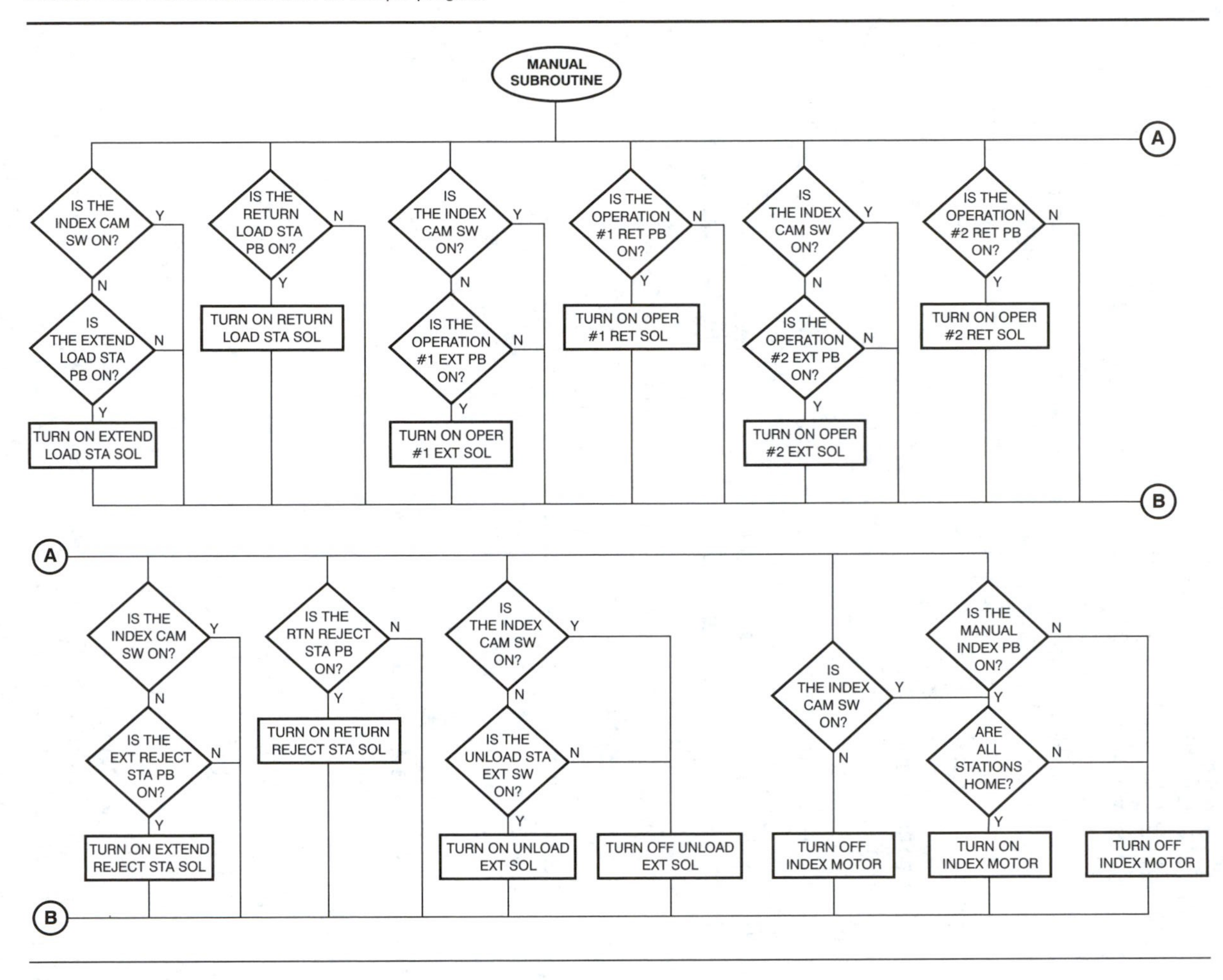

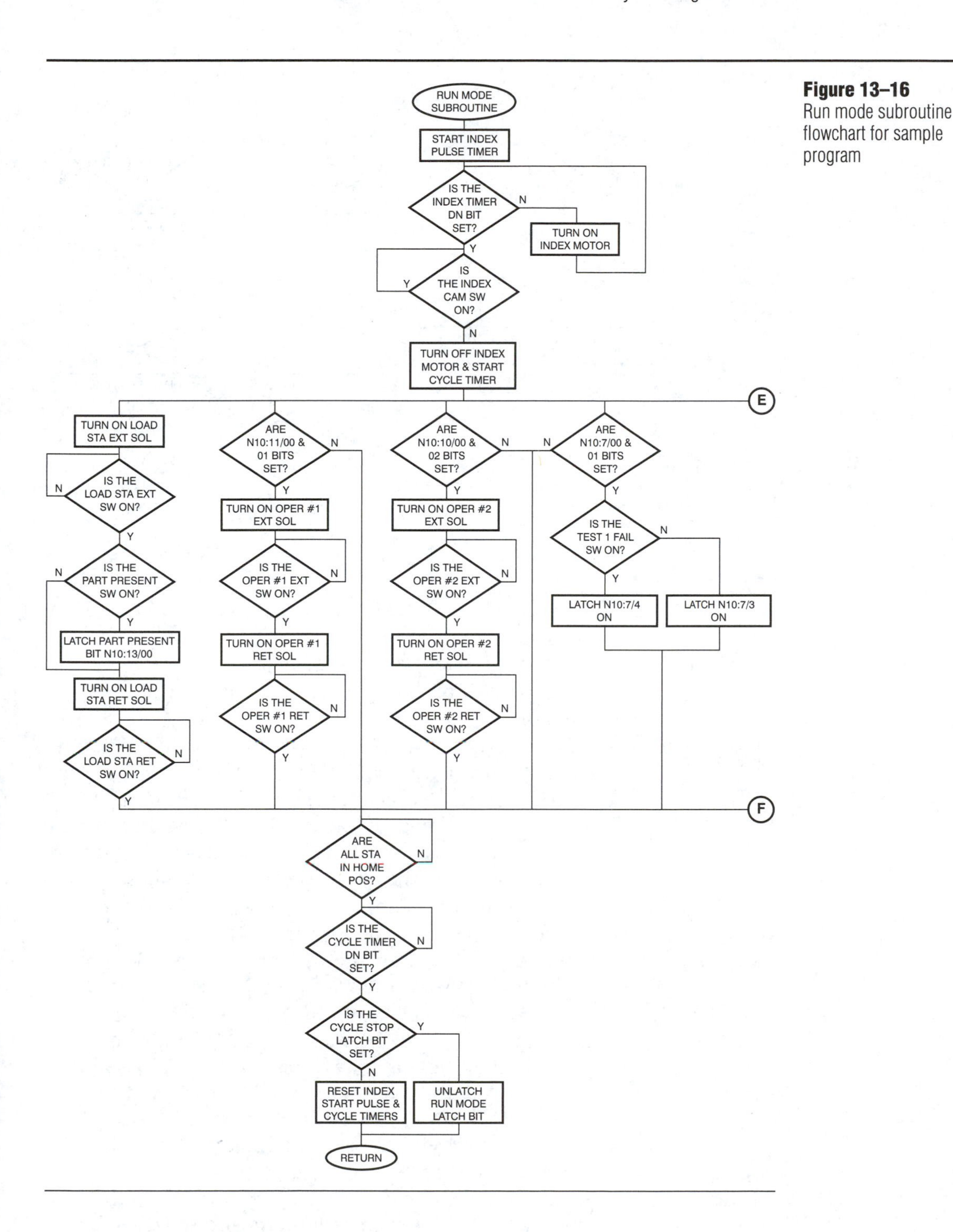

Figure 13–16
Run mode subroutine flowchart for sample program

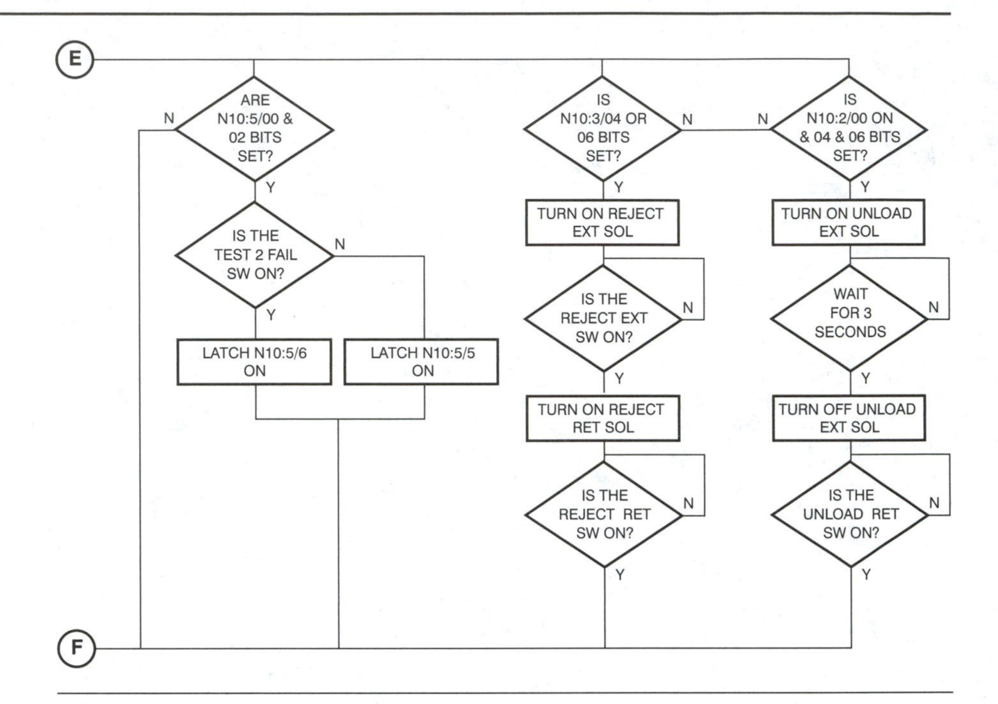

Figure 13–17 Run mode subroutine flowchart for sample program, *continued*

Rung 4 The second flash timer function.

Rung 5 Tests to see if all the cylinders are in their home positions.

Rung 6 Contains the three conditions that can unlatch the run mode. Branch 1 will unlatch the run mode if the selector switch is changed to manual mode. Branch 2 will unlatch the run mode if the "too long" cycle timer times out. Branch 3 will unlatch the run mode if the cycle stop latch bit is set and the cycle done pulse activates.

Rung 7 Our cycle start function. If the cycle start PB is pressed, the selector switch is set for run mode, all cylinders are home, and the index cam switch is off, the timer will begin timing. After 3 seconds the DN bit will be set.

Rung 8 The "too long" cycle timer function. Any time the line is in run mode, this timer runs while the index cam is in dwell. If the line indexes normally at 10-second intervals this timer is reset. If the line jams and does not index within 15 seconds the timer will time out and the DN bit will cause the run mode to be unlatched in rung 6.

Rung 9 Starts the index motor and gets the cam up on the lobe. When the line is in run mode and the done bit is

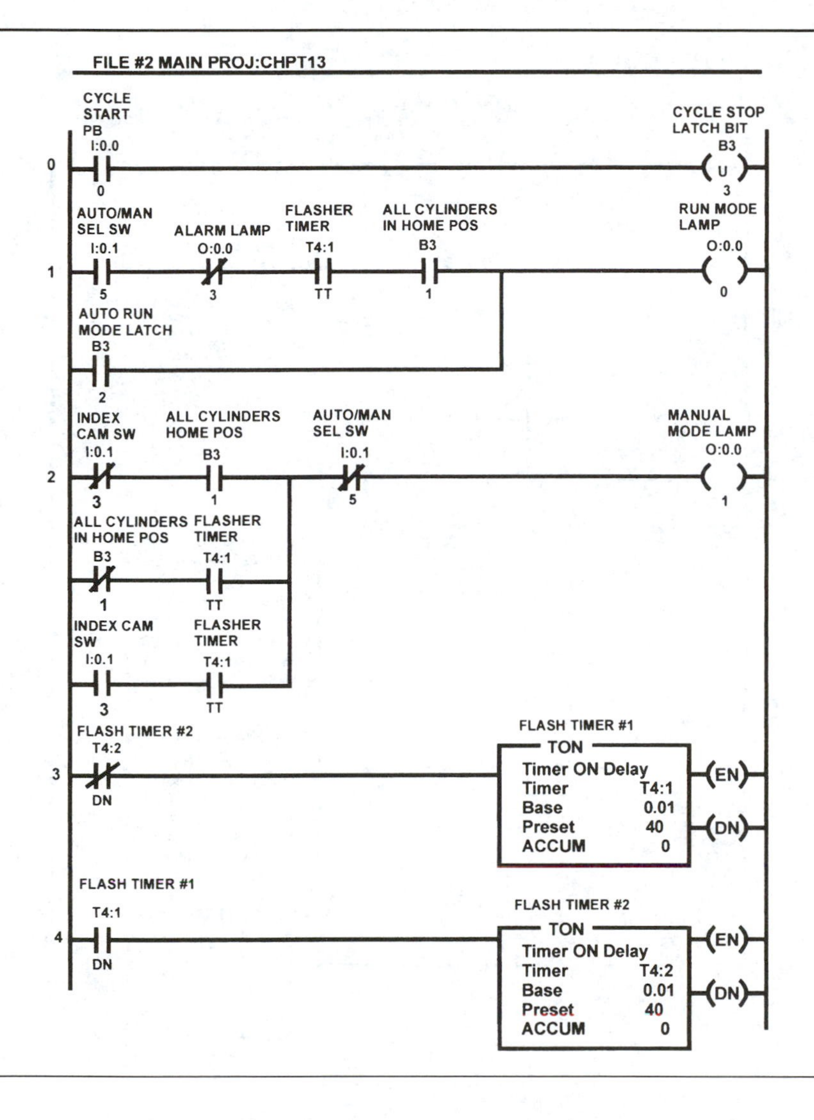

Figure 13–18
Main program ladder file 2

not set, this timer runs. The TT bit is used in rung 11 to start the index function.

Rung 10 Latches the run mode latch bit B3/2 when the cycle start timer is done.

Rung 11 Controls the line index function. Note that this output is turned off if any cylinder is moved out of home position. Branch 1 starts the index in manual mode. Branch 2 starts the index in run mode. Branch 3 keeps the motor turning until the cam falls back into the dwell slot.

Figure 13–19
Main program ladder file 2, *continued*

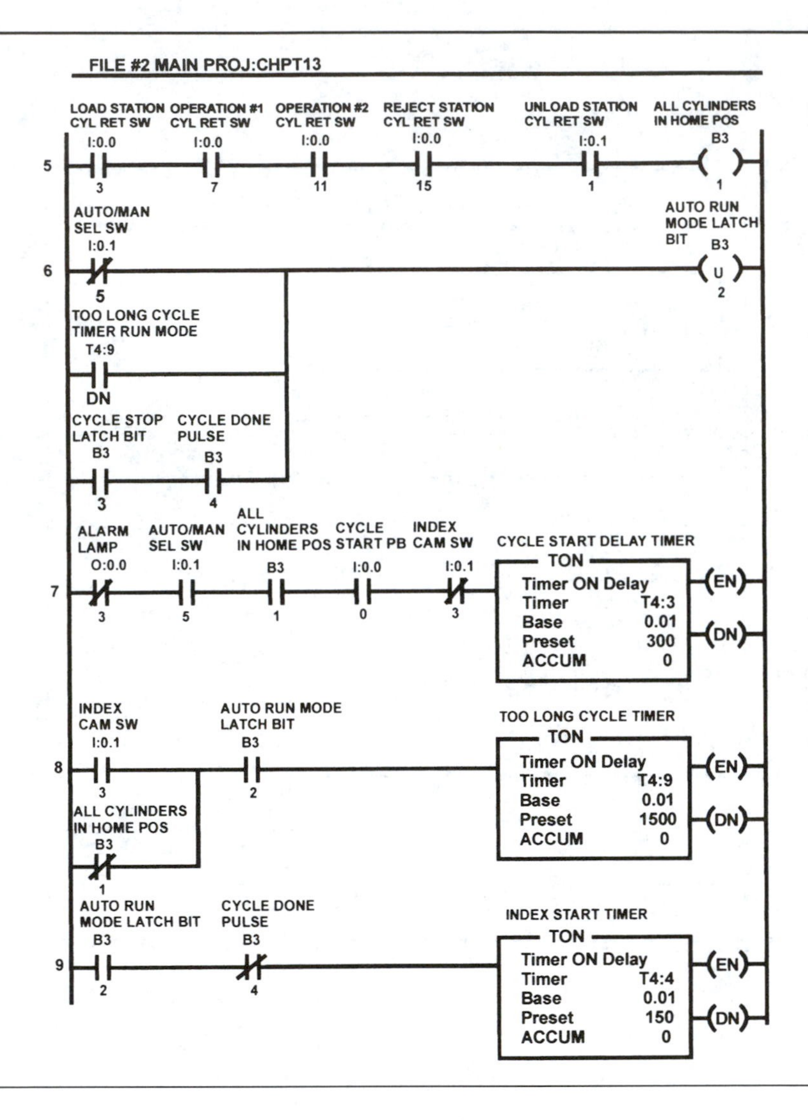

Rung 12 Our shift register control function. We use a copy instruction to move the data in our shift register. The copy instruction will copy data every scan unless we limit it with a one-shot instruction. Each time the index cam switch transitions from off to on, the copy instruction will move our shift register data one position.

Rung 13 Latches the cycle stop bit B3/3 when the cycle stop PB is pressed.

Rung 14 Jumps to the manual subroutine if the selector switch is set for manual mode.

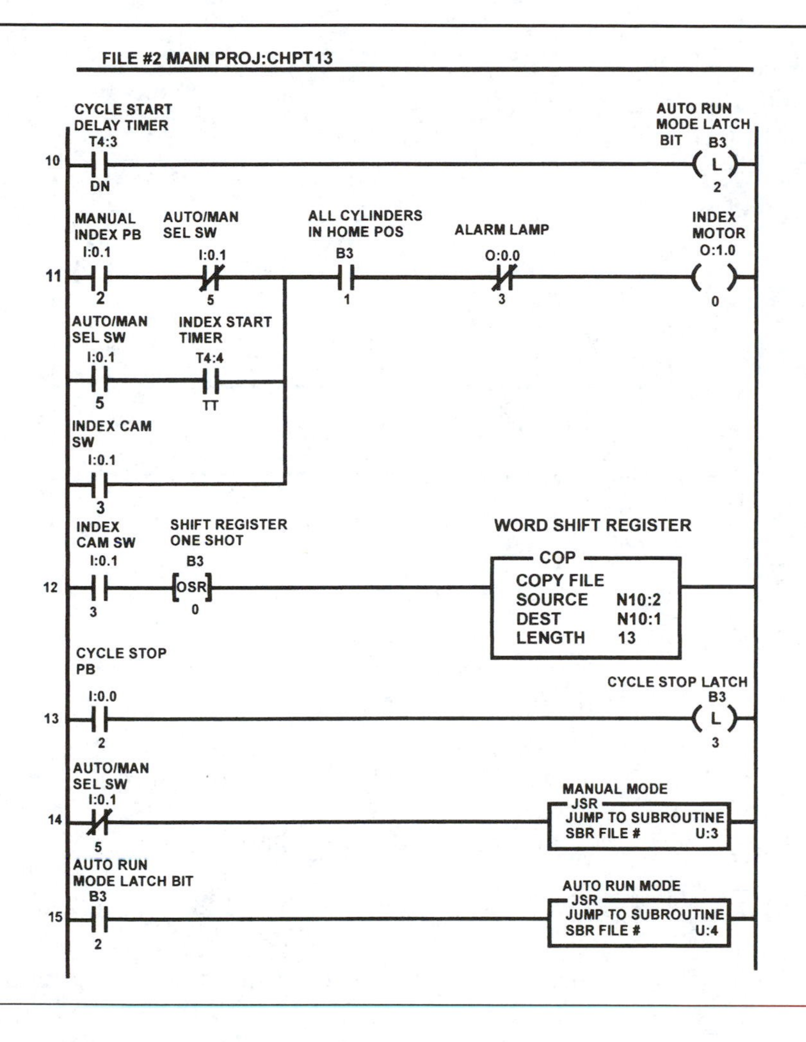

Figure 13–20 Main program ladder file 2, *continued*

Rung 15	Jumps to the run mode subroutine if the run mode latch bit is set.
Rung 16	Turns on the warning lamp during the cycle start safety delay time.
Rung 17	Latches the alarm bit if the "too long" timer times out.
Rung 18	Resets the alarm latch when the selector switch is set to manual.
Rung 19	End of scan.

Manual Mode

Rung 0	Extends the load station cylinder when button is pressed.
Rung 1	Returns the load station cylinder when the button is released.

Figure 13–21
Main program ladder file 2, *continued*

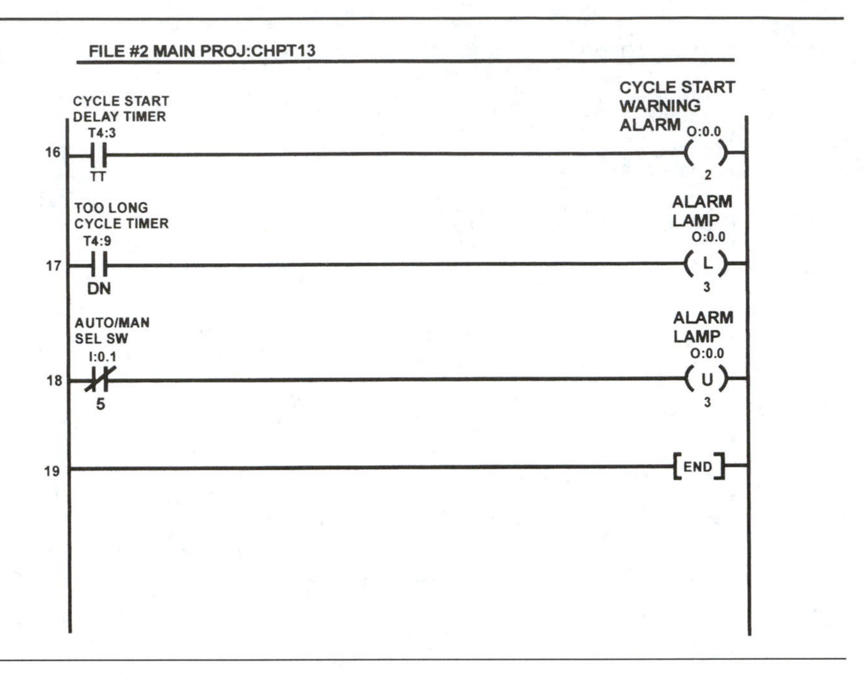

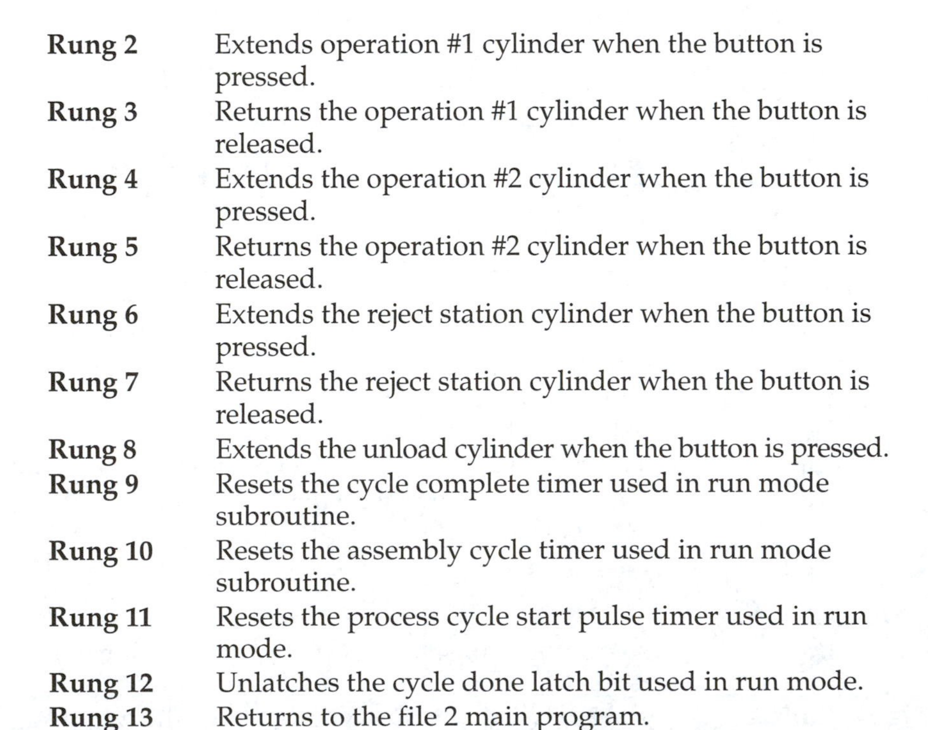

Rung 2	Extends operation #1 cylinder when the button is pressed.
Rung 3	Returns the operation #1 cylinder when the button is released.
Rung 4	Extends the operation #2 cylinder when the button is pressed.
Rung 5	Returns the operation #2 cylinder when the button is released.
Rung 6	Extends the reject station cylinder when the button is pressed.
Rung 7	Returns the reject station cylinder when the button is released.
Rung 8	Extends the unload cylinder when the button is pressed.
Rung 9	Resets the cycle complete timer used in run mode subroutine.
Rung 10	Resets the assembly cycle timer used in run mode subroutine.
Rung 11	Resets the process cycle start pulse timer used in run mode.
Rung 12	Unlatches the cycle done latch bit used in run mode.
Rung 13	Returns to the file 2 main program.

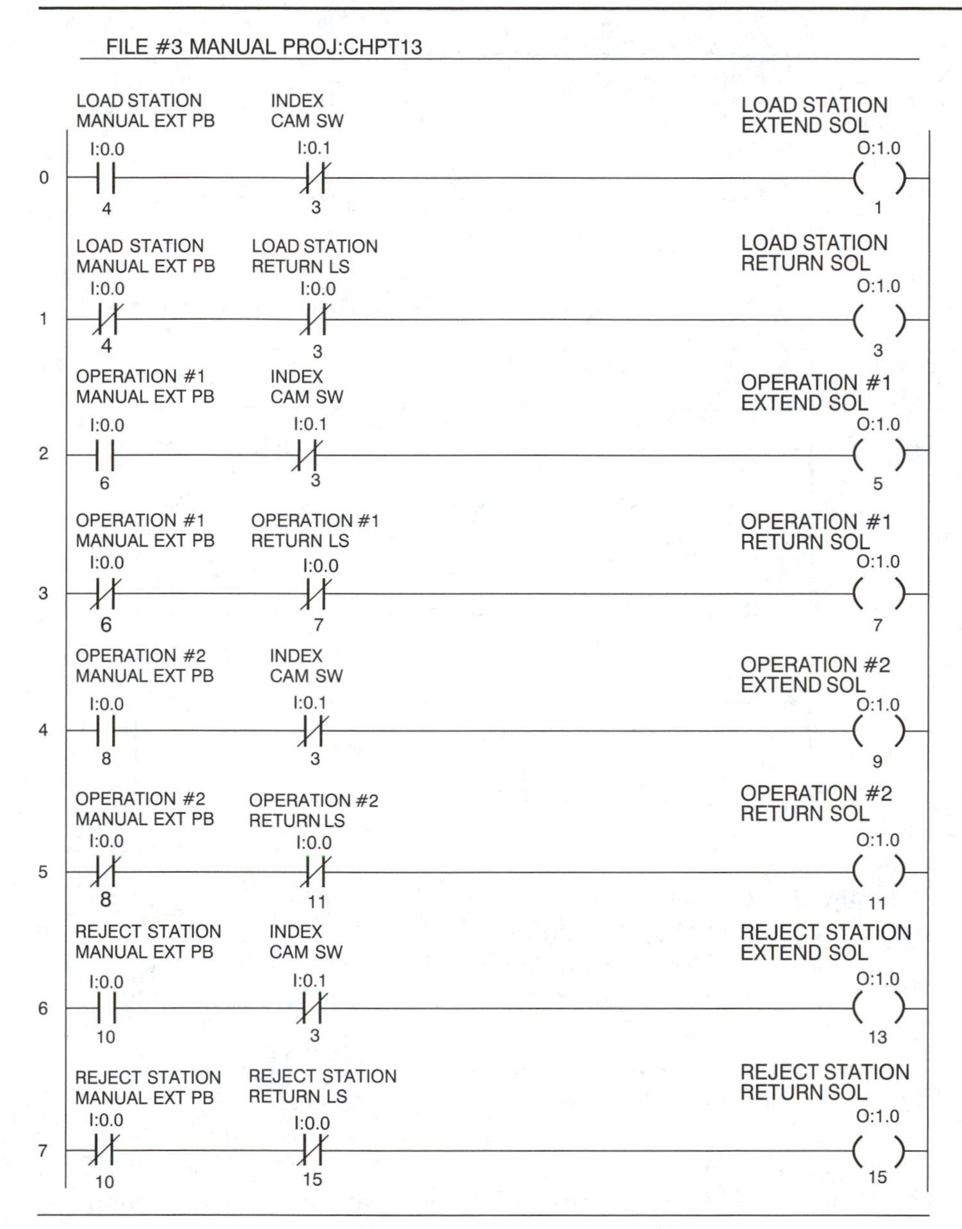

Figure 13–22
Manual mode ladder file 3

Run Mode

Rung 0 Unlatches the cycle done latch bit when the cycle done timer times out.

Rung 1 Starts the cycle done delay timer to set the length of the cycle done pulse.

Figure 13–23
Manual mode ladder file 3, *continued*

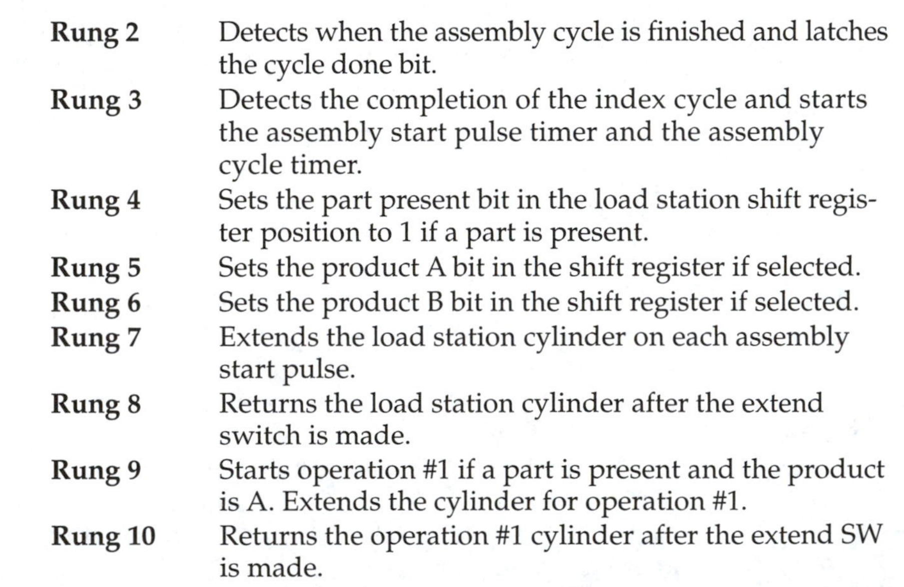

Rung 2 Detects when the assembly cycle is finished and latches the cycle done bit.

Rung 3 Detects the completion of the index cycle and starts the assembly start pulse timer and the assembly cycle timer.

Rung 4 Sets the part present bit in the load station shift register position to 1 if a part is present.

Rung 5 Sets the product A bit in the shift register if selected.

Rung 6 Sets the product B bit in the shift register if selected.

Rung 7 Extends the load station cylinder on each assembly start pulse.

Rung 8 Returns the load station cylinder after the extend switch is made.

Rung 9 Starts operation #1 if a part is present and the product is A. Extends the cylinder for operation #1.

Rung 10 Returns the operation #1 cylinder after the extend SW is made.

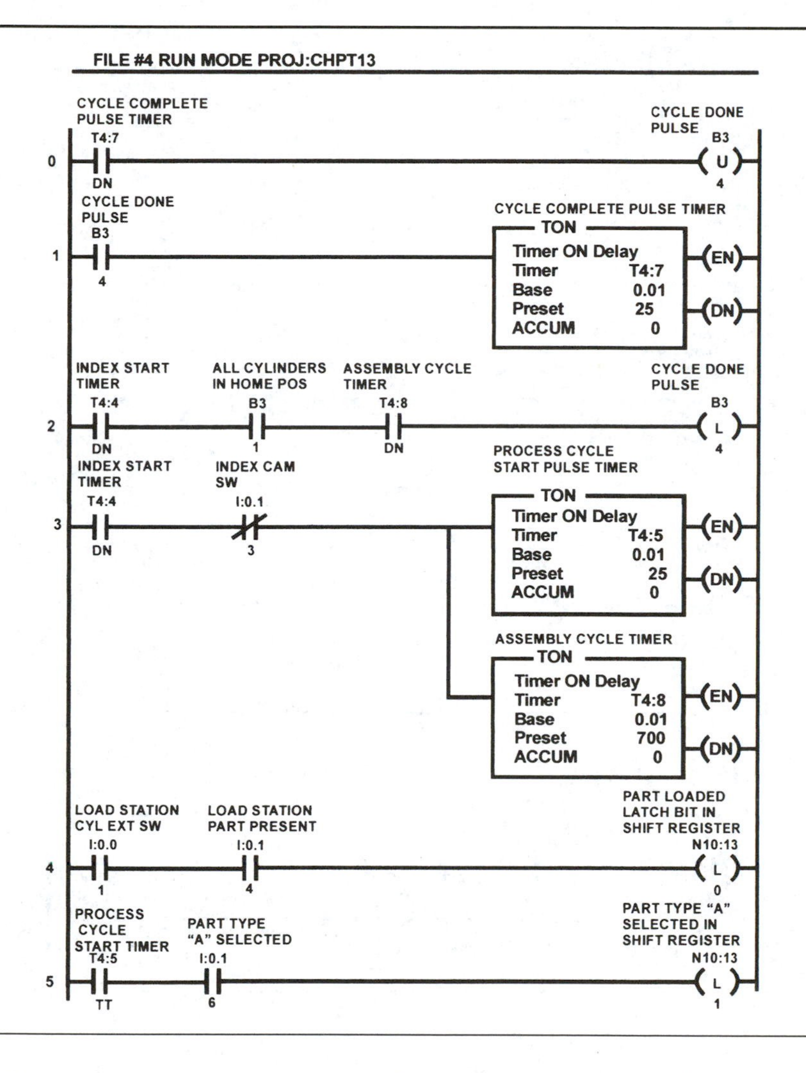

Figure 13–24
Run mode ladder file 4

Rung 11 Starts operation #2 if a part is present and the product is B. Extends the cylinder for operation #2.

Rung 12 Returns the operation #2 cylinder after the extend SW is made.

Rung 13 Performs test #1 if a part is present and it is product A. Sets the pass/fail bits in the shift register depending on the test results.

Rung 14 Performs test #2 if a part is present and it is product B. Sets the pass/fail bits in the shift register depending on the test results.

Rung 15 Starts the reject station if a part is present and either fail bit is set.

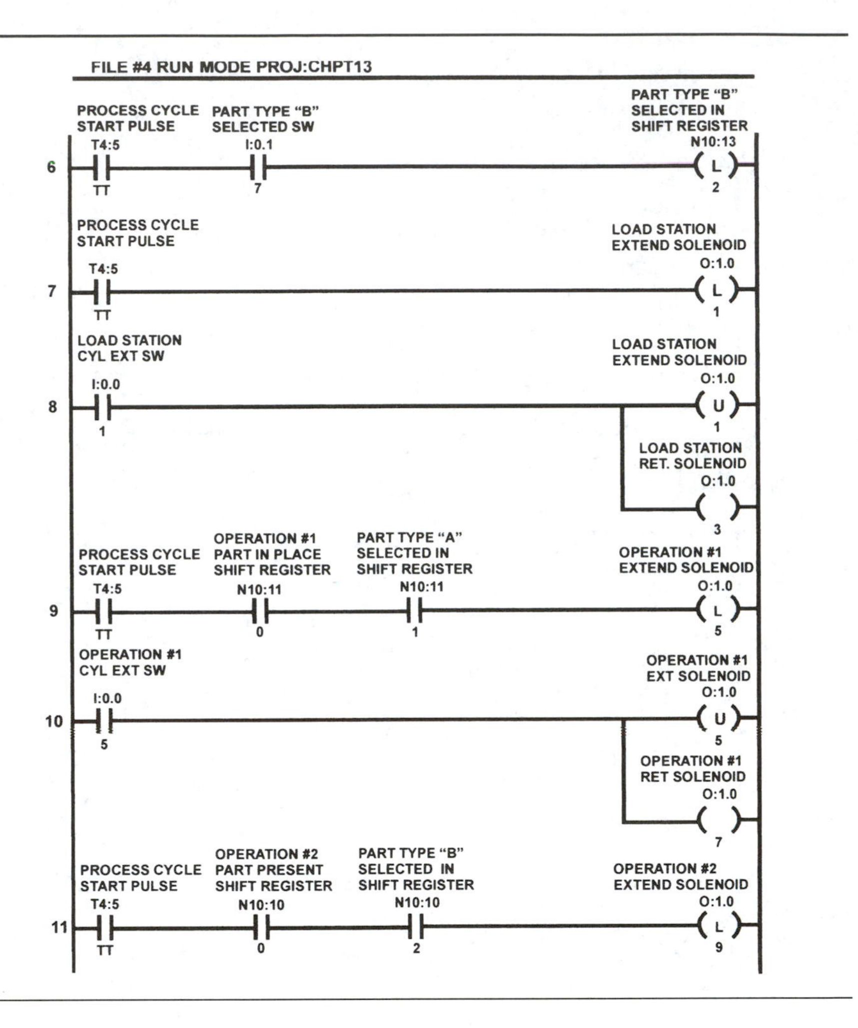

Figure 13–25
Run mode ladder file 4, *continued*

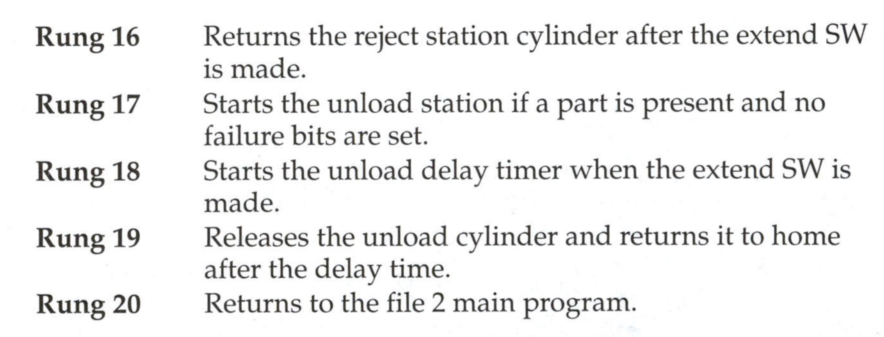

Rung 16	Returns the reject station cylinder after the extend SW is made.
Rung 17	Starts the unload station if a part is present and no failure bits are set.
Rung 18	Starts the unload delay timer when the extend SW is made.
Rung 19	Releases the unload cylinder and returns it to home after the delay time.
Rung 20	Returns to the file 2 main program.

At this point you should get a copy of the program from the instructor and run it on the teaching kit to verify for yourself that it works. Change

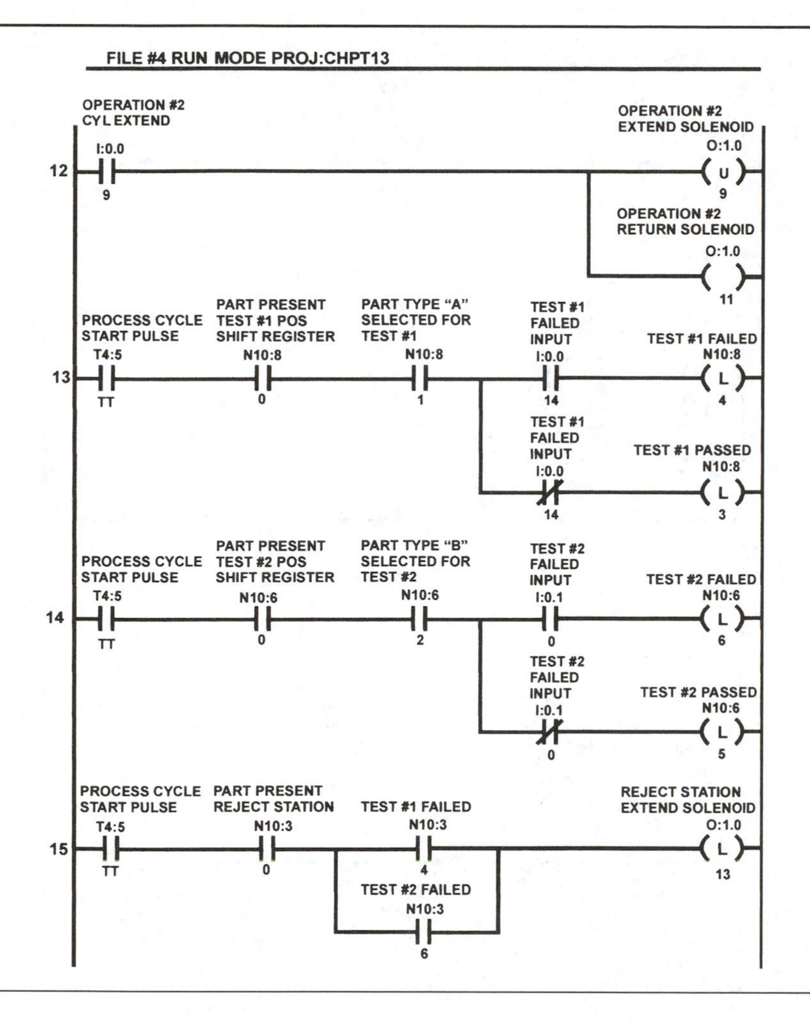

Figure 13–26
Run mode ladder file 4, *continued*

the product switch settings and verify that the actions of the four stations reflect the different product mix. Also pass and fail some units and verify that they get scrapped or unloaded based on their test results.

Note that while we have shown this example as a linear motion assembly line, it could just as easily have been a rotary table. All of the concepts would still have applied. Real assembly line programs will be considerably larger but the basic concepts are the same. In large programs each station on the line may be written in a separate subroutine called from the run subroutine. This separation will help if we have a number of different products to run down the line and not all of the workstations operate on every product. We can load a bit pattern into the shift register as the product starts down the line and use it to activate or deactivate each

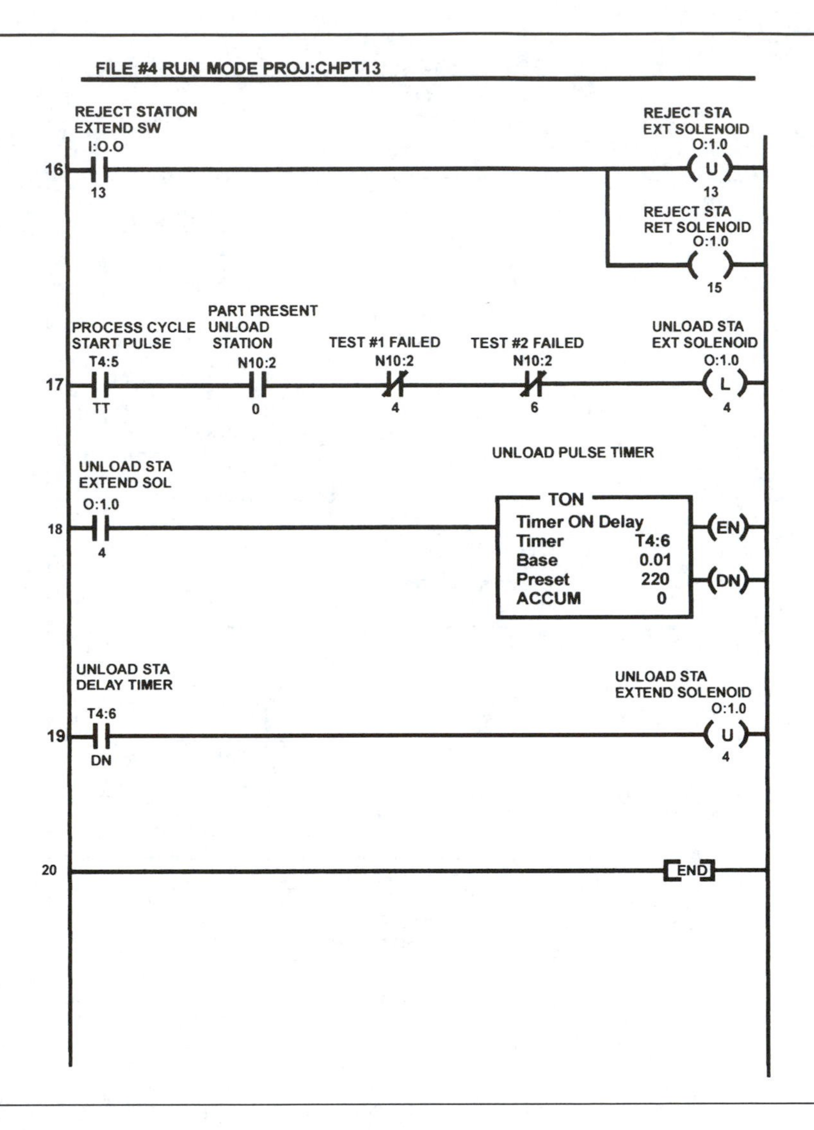

Figure 13–27 Run mode ladder file 4, continued

workstation's subroutine as the product moves into position. Because only the subroutine's calling rung needs to be controlled, it may save a lot of memory words.

In larger programs in which each workstation has numerous moving elements, our main file's rung 5, which indicates all elements are home, should be separated into one rung for each station. This will be important later when we talk about diagnostic systems. If the operator presses the cycle start PB and our program will not enter run mode, we need to indicate to the operator which station is out of home position. If all station home elements are combined into one rung, it will be difficult to tell which station is not

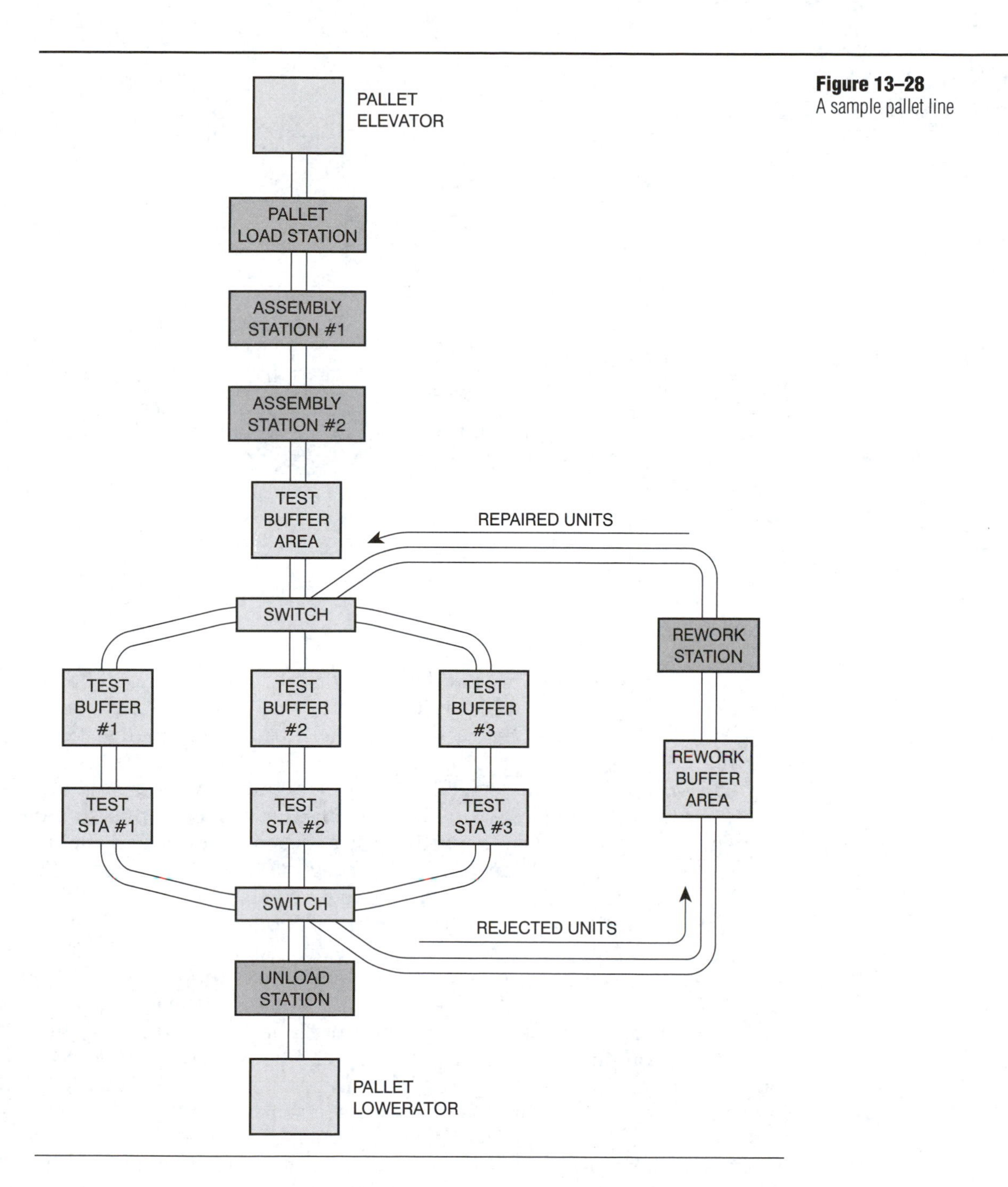

Figure 13–28
A sample pallet line

Figure 13–29
Sample pallet data buses

PALLET 1	PRODUCT TYPE DATA	N7:40
	TEST DATA	N7:41
	TEST DATA	N7:42
	TEST DATA	N7:43
PALLET 2	PRODUCT TYPE DATA	N7:44
	TEST DATA	N7:45
	TEST DATA	N7:46
	TEST DATA	N7:47
PALLET 3	PRODUCT TYPE DATA	N7:48
	TEST DATA	N7:49
	TEST DATA	N7:50
	TEST DATA	N7:51

home. By using a separate rung and B3 memory bit for each station's home elements, we can then use these bits to indicate which station is out of position. We can then sum all the B3 bits together in one rung for all stations home.

Earlier in this chapter we talked a little about how we needed to develop a timing sequence to run this machine. Let us look at this timing sequence in detail and explain which rungs are used to develop it. The sequence starts with file 2, rung 7. In this rung we look at the machine's start-up conditions. If they are correct and the cycle start PB is on, we begin the 3-second safety timer delay. This timer's DN bit is used in rung 10 to latch the run mode latch bit B3/2. Rung 9 begins the run mode index start pulse sequence when the run mode latch bit is set on and the cycle done bit B3/4 is off.

We will use the cycle done bit to kick the index timer on again after the cycle is complete. The TT output of timer T4:4 is used in the middle branch of rung 11 to start the index cycle. Once the index is complete, we move to file 4, rung 3. This rung looks for the conditions that tell us the index has completed and if so starts two timers. The process cycle start timer's TT bit will be used to start all the assembly operations. The assembly cycle timer will give us the 7-second delay required for our workers to complete their tasks. The next step is controlled by file 4, rung 2. When all the stations are back home and the assembly cycle timer T4:8

has timed out, we latch the cycle done B3/4 bit on. This bit is used in three places. In file 4, rung 2 starts the cycle complete delay timer. In file 2, rung 8, this bit is ANDed with the stop bit and if both are on the run mode is deactivated.

At the end of 0.25 second, the T4:7/DN bit is used in file 4, rung 0, to reset the cycle done bit. Note that this same bit is now used in file 2, rung 9, to restart the index start time for another index cycle. This completes the sequence because we will continue to cycle until the stop bit is detected and run mode is deactivated.

In future chapters we will be writing a diagnostic program to add to this sample program. For now we will not elaborate further than to say that the "too long" cycle timer T4:9 is used as a safety tool. Timer T4:9's DN bit is used in file 2, rung 6, to shut down run mode if the cycle does complete within the 15-second window. This function is also called a watch dog timer and is a common safety element in most control system designs.

ASYNCHRONOUS PALLET PULL ASSEMBLY LINES

Unlike indexed assembly lines, the pallet pull system is complicated by the fact that each station is independent of all others and that all the pallets can circulate in a different order each time through the line. This independent movement is one of the main advantages of a pallet pull line, but also the one that can create problems for a programmer. Because we can have multiple branches and loops in our pallet line, the order in which the pallets enter the line will often not be the order in which they leave the line. We must find a different way to store and retrieve pallet data than the method we used for the indexed line. Shift registers and FIFOs will only work if our pallets all move together and we have already said they will not.

We do have available, however, two data collection methods that do work on this type of line. One is the *RF tag system* and the other is a *bar code tag and reader system*. Because the order of our pallets is always changing, we will code each pallet with a different number. This can be accomplished with a bar code tag or RF tag number that is physically attached to each pallet. In Figure 13–28 we have diagrammed a sample pallet line.

In an indexed assembly line, the line speed is limited by the slowest station. In a pallet system we can get around this by branching the line and using multiple stations where the cycle time is long. In the example shown in Figure 13–28 we have determined that the test function requires considerably more time than the other functions; therefore, we have designed three test functions into the system.

Our simple line starts with a pallet elevator that raises an empty returned pallet from below. The first station loads a base plate on the pallet. We then have two assembly stations that perform some operation on the base plate. Pallets are released from each station when an empty position is created in front of it. When they come to the test branch, the pallets are switched to whichever path has an empty position in its buffer. If one test line is experiencing problems, the pallets will be passed through the other two branches. This is the kind of process flexibility that is available with pallet lines.

All rejects are switched through the repair station and then back to be retested again. If the product passes the test, it is unloaded at the unload station and the empty pallet is returned by the lower path to the front of the line. As each pallet enters a station, a reading device reads the pallet number. To keep track of pallet data we must set up a section of memory into multiple word data locations for each pallet number. When each pallet is raised at the beginning of the line, its memory locations are cleared and new product data are entered into the first word. As the pallet moves through the assembly process, data are collected about the product and stored in its memory locations (see Figure 13–29). Because each pallet is coded separately, the order is not important.

If a product fails the test and is routed to the repair station, the repair technician can read the test data from the pallet's memory locations and use it to decide what needs to be repaired. When the pallet returns to the test station, the old data are rewritten with the new test data. Our trainer kits are too limited in I/Os to simulate this type of assembly line, and the kit has no bar code or RF readers. Therefore we will not write a simulated program for this type of line because it cannot be run on the kit.

CAM AND ENCODER DRIVEN MACHINES

Another type of continuously running machine is driven by a cam and lever system. Many packaging machines and connector assembly machines are of this type. They are usually driven by a variable-speed motor drive that turns the cams that activate main assembly mechanisms. Most of the mechanisms are coupled to the main driveshaft by a series of cams and levers riding on the cams. One turn of the main cam section is usually one assembly cycle.

We often have additional equipment and mechanisms that must perform their function in time with the main cam drive assembly. By adding a rotary encoder device to the motor drive we can develop a method of timing the start of these additional mechanisms. If, for example, we used a 720-pulse-per-turn encode to the motor drive, we could time our cycle in ½-degree increments. Our ladder rungs would now all begin with an equal statement that would compare the encoder count value with a degree of rotation number to begin each mechanism.

One major advantage of designing machines in this manner is they can be started slowly and ramped up to full speed gradually. The encoder rotation number that we code to start certain mechanisms can be changed as the motor speed increases. This ability to adjust our timing based on motor speed gives the mechanical designer much greater flexibility in his design limits.

CONCLUSIONS

Each of these assembly line types presents certain advantages and disadvantages to the engineer. Manufacturing engineers must look at the products to be assembled and decide which type of assembly line best suits the product mix being assembled. Whichever line type is chosen, the designers's main job is to build a safe and reliable line that is easy to maintain and work on. We have covered a number of the safety options and operator interface options for assembly lines but these are just the basic principles. Each line must be carefully designed for safe operation.

REVIEW QUESTIONS

1. In the three-state control system, which state is considered the safe state?
2. When starting up a continuously running machine, what must be done to provide a safe start-up sequence?
3. Why must the cycle stop input be latched for proper machine operation?
4. When setting up a product data shift register, why must each line position be assigned to a memory address even if no action is performed at that position?
5. Why must the memory addresses assigned to the line positions begin with the highest address attached to the first position on the line?
6. Why is a one-shot instruction used in the control rung of the shift register copy instruction?
7. Why should each station mechanism have its own rung and memory bit to indicate if it is in the home position?
8. What are some of the main advantages of a pallet line over a synchronous indexed line?
9. Why does the normal shift register function not work on a pallet line?
10. How do we make sure the line does not become a trap if some part of a mechanism fails to complete its cycle?
11. What are the main timing functions that must be generated to create a continuously running program?
12. What functions must you always reset in manual mode to ensure that the auto mode will restart correctly after a jam-up or E stop condition?
13. What are some advantages of cam-driven assembly equipment?
14. How do we stop a continuously running machine when the cycle stop push button is pressed?
15. How is the "too long" cycle timer or watch dog timer used to stop the machine if a jam-up occurs?

LAB ASSIGNMENT

Objective

In this lab you will write a continuously running machine program to accomplish the following specs. You must use two separate subroutines to control manual mode and auto mode. The program must conform to the three-state program configuration. Use the system timing and program rungs shown in Figures 13–4, 13–5, and 13–6 as a guide to writing your own program. Start this lab by generating a flowchart or a sequence of state diagrams for each major portion of the program. Show your diagrams to the instructor before you write your ladder code.

Program Specifications

Manual Mode

All the cylinders must be moveable with the push buttons provided. An additional limitation is that cylinder #2 cannot be extended unless cylinder #1 is extended first. Cylinder #2 must be returned before cylinder #1 can be returned. The index operation must not start if any cylinder is not retracted and must stop immediately if any cylinder moves out of its retracted position. All latched bits used in the auto mode subroutine must be reset when the manual mode is activated.

Auto Mode

To start auto run mode, all cylinders must be home and the cycle start push button must be held in for three seconds. Once running, the cycle stop push button can be pressed and released at any time during the cycle. Only when the cycle finishes will the program drop out of auto run mode and return to the auto stop condition. The auto run mode sequence is as follows:

1. Index motor cycles one turn.
2. Cylinder #1 extends.
3. Once fully extended, delay for 2 seconds, then extend cylinder #2.
4. Once cylinder #2 extends fully, extend cylinders #3 and #4 at the same time.
5. When all four cylinders are fully extended, extend cylinder #5 and hold for 4 seconds, then return.
6. When cylinder #5 is returned, return cylinder #4.
7. When cylinder #4 is returned, return cylinder #3.
8. Once cylinder #3 is returned, return cylinder #2.
9. Once cylinder #2 is returned, return cylinder #1.
10. When all cylinders are returned, the cycle will repeat if the cycle stop bit is not latched.

You should have a "too long" cycle timer to shut down the auto run mode if the cycle takes longer than normal. This should also light the alarm lamp. The operator should reset the alarm lamp when switching to manual mode.

I/O List

Inputs:

I:0.0/00	CYCLE START PB
I:0.0/01	CYLINDER #1 DOWN L.S.
I:0.0/02	CYCLE STOP PB
I:0.0/03	CYLINDER #1 UP L.S.
I:0.0/04	CYLINDER #1 DOWN MANUAL PB
I:0.0/05	CYLINDER #2 DOWN L.S.
I:0.0/06	CYLINDER #1 UP MANUAL PB
I:0.0/07	CYLINDER #2 UP L.S.
I:0.0/08	CYLINDER #2 DOWN MANUAL PB
I:0.0/09	CYLINDER #3 DOWN L.S.
I:0.0/10	CYLINDER #2 UP MANUAL PB
I:0.0/11	CYLINDER #3 UP L.S.
I:0.0/12	CYLINDER #3 DOWN MANUAL PB
I:0.0/13	CYLINDER #4 DOWN L.S.
I:0.0/14	CYLINDER #3 UP MANUAL PB
I:0.0/15	CYLINDER #4 UP L.S.
I:0.1/00	MANUAL INDEX PB
I:0.1/01	CYLINDER #5 UP L.S.
I:0.1/02	CYLINDER #5 DOWN MANUAL PB
I:0.1/03	INDEX CAM L.S.
I:0.1/04	AUTO / MANUAL SELECTOR SW
I:0.1/05	CYLINDER #4 DOWN / UP MANUAL SELECTOR SW
I:0.1/06	NOT USED
I:0.1/07	NOT USED

Outputs:

0:0.0/00	AUTO MODE LAMP
0:0.0/01	MANUAL MODE LAMP
0:0.0/02	NOT USED
0:0.0/03	ALARM LAMP
0:0.0/04	INDEX IN DWELL POSITION LAMP
0:0.0/05	NOT USED
0:0.0/06	NOT USED
0:0.0/07	NOT USED
0:0.0/08	NOT USED
0:0.0/09	NOT USED
0:0.0/10	NOT USED
0:0.0/11	NOT USED
0:0.0/12	NOT USED
0:0.0/13	NOT USED
0:0.0/14	NOT USED
0:0.0/15	NOT USED

0:1.0/00	INDEX MOTOR
0:1.0/01	CYLINDER #1 DOWN SOL
0:1.0/02	NOT USED
0:1.0/03	CYLINDER #1 UP SOL
0:1.0/04	CYLINDER #5 DOWN SOL
0:1.0/05	CYLINDER #2 DOWN SOL
0:1.0/06	NOT USED
0:1.0/07	CYLINDER #2 UP SOL
0:1.0/08	NOT USED
0:1.0/09	CYLINDER #3 DOWN SOL
0:1.0/10	NOT USED
0:1.0/11	CYLINDER #3 UP SOL
0:1.0/12	NOT USED
0:1.0/13	CYLINDER #4 DOWN SOL
0:1.0/14	NOT USED
0:1.0/15	CYLINDER #4 UP SOL

Program Checkout

Once the program runs properly as defined in the program specs, verify that the following conditions operate correctly.

1. Switch to auto run mode and press the cycle start push button. Release the button before the three-second safety timeout period is up. Verify the machine does not start. Verify that you have to hold the cycle start push button again for the full three-second safety period before the machine will start.
2. Switch the mode from auto to manual in the middle of the cycle. Reset all the cylinders to the home position, then restart the auto run mode. The machine should start up correctly without any erroneous movements.
3. Run the machine in auto run mode and press the cycle stop button. The machine should stop when the cycle is complete and not before. Restart the auto run mode and verify that you have no start-up problems.
4. Run the machine in auto run mode. After several cycles hold one of the cylinder rods so it cannot go home. Verify that the watch dog timer shuts down the run mode after the proper amount of time and turns on the alarm lamp. Switch to manual mode and reset all the cylinders to their home positions. Verify that the alarm lamp is reset. Switch to run mode and restart the auto run mode. Verify that the machine restarts correctly without any erroneous movements.

Once you have the program running within specification, print out a copy of your ladder code and then demonstrate the program to the instructor.

CHAPTER

14

Shift Register, FIFO, and LIFO Instructions

LEARNING OBJECTIVES

After completing this chapter the reader should understand:

1. The programming and use of bit shift registers.
2. Setup and programming of FIFO instructions.
3. Setup and programming of LIFO instructions.

In Chapter 13 we demonstrated a word wide shift register. In this chapter, we look at single-bit shift registers. We also look at FIFO and LIFO instructions. For those who are not familiar with these computer terms, FIFO stands for first in/first out and LIFO stands for last in/first out (also sometimes called a *stacking operation*). We start with a discussion of shift register instructions because they are easier to use and understand.

BIT SHIFT REGISTER

In the SLC-500 control system, we have two bit shift instructions: BSL, which stands for *bit shift left*, and BSR, which stands for *bit shift right*. Both are output instructions that have DN and EN bits for control.

BSL Instruction

Let's look first at the BSL instruction, as shown in Figure 14–1.

File

The file address is the address of the bit array where data are shifted. You can use bit file B3 but you must be very careful that new additions to the B3 file do not overwrite the shift register locations. It is usually better to use a separate bit register for shift register instructions.

As an example, if you had used B3:2 as a shift register file with a length of 16 bits, you would have used B3/32 through B3/47. If you make program changes and add a few new bit locations to B3 you might not remember that B3/32 has been used. If you assign a new output instruction to B3/32, the program will not know it is already used in a shift register instruction. Any time you activate the new rung you will be changing shift register data without realizing it.

Control

The control address is the location in our control register where this instruction will store its status bits and other instruction data. Each shift instruction uses three memory locations to hold data as shown in Figure 14–2.

EN (bit 15) The enable bit, which is set on when the rung condition is true.

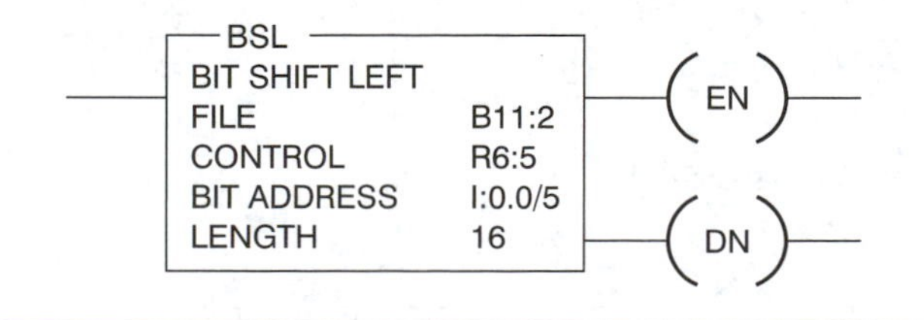

Figure 14–1
BSL instruction

Figure 14–2
Memory allocations for bit shift register

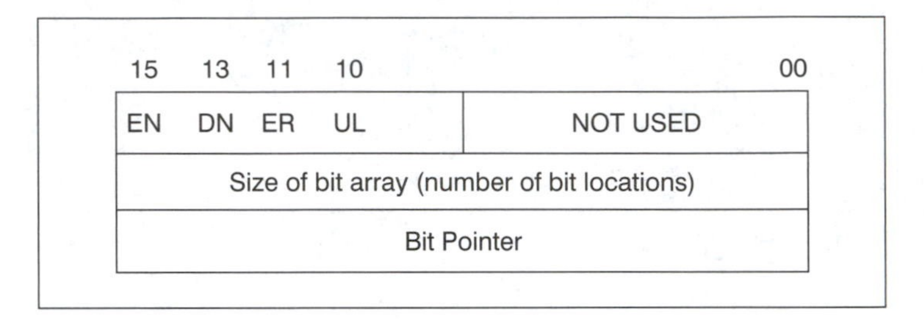

DN (bit 13) The done bit, which is set on when the shift operation has been completed on all bits in the file.

ER (bit 11) The error bit, which is set on if an error is detected in the instruction execution.

UL (bit 10) The unload bit. It is where the instruction places the data bit that exited the last shift register position when the file was shifted.

Note that bits 15, 13, and 11 are all cleared when the instruction rung goes FALSE. Bits beyond the last shift register bit and the next word boundary are held invalid by the processor.

Bit Address

The bit address is the address of the bit that will be loaded into the shift register when the instruction is activated.

Length

The length value is the number of bits contained in the shift register. Any single array has a limit of 2047 locations. A length value of 0 will cause the address bit to be loaded directly into the unload bit location. If more than 2047 locations are needed, two shift registers can be concatenated together for more locations.

In the example instruction given earlier, the data would flow as shown in Figure 14–3.

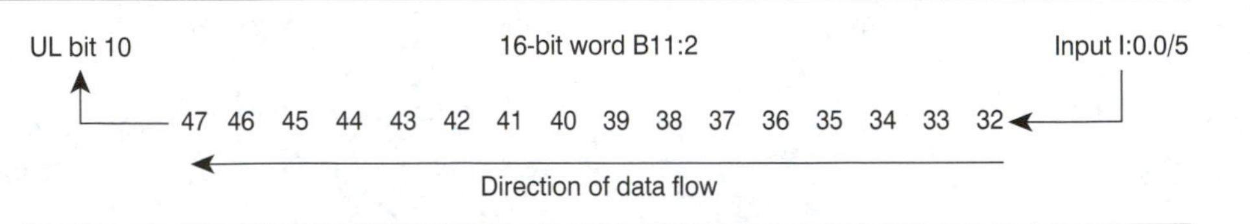

Figure 14–3
Example of data flow for BSL instruction

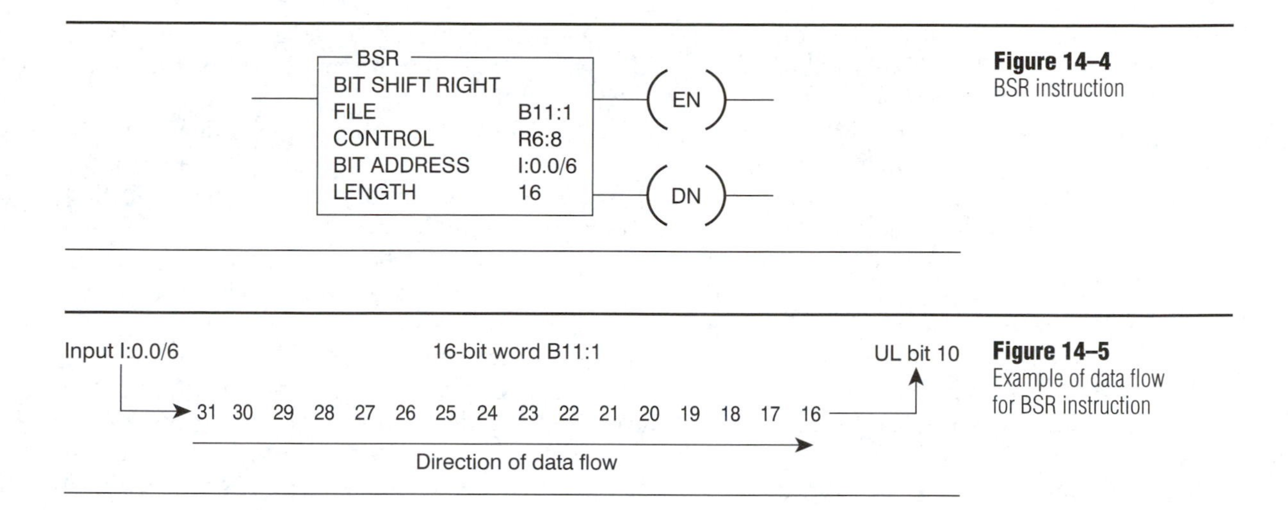

Figure 14–4 BSR instruction

Figure 14–5 Example of data flow for BSR instruction

BSR Instruction

The bit shift right instruction is the same as the BSL instruction except the data are shifted in the opposite direction. The instruction shown in Figure 14–4 is a typical bit shift right instruction using B11:1 as a word address and a length of 16 bits. We have also changed the input word to I:0.0/6. Figure 14–5 shows the direction of data flow. Note that except for the direction of data flow, the BSR instruction is exactly the same as the BSL instruction.

Shift registers can be used in control programs any time data must flow with the product and the number of data bits is small. For more than four data bits, a word shift register is a better choice. One advantage of using shift registers is that they are self-clearing. For example, if a line person pulls a product out of a carrier on an indexed line, the part in place bit will be wrong for that position. This may cause some problems for that position as it travels down the line, and then when it reaches the end of the assembly line it will fall out of the shift register and be cleared. As we will see in our explanation of FIFO, that is not the case. Without this automatic clearing feature, we may experience major problems.

FIFO INSTRUCTIONS

Two instructions are used in FIFO operations and they are always used in pairs. One is FFL, which stands for FIFO load; the other is FFU, which stands for FIFO unload. We should begin, however, by defining what FIFO means.

In a FIFO register array, we have a set number of locations in which to store data—just as we had in shift registers—and just as in our last chapter this FIFO array is a word wide array. The main difference between a shift register and a FIFO is that the data word loaded into a FIFO is moved to the end of the register and is available to be unloaded right away.

Because we have a load and unload instruction, we can load several words into the FIFO before we unload any. We do not need to synchronize the data as must be done in a shift register. This allows data to be handled in a non-synchronized manner such as will occur in pallet and free type operations. As each pallet enters a buffer area, the word containing its data is loaded into the FIFO for that area. As the pallets are released from the gate at the end of the work area, data words are removed from the FIFO. This register function works as long as no one removes any pallets from the area.

Because words in the FIFO are removed only when pallets are released, the order will be off if one pallet was removed and one word will always be left in the register. As we stated before, FIFOs are not self-clearing. Once the order is corrupted, only clearing the whole FIFO will fix the problem.

FFL Instruction

The FIFO load instruction is shown in Figure 14–6.

Source

The source address is the address location whose contents will be loaded into the FIFO's first empty location. Source can be a word address or a constant value.

FIFO

The FIFO address is the first address of the stack. It must be an indexed word address in the input, output, status, bit, or integer files. The programmer must be careful to use the same address in both the load and unload instructions.

Control

The control address is the address where the instruction will be located in the control file. Again the programmer must be sure to use the same

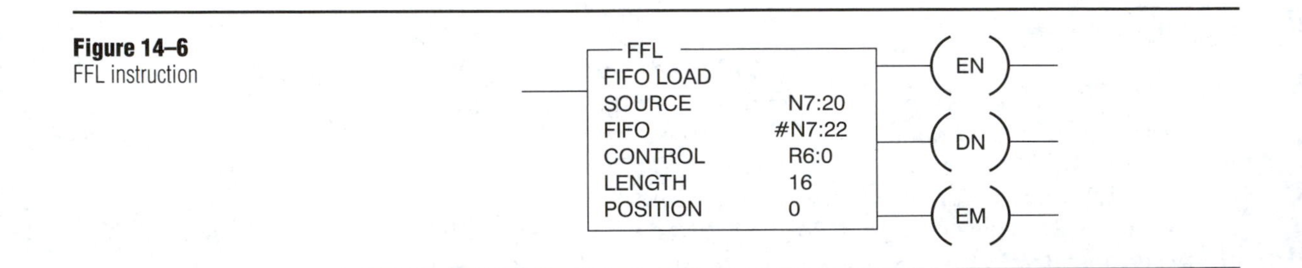

Figure 14–6
FFL instruction

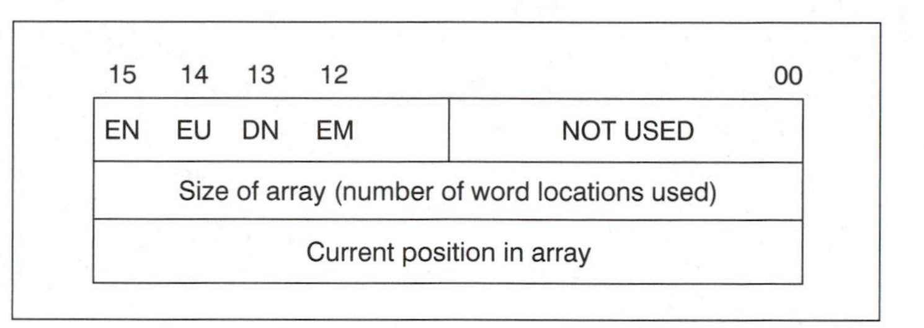

Figure 14–7 Memory allocations for FFL instruction

address in both the load and unload instructions. Each instruction uses three words in the control register as shown in Figure 14–7.

EN (bit 15)	The FFL instruction enable bit. The bit is set when the control rung goes TRUE and is reset when the rung goes FALSE.
EU (bit 14)	The FFU instruction enable bit. This bit is set when the control rung goes TRUE and is reset when the rung goes FALSE.
DN (bit 13)	This bit is called the done bit but really indicates when the stack register is full. It is set by the FFL instruction and when set, inhibits additional loading of the stack register.
EM (bit 12)	The empty bit. It is set by the FFU instruction to indicate the stack is empty.

Length

The length variable is the number of words in the array. A maximum of 128 words can be allocated.

Position

The position variable is the position of the next available empty location in the array. The same number will appear in both the FFL and FFU instruction boxes.

FFU Instruction

The FIFO unload (FFU) instruction (Figure 14–8) is the same as the FFL instruction except for the Dest address. This is the address that will contain the unloaded word from the FIFO. The FFU instruction unloads the last word in the FIFO array, places it in the Dest address, and moves all the other FIFO words forward one position.

As we explained before, the main use of FIFO registers is to collect and hold data that will not be inserted or removed in a synchronous manner. As an example, we look at a section of a pallet and free line with two

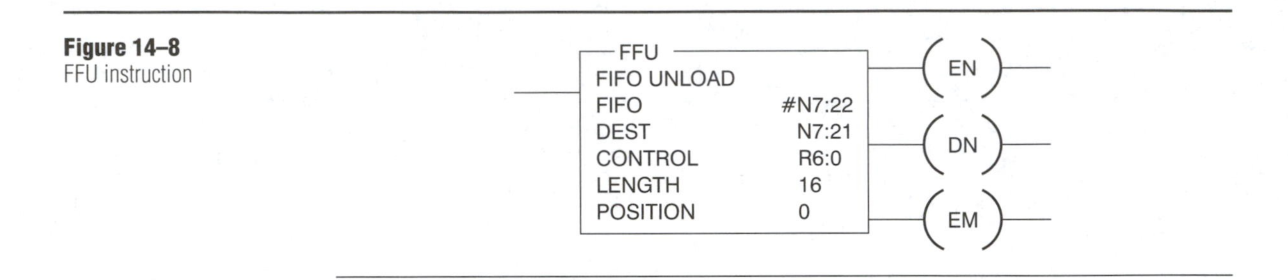

Figure 14–8
FFU instruction

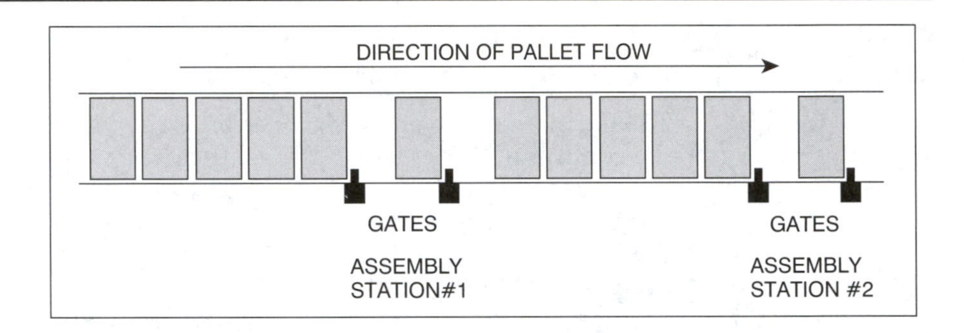

Figure 14–9
Sample assembly line

manual assembly stations that are in series with each other (Figure 14–9). Each station has a buffer gate and a release gate.

We would assign a FIFO array to each station with an array length of 6. When station #2 shows a full FIFO, we should inhibit station #1 from releasing any more pallets. Once station #2 releases a pallet, we reactivate station #1 gates again. In this manner we can move the accumulated data with each pallet as it is passed down the line. Each station can work independently of any other station in releasing pallets as long as the next station has room to accept the released pallet.

Once again, note that the whole plan will be corrupted if any pallet is removed from a station or buffer by hand. The FIFO will be out of sync and the only way to clear it will be to empty all the pallets from the buffer area and station and then clear all FIFO locations. The real danger in this arrangement is that if a pallet is removed without anyone's knowledge, the system will continue to function without an error message, even though the data are skewed by one pallet location.

LIFO INSTRUCTIONS

Two instructions are used for LIFO operations and they are always used in pairs. One is LFL, which stands for LIFO load; the other is LFU, which stands for LIFO unload. LIFO operations are just like FIFO instructions

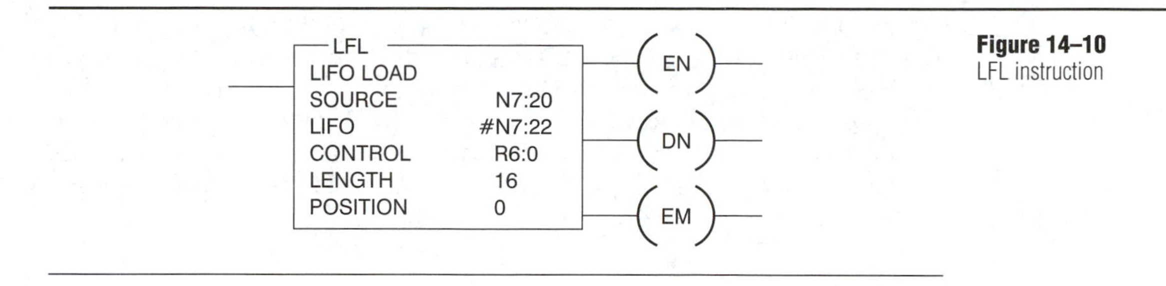

Figure 14–10
LFL instruction

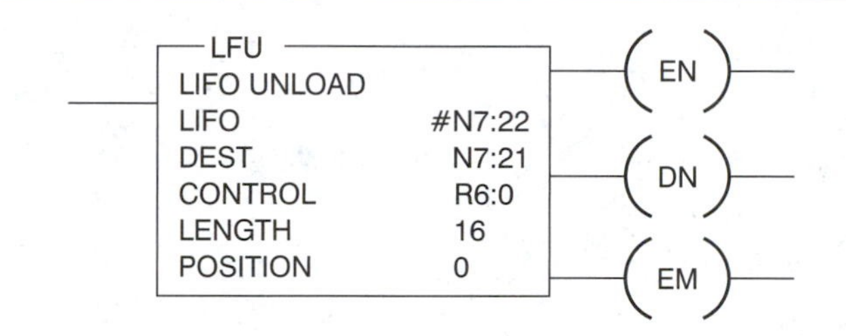

Figure 14–11
LFU instruction

except the data words are pushed and pulled from the same end of the array. This is a typical computer stacking operation. Data words are pushed down on the array by the load instruction and unloaded off the top by the unload instruction. From this action we get the name last in/first out. The LFL and LFU instructions are shown in Figures 14–10 and 14–11, respectively.

All status bits and word locations are the same as those used in the FIFO instructions. One of the main uses of LIFO instructions is for the temporary storage of data words between functions. One program can save a block of data on the array and then allow a second program to retrieve the data when control is passed to the other program.

CONCLUSIONS

In conclusion, we can say that bit shift instructions should be used any time we have a synchronous system and need to move data with the parts being processed. If the total number of data bits to keep track of is four or fewer, the bit shift instruction is a good choice; for larger requirements, the word shift register introduced in the previous chapter is a better choice.

These types of instructions are most commonly used on index tables or indexer driven lines with all carriers fastened to a chain or belt.

For asynchronous type lines, such as pallet lines in which individual pallets are stopped and released in zones, the FIFO instructions are a better choice for a data tracking system than the LIFO instructions. The designer must be careful to never let the FIFO get out of sync due to pallets being removed from a zone. As we have stated, this condition is not self-clearing and will cause data to be skewed forever until the FIFO is cleared.

REVIEW QUESTIONS

1. What function does the control register play in the bit shift instructions?
2. Why is it not a good idea to use the B3 file for bit shift instructions?
3. What happens to the bit that exits the last position in a bit shift register?
4. What happens to the first data word loaded into a FIFO register?
5. We said FIFOs are not self-clearing. Why is this a problem and when might it happen?

LAB ASSIGNMENT

Objective

To gain an understanding of bit shift registers and how to use them.

1. Enter the program shown in Figure 14–12 and save it as Lab 14.
2. The program is set up to use PB1 as the shift control and SEL Sw1 as the input condition. The output for the 8 bits is shown on lamps #1 to #8.
3. Note that the BSL instruction uses the second word in the B3 file for the actual shift register. These bits are then assigned to the real outputs. Each time the PB1 button is pressed the status of SEL Sw1 is shifted into B3:1.0 and the other seven bits are moved down by one.
4. Run the program until you feel comfortable with its operation.
5. Open the B3 file and observe the shift motion on the second word.
6. Add an additional BSR instruction that will move the data to the right using PB2 and SEL Sw2 as a data input. Be sure to use a different R6 location for this instruction.
7. Design a pulse generator by using two timers and use it to feed the BSL instruction so that the data is continuously shifted every 0.5 second.
8. Add the push button #3 input so that it will clear all the shift register bits.

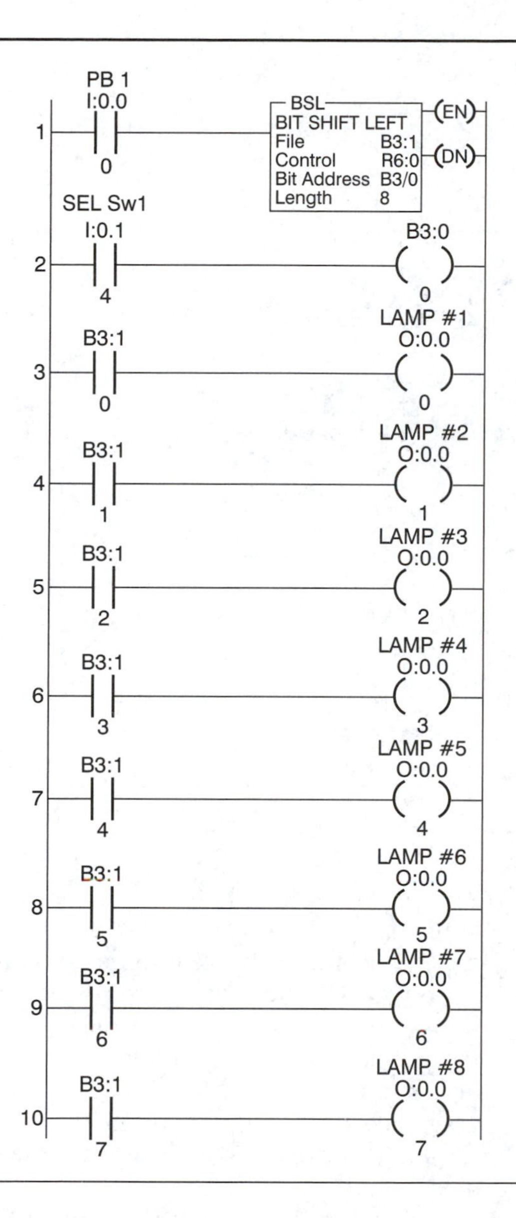

Figure 14–12
Lab ladder program

CHAPTER

15

Sequencer Functions

LEARNING OBJECTIVES

After completing this chapter the reader should understand:

1. The programming and use of sequencers.
2. The setup and layout of sequencer data files.
3. Setting up the sequencer mask word.
4. Using the combination of sequencer output and compare instructions.
5. Use of the sequencer load instruction.

Many machines designed and built from the early 1900s until the late 1960s used programming devices called *drum sequencers*. These devices consisted of a rotating cylinder with pins or raised cams attached to it. Riding on the cylinder was a multitude of micro switches. By properly placing the cams on and around the cylinder, the designer could program a sequence into the micro switches as the drum turned. Each of these switches was in turn attached to an output device that controlled part of the machine. As the drum turned, each device was turned on and off in the sequence programmed by the pin locations (see Figure 15–1). This concept is similar to the paper rolls used to control musical instruments such as player pianos.

We have sequencer instructions available to us in the SLC-500 controller. The instruction labels are SQO for sequencer output, SQC for sequencer compare, and SQL for sequencer load.

Before we get into the explanations of the actual sequencer instructions let us explore how the sequencer function works. The main concept of the sequencer function is to enable the programmer to store a sequence of I/O states in a file, and then recall the sequence and output each word of the stored data to the real I/O addresses. Because we always store 16-bit words of data, there can be problems if we only want some

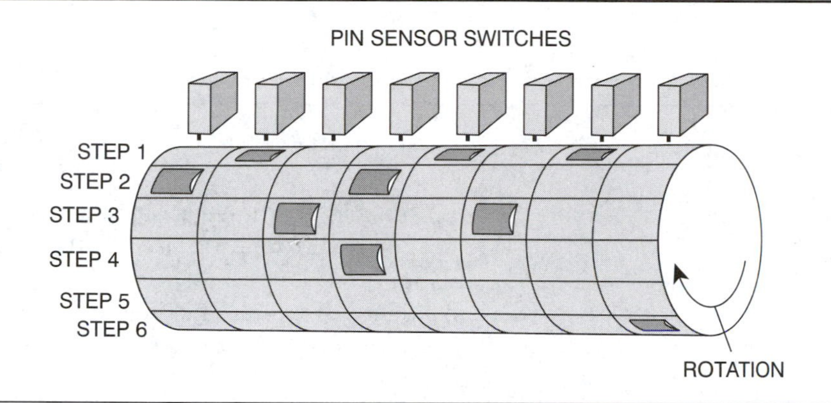

Figure 15–1
Eight-bit sequencer drum

portion of the word. For example, we might want to control only the lower 8 bits of a 16-bit output word. If we moved the complete 16-bit word of stored data to the output register, even if we have zeros in the unused positions, we would overwrite 8 bits of data in the output word that are not part of our control problem. To get around this problem we will pass the stored data through a mask before presenting it to the output word. The mask will allow us to hold back any bit locations that are not part of our control problem. Let us now look at the first sequencer command.

SQO INSTRUCTION

Allen-Bradley's sequencer output instruction requires three words of memory in control file 6 (see Figure 15–2).

As you can see from the figure, the first word of the memory block is used to hold the status bits for this instruction. Let us look at the function of each of these status bits.

EN The enable bit, which is on any time the rung conditions are TRUE.

DN The done bit, which is set on when the sequencer steps to the last position in the file. The next FALSE-to-TRUE transition of the rung clears this bit and resets the position

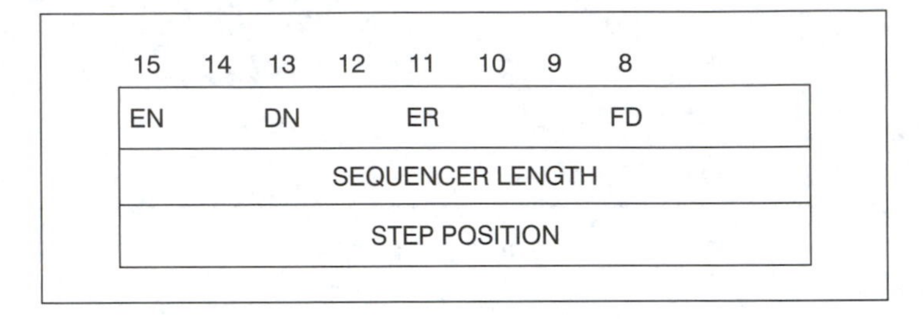

Figure 15–2
Memory locations for SQO instruction

counter to position 1. Note that position 0 is only encountered when the PLC is switched from program to run mode.

ER The error bit, which is set on if the processor sees a negative position value or no length value. This is only encountered if you are changing the sequencer values as part of the ladder program.

FD This bit is only used in SQC instructions.

Instruction Elements

Figure 15–3 shows a sample instruction rung format. As in any instruction the input can have multiple elements. We now describe each element of the instruction.

File

The file address points to the first word of the processor file that contains the output data to be sequenced by the instruction. It should be a bit file; we recommend using a separate file above #9. (You must use the # character in front of this address to indicate it is a file address.) Using the assigned B3 file can cause problems if the programmer is not careful when allocating new bit functions. There is no write protection against overwriting words assigned to the sequencer.

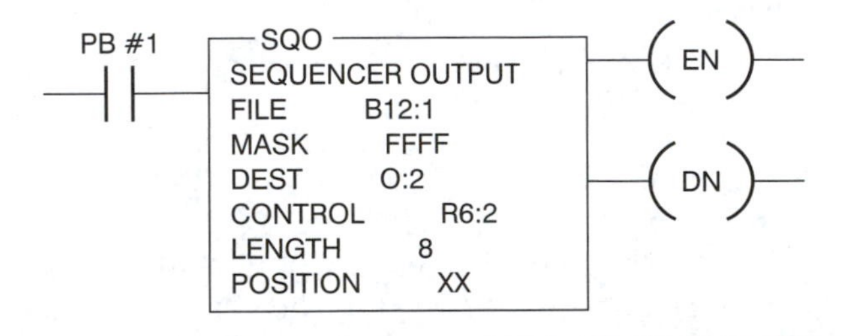

Figure 15–3
Sample rung format for SQO instruction

Mask

All data transferred from the file address to the output word will be passed through this mask. The mask can be a hex value as shown in Figure 15–3 or an additional file address. In any case, each position set to a 1 will pass data; each position set to a 0 will block data. If the mask is a file, the file length will be the same as the instruction file length and the position will track in sequence. Note that when entering a hex number higher than 9FFF you must precede the number with a 0. To set the mask at FFFF you would have to enter 0FFFF.

Dest

This is the output destination for the SQO instruction. Each time the rung conditions go from FALSE to TRUE, the contents of the next position in the bit file are transferred through the mask to this location. This address can be a real output word or an internal address.

Control

This is the position of the three status words allocated in the control file that are used to keep track of status bits and position counts for this instruction. The programmer must be careful to keep track of the control addresses used in his program. If you assign the same control address to more than one instruction, unpredictable operation will result.

Length

Contains the number of steps assigned to the instruction beginning at step 1. All SQO instructions start at position 0 when the PLC is switched from program mode to run mode. Due to this extra state, all sequencer files use file length +1 number of words. The sequencer resets to position 1 after reaching the last step and not to position 0.

Position

The position value is the count of how far we have moved in the sequence file. The counter will start at 1 after being reset or after rolling over from the max count number. The count will increment once for each FALSE-to-TRUE transition of the input rung.

Sample Application

The best way to gain an understanding of sequencers is to use one in a real-world application. We often encounter the situation in manual mode machine control operations in which the sequence of events required to operate a mechanical device is very complex. If it is difficult for the designer to remember the correct sequence of moves, it will most likely be impossible for the machine operator. In these situations it is often better to give the operator a single step button to cycle the mechanism instead of 12 push buttons with confusing labels.

In our example we are going to control a mechanism that has six air cylinders and therefore 12 outputs. If we were to control this mechanism with the old style manual station, we would need 12 push buttons. We will simplify the station by having only one button.

One of the first design parameters we must deal with in setting up a sequencer function is the type of output word to use. If all the I/O bits we need for the application are contained in one 16-bit output module we can use that output module as our output word. If, as is the usual case, the I/Os are spread over several output modules, we must use an internal word of memory as our output word. We then use additional ladder rungs to take each of these internal memory bits and drive the real output bit from the rung.

Let's assume we have all the outputs for this station connected to one output module. This is a rare case in the real world but for our first example it makes things simpler. Note that four addresses of our I/O module are assigned to a different mechanism and therefore will not be controlled by this sequencer.

Sample Outputs:

O:2.0/15	CYLINDER 6 EXT SOL
O:2.0/14	CYLINDER 6 RET SOL
O:2.0/13	DIFFERENT STATION'S I/O
O:2.0/12	CYLINDER 5 EXT SOL
O:2.0/11	CYLINDER 5 RET SOL
O:2.0/10	CYLINDER 4 EXT SOL
O:2.0/9	CYLINDER 4 RET SOL
O:2.0/8	DIFFERENT STATION'S I/O
O:2.0/7	DIFFERENT STATION'S I/O
O:2.0/6	DIFFERENT STATION'S I/O
O:2.0/5	CYLINDER 3 EXT SOL
O:2.0/4	CYLINDER 3 RET SOL
O:2.0/3	CYLINDER 2 EXT SOL
O:2.0/2	CYLINDER 2 RET SOL
O:2.0/1	CYLINDER 1 EXT SOL
O:2.0/0	CYLINDER 1 RET SOL

As previously noted, four of the addresses are not part of this station. We must therefore mask off these addresses (see Figure 15–4). Our mask word would be DE3F hex.

The sequence of operation for this mechanism is as follows:

1. Extend cylinder #1.
2. Extend cylinders #3 and #4.
3. Extend cylinder #5.

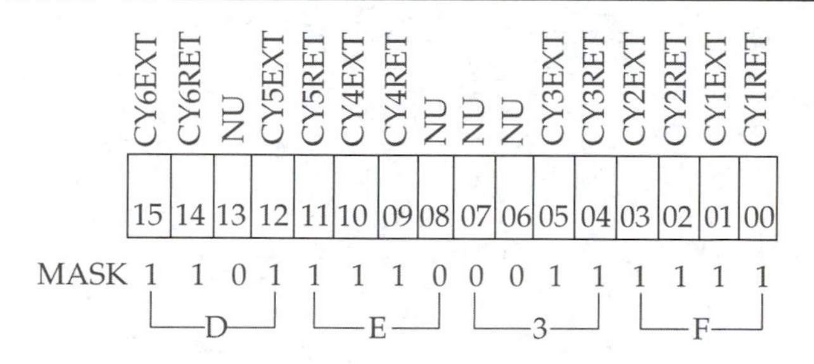

Figure 15–4
Output word assignment and mask generation

4. Extend cylinders #2 and #6.
5. Return cylinder #6.
6. Return cylinder #2.
7. Return cylinder #5.
8. Return cylinders #3 and #4.
9. Return cylinder #1.

We will use a single push button to step the mechanism through its cycle. See the rung layout shown in Figure 15–5. We show nine states for the mechanism to complete its cycle, but we will require one additional state where all outputs are off.

Let us look at how this sequencer will function from a file transfer state. Figure 15–6 shows the memory layout, mask word, and output word for this sequencer function. This is based on the data contained in the SQO instruction shown in Figure 15–5.

Each time the push button is pressed in the sequencer rung, the sequencer counter will step one position. The data from this next position will be fetched, passed through the mask, and stored at the actual output word. Figure 15–6 shows the procedure for the third step of the sequence. We must now code which output bits we want on for each step in the mechanical sequence. The best way to do this is to create a chart with the output functions listed across the top and the memory locations listed below as shown in Figure 15–4. For our sequencer the complete file is shown in Figure 15–7.

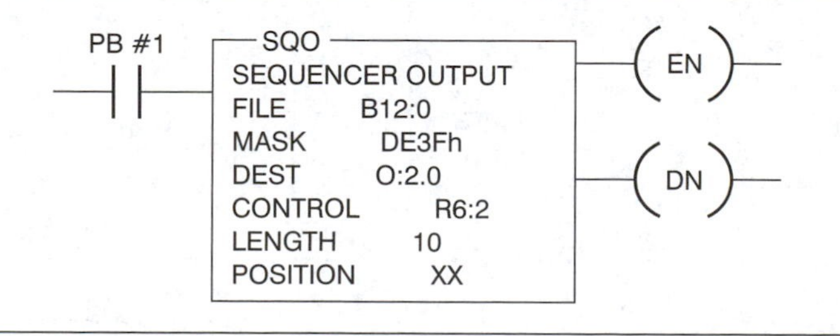

Figure 15–5
Rung layout for sample application

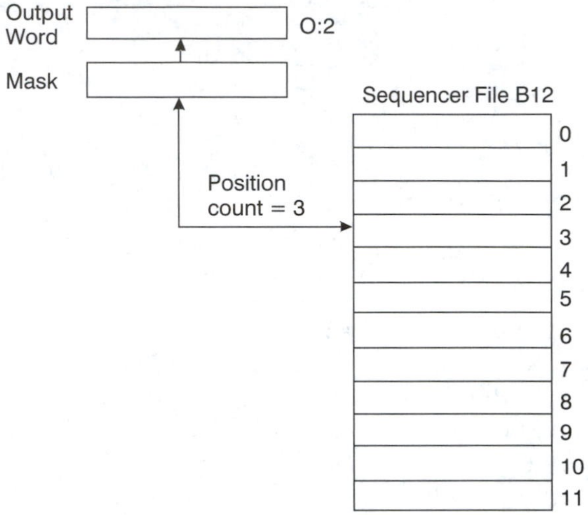

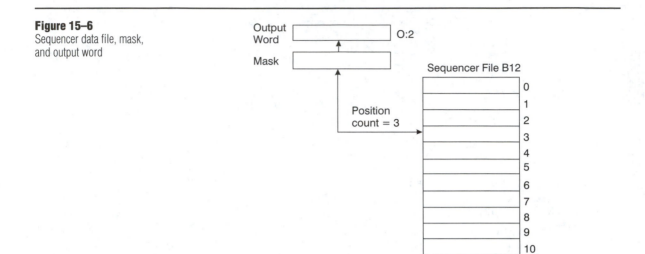

Figure 15–6
Sequencer data file, mask, and output word

Figure 15–7
Data file bit allocations for sequencer

	CY6EXT	CY6RET	NU	CY5EXT	CY5RET	CY4EXT	CY4RET	NU	NU	NU	CY3EXT	CY3RET	CY2EXT	CY2RET	CY1EXT	CY1RET
Output Word	15	14	13	12	11	10	09	08	07	06	05	04	03	02	01	00

Step	Bit Pattern																File B12 Word
0	0	0	0	0	0	0	0	0	0	0	0	0	0	0	0	0	0 = 0000h
1	0	0	0	0	0	0	0	0	0	0	0	0	0	0	0	0	1 = 0000h
2	0	0	0	0	0	0	0	0	0	0	0	0	0	0	1	0	2 = 0002h
3	0	0	0	0	0	1	0	0	0	0	1	0	0	0	0	0	3 = 0420h
4	0	0	0	1	0	0	0	0	0	0	0	0	0	0	0	0	4 = 1000h
5	1	0	0	0	0	0	0	0	0	0	0	0	1	0	0	0	5 = 8008h
6	0	1	0	0	0	0	0	0	0	0	0	0	0	0	0	0	6 = 4000h
7	0	0	0	0	0	0	0	0	0	0	0	0	0	1	0	0	7 = 0004h
8	0	0	0	0	1	0	0	0	0	0	0	0	0	0	0	0	8 = 0800h
9	0	0	0	0	0	0	1	0	0	0	0	1	0	0	0	0	9 = 0210h
10	0	0	0	0	0	0	0	0	0	0	0	0	0	0	0	1	10 = 0001h
																	11

Notice that we have calculated the hex value for each word in memory. You can enter the data into the file addresses as bits or as hex data. You're less likely to make a mistake entering the data as a hex value than entering each individual bit location.

Before we can load the sequence data we must create bit file 12 and open at least the first 11 words of the file. To open a new data file you first right click the data files folder on the left edge of the screen, then click on the NEW option. This will open the pop-up window shown in Figure 15–8.

Enter 12 in the File box to create file 12, and make sure the Type box is set for Binary. Enter a name and description in the Name and Desc boxes. The last variable you need to enter is the Last position address. Because we need 10 words for our program, the file was opened up to B12:12 or the first 13 words, because the file starts at B12:0. This gives us a few extra locations in case we decide to add steps to our program at a later date. You can always open this window and extend the Last variable if you need additional space.

As stated earlier, it is an unusual situation in I/O assignments when all the addresses for a given mechanism are contained within a single output or input module. This being the case, we can work around the problem by using a bit address file for our SQO output word. We then will need a number of rungs to reassign the bit addresses to real output addresses. The main advantage of using an internal word is that we can use all 16 bits and we will not have to mask off any locations. In our example we could have used one of the masked bits to activate a light that would have indicated to the operator that he was in step 1. In this way

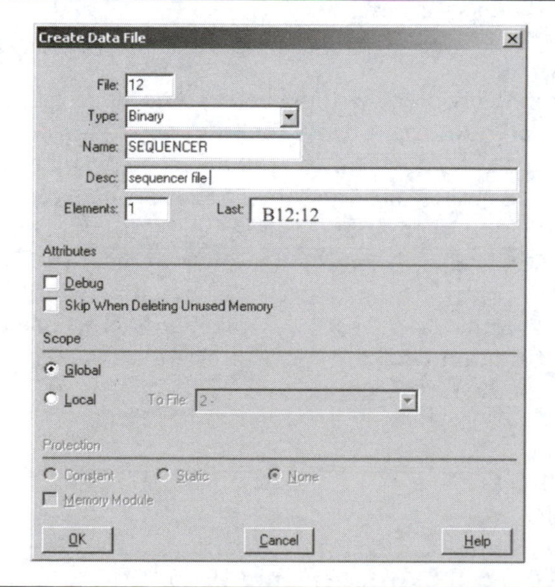

Figure 15–8
Pop-up window for entering a new bit file

the operator would always know when he had completed the cycle and the mechanism was in home position.

It is important to point out that the sequencer will step each time the button is pressed even if the previous step was not completed. In a manual mode type of operation, this is a minor problem because an operator will be present to watch the operation. If, however, you were going to use a sequencer output file to control a mechanism in auto mode it becomes a major problem.

A common mistake many beginning programmers make is attempting to use SQO instructions to control the automatic machine cycle. At first glance it seems to be so much easier than writing all those rungs of code we would normally need if we were using standard ladder logic. We could just use two timers tied together for step timing and run the whole machine in auto cycle with just one sequencer instruction. Although this seems like a good idea, it has some major problems.

The first major control problem is the lack of any mechanical position feedback. In our standard ladder programming we use the position switches on the mechanisms to trigger the next movement. If the previous output movement is not sensed to be complete, the sequence stops. In this SQO program the sequence will step regardless of the completion of the previous step. If a mechanism becomes jammed, the sequence will continue anyway and may cause extensive damage to our equipment.

The second problem with an SQO-only design is that of timing. Because we only have one set of step timers running, we must choose a time delay between steps that is long enough for the slowest mechanism in the machine. This means most of the other mechanisms will have completed their movement and be waiting for the next step time-out. Due to this timing limitation, the machine will cycle at a much slower rate than it is capable of.

An additional problem will occur if some variable in the machine design changes. If the air pressure is set lower or some of the cylinder seals become worn, the machine will experience repeated jam-ups because it cannot sense the changes and the timer will assume everything takes the same amount of time to complete as before.

For the reasons just given, the use of sequencer output instructions by themselves to control the auto cycle of a machine is not recommended. The way to overcome these feedback problems is to use a combination SQC and SQO function working together. We show a sample of this combined function in the next section.

SQC INSTRUCTION

Allen-Bradley's sequencer compare instruction requires three words of memory in the control file 6 (see Figure 15–9).

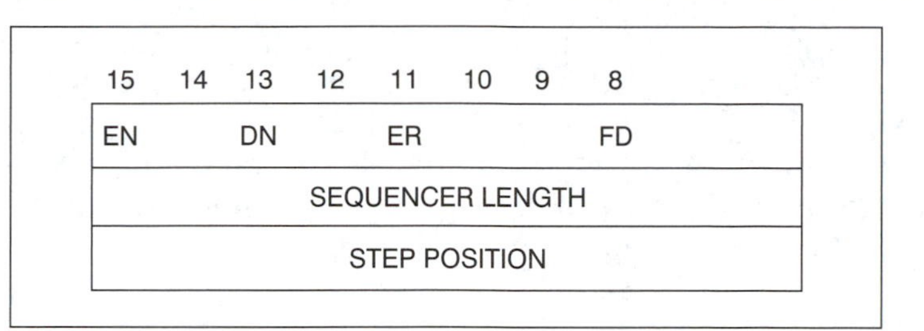

Figure 15–9
Memory locations for SQC instruction

EN The enable bit, which is on any time the rung conditions are TRUE.

DN The done bit, which is set on when the sequencer steps to the last position in the file. The next FALSE-to-TRUE transition of the rung clears this bit and resets the position counter to position 1. Note that position 0 is only encountered when the PLC is switched from program to run mode.

ER The error bit, which is set on if the processor sees a negative position value or no length value. This is only encountered if you are changing the sequencer values as part of the ladder program.

FD The found bit, which is set on when a match is found between the contents of the input word and the contents of the current position in the sequencer file. The contents of the input word are passed through the mask before the compare function is performed.

Instruction Elements

Figure 15–10 shows the instruction rung format. As in any instruction the input can have multiple elements.

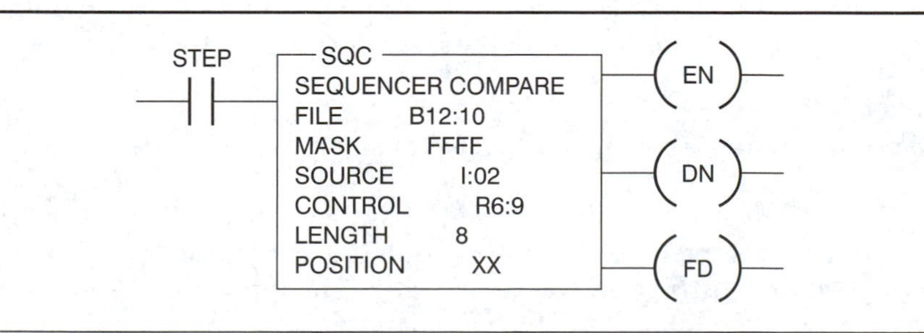

Figure 15–10
Sample rung format for SQC instruction

File

The file address is the processor file address that contains the compare data to be sequenced by the instruction. It should be a bit type file and we recommend using a separate file above #9. (You must use the # character before the file number to indicate that it is a file address.) Using the assigned B3 file can cause problems if the programmer is not careful when allocating new bit functions.

Mask

All data compared to the file address word will be passed through this mask. The mask can be a hex value as shown in Figure 15–10 or an additional file address. In any case each position set to a 1 will pass data; each position set to a 0 will block data. If the mask is a file, the file length will be the same as the instruction file length and the position will track in sequence. Note that when entering a hex number higher than 9FFF you must precede the number with a 0. To set the mask at FFFF you would have to enter 0FFFF.

Source

The source address is the input address for the SQC instruction. Each time the rung conditions go from FALSE to TRUE, the contents of this address are compared through the mask to the contents of the file word pointed to by the position counter. If the two words are equal, the FD bit will be set. The source can be a real input word or an internal word.

Control

This is the position of the three status words allocated in the control file that are used to keep track of status bits and position counts for this instruction. The programmer must be careful to keep track of the control addresses used in his program. If you assign the same control address to more than one instruction, unpredictable operation will result.

Length

The length address contains the number of steps assigned to the instruction beginning at step 1. All SQC instructions start at position 0 when the PLC is switched from program mode to run mode. Due to this extra state, all sequencer files use file length + 1 number of words. The sequencer resets to position 1 after reaching the last step and not to position 0.

Position

The position address contains the actual position in our bit file that is being compared to the input word.

The SQC instruction can be used in parallel with an SQO instruction. When used together, the SQC instruction verifies that the function called for by the SQO instruction was completed. This prevents the sequencer from continuing on if any step was not completed.

SAMPLE PROGRAM

We will now write a sample program using a pair of SQO and SQC instructions connected together. The mechanical sequence for this sample program has been kept simple for this illustration. For this example we will need to control three air cylinders, four lamps, and two timers. Because our outputs will be on two different I/O modules and we have to activate two timers, we will have to use an internal output word. While all our inputs come from one input module, we must also look at the status bits from the two timers. This means that we will also have to use an internal word as our input word. Let's take a look at how our two instructions will interface.

Figure 15–11 gives us an idea how the two instructions should relate to each other. Note that we now have two different sequencer data tables, one for the output sequence and one for the input compare sequence. The two sequencer instructions will use the SQC's FD bit and the SQO's EN bit to handshake back and forth as they move in sequence down the data files. We will set up the program so that the SQO is enabled when the FD bit is off. This will cause the SQO to step as soon as we activate the run mode. Once the SQO steps it will enable the SQC instruction. The SQC will then step and compare the word at the input address with the word pointed to in the SQC data table. If the contents of the two words match, the FD bit will come on. This will drop out the SQO instruction, which in turn drops out the SQC instruction. When the SQC FD bit drops out, the SQO instruction reactivates and we step to the next output state. The sequence continues as long as each output step is followed by the correct input states for that step. When the sequence reaches the end of the data table, both sequence counters return to step 1.

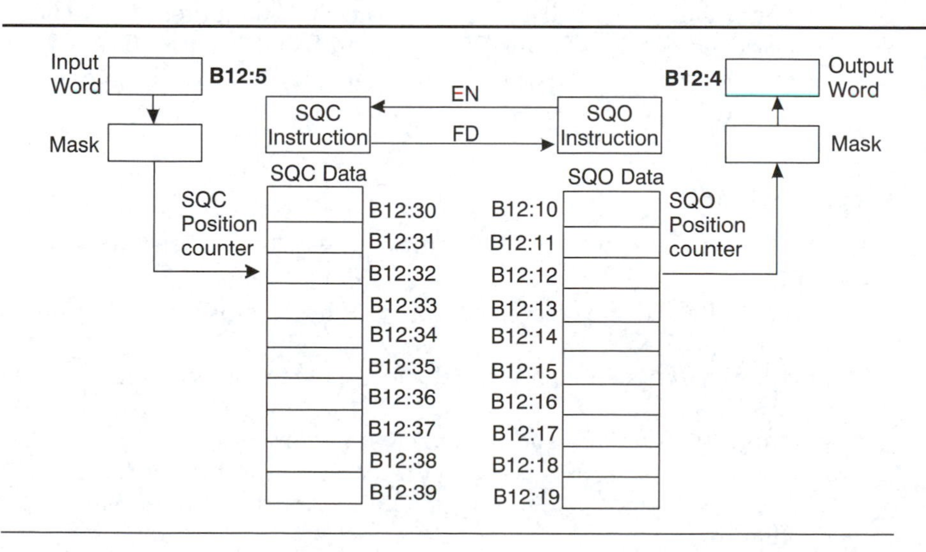

Figure 15–11 Operational chart for SQC and SQO instructions working as a pair

Figure 15–12
I/O assignments for SQO and SQC input and output words

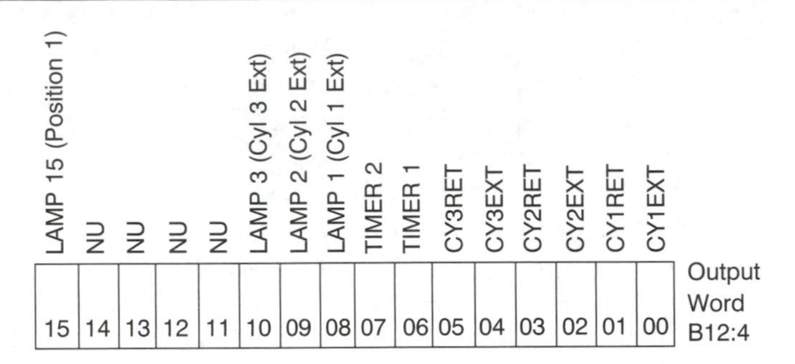

NU
NU
NU
NU
NU
NU
NU
START PB
TIMER 2 DN Bit
TIMER 1 DN Bit
CY3RET LS
CY3EXT LS
CY2RET LS
CY2EXT LS
CY1RET LS
CY1EXT LS
15 14 13 12 11 10 09 08 07 06 05 04 03 02 01 00
Input
Word
B12:5

Let us now look at how we will allocate our I/O to the input and output words. Figure 15–12 shows how we have allocated our output word.

We have assigned B12:4 as our output word and B12:5 as our input word. There is nothing special about these addresses; they were picked because they are at the top of the B12 file. We will use B12:10 as the top of our SQO data file and B12:30 as the top of our SQC data file. We will need only nine addresses for each data file to accomplish the required sequence, but you should leave some extra room in your file structure for expansion in case you decide to add steps at a later date.

Program Sequence

1. When switched to run mode, light lamp 15 and wait for the start PB.
2. When the start PB closes, extend cylinder 1, light lamp 1, and turn off lamp 15.
3. When cylinder 1 extend switch closes, extend cylinder 2 and light lamp 2.
4. When cylinder 2 extend switch closes, wait 5 seconds.
5. When timer is done, extend cylinder 3 and light lamp 3.
6. When cylinder 3 extend switch closes, return cylinder 3 and turn off lamp 3.

7. When cylinder 3 return switch closes, wait 3 seconds.
8. When timer is done, return cylinder 2 and turn off lamp 2.
9. When cylinder 2 return switch is closed, return cylinder 1 and turn off lamp 1.
10. When cylinder 1 return switch closes, move to step 1.

I/O Allocations

I:0.0/00	START PB
I:0.0/01	CYLINDER 1 EXTEND LS
I:0.0/02	CYLINDER 1 MANUAL RETURN PB
I:0.0/03	CYLINDER 1 RETURN LS
I:0.0/04	CYLINDER 2 MANUAL RETURN PB
I:0.0/05	CYLINDER 2 EXTEND LS
I:0.0/06	CYLINDER 3 MANUAL RETURN PB
I:0.0/07	CYLINDER 2 RETURN LS
I:0.0/08	SEQUENCER RESET PB
I:0.0/09	CYLINDER 3 EXTEND LS
I:0.0/11	CYLINDER 3 RETURN LS
I:0.1/4	AUTO/MANUAL SEL SW
O:1.0/01	CYLINDER 1 EXTEND SOL.
O:1.0/03	CYLINDER 1 RETURN SOL.
O:1.0/05	CYLINDER 2 EXTEND SOL.
O:1.0/07	CYLINDER 2 RETURN SOL.
O:1.0/09	CYLINDER 3 EXTEND SOL.
O:1.0/11	CYLINDER 3 RETURN SOL.
O:0.0/00	LAMP 1 CYLINDER 1 EXTENDED
O:0.0/01	LAMP 2 CYLINDER 2 EXTENDED
O:0.0/02	LAMP 3 CYLINDER 3 EXTENDED
O:0.0/14	LAMP 15 STEP 1 POSITION
O:0.0/15	LAMP 16 ALARM LAMP

Let us now set up our output sequence. We start with our output word labeled with the position allocations. Once we have assigned each bit we can begin filling in the data file. Our completed data file is shown in Figure 15–13.

If we work down our data file we see the following:

Step 0 has all outputs off.

Step 1 turns on the step 1 lamp.

Step 2 turns off the step 1 lamp and turns on cylinder 1 extend solenoid and lamp 1.

Step 3 turns on cylinder 2 extend solenoid and lamp 2.

Figure 15–13
Output data table

Step	LAMP 15 (Position 1)	NU	NU	NU	NU	LAMP 3 (Cyl 3 Ext)	LAMP 2 (Cyl 2 Ext)	LAMP 1 (Cyl 1 Ext)	TIMER 2	TIMER 1	CY3RET	CY3EXT	CY2RET	CY2EXT	CY1RET	CY1EXT	Output Word B12:4
	15	14	13	12	11	10	09	08	07	06	05	04	03	02	01	00	
Step	Bit Pattern																File B12
0	0	0	0	0	0	0	0	0	0	0	0	0	0	0	0	0	10 = 0000h
1	1	0	0	0	0	0	0	0	0	0	0	0	0	0	0	0	11 = 8000h
2	0	0	0	0	0	0	0	1	0	0	0	0	0	0	0	1	12 = 0101h
3	0	0	0	0	0	0	1	1	0	0	0	0	0	1	0	1	13 = 0305h
4	0	0	0	0	0	0	1	1	0	1	0	0	0	1	0	1	14 = 0345h
5	0	0	0	0	0	1	1	1	0	0	0	1	0	1	0	1	15 = 0715h
6	0	0	0	0	0	0	1	1	0	0	1	0	0	1	0	1	16 = 0325h
7	0	0	0	0	0	0	1	1	1	0	1	0	0	1	0	1	17 = 03A5h
8	0	0	0	0	0	0	0	1	0	0	1	0	1	0	0	1	18 = 0129h
9	0	0	0	0	0	0	0	0	0	0	1	0	1	0	1	0	19 = 002Ah
10	0	0	0	0	0	0	0	0	0	0	0	0	0	0	0	0	20 = 0000h
																	21

Step 4 turns on timer 1.

Step 5 turns on cylinder 3 extend solenoid and lamp 3.

Step 6 turns off lamp 3 and returns cylinder 3.

Step 7 turns on timer 2.

Step 8 turns off lamp 2 and returns cylinder 2.

Step 9 turns off lamp 1 and returns cylinder 1.

Now that we have programmed our output steps we need to calculate the input pattern we should see for each output step. We again set up a chart, but this time we need our input bit assignments listed on top (see Figure 15–14) .

We start with all inputs off except the three cylinder return switches.

Step 1 looks for the three return switches to be on and the start PB.

Step 2 looks for the cylinder 2 and 3 return switches and the cylinder 1 extend switch.

Figure 15–14
Input data table

Step	NU	NU	NU	NU	NU	NU	NU	START PB	TIMER 2 DN Bit	TIMER 1 DN Bit	CY3RET LS	CY3EXT LS	CY2RET LS	CY2EXT LS	CY1RET LS	CY1EXT LS	
	15	14	13	12	11	10	09	08	07	06	05	04	03	02	01	00	Input Word B12:5
Step	**Bit Pattern**																**File B12**
0	0	0	0	0	0	0	0	0	0	0	1	0	1	0	1	0	30 = 002Ah
1	0	0	0	0	0	0	0	1	0	0	1	0	1	0	1	0	31 = 012Ah
2	0	0	0	0	0	0	0	0	0	0	1	0	1	0	0	1	32 = 0029h
3	0	0	0	0	0	0	0	0	0	0	1	0	0	1	0	1	33 = 0025h
4	0	0	0	0	0	0	0	0	0	1	1	0	0	1	0	1	34 = 0065h
5	0	0	0	0	0	0	0	0	0	0	0	1	0	1	0	1	35 = 0015h
6	0	0	0	0	0	0	0	0	0	0	1	0	0	1	0	1	36 = 0025h
7	0	0	0	0	0	0	0	0	1	0	1	0	0	1	0	1	37 = 00A5h
8	0	0	0	0	0	0	0	0	0	0	1	0	1	0	0	1	38 = 0029h
9	0	0	0	0	0	0	0	0	0	0	1	0	1	0	1	0	39 = 002Ah
10	0	0	0	0	0	0	0	0	0	0	0	0	0	0	0	0	40 = 0000h
																	41

Step 3 looks for the cylinder 1 and 2 extend switches and the cylinder 3 return switch.

Step 4 looks for the same switch setup but with the addition of the timer 1 done bit.

Step 5 looks for all three cylinders to be in the extend position.

Step 6 looks for cylinders 1 and 2 extended and cylinder 3 returned.

Step 7 looks for the same switch setup but with the timer 2 done bit.

Step 8 looks for cylinders 2 and 3 returned and cylinder 1 extended.

Step 9 looks for all three cylinders to be in the returned position.

Sequencer Program Generation

The first step in generating our ladder program will be to take each bit of the sequencer output word and drive the real output. The program rungs for this sample program are shown in Figures 15–15, 15–16, 15–17, and 15–18.

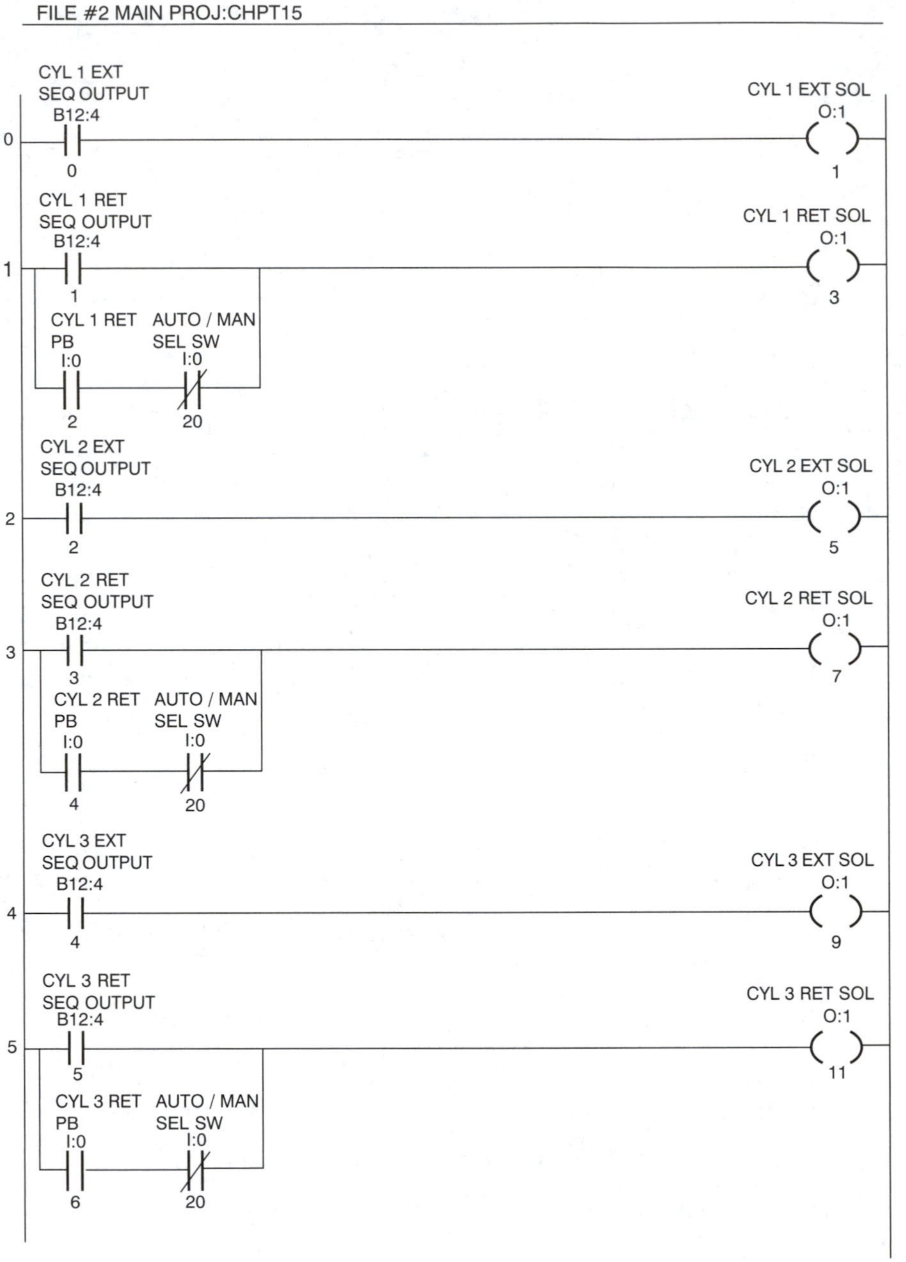

Figure 15–15
Sample sequencer program using SQC and SQO instructions

FILE #2 MAIN PROJ:CHPT15

Rung 0: CYL 1 EXT SEQ OUTPUT B12:4 / 0 — CYL 1 EXT SOL O:1 / 1

Rung 1: CYL 1 RET SEQ OUTPUT B12:4 / 1 — CYL 1 RET SOL O:1 / 3
CYL 1 RET PB I:0 / 2 — AUTO / MAN SEL SW I:0 / 20

Rung 2: CYL 2 EXT SEQ OUTPUT B12:4 / 2 — CYL 2 EXT SOL O:1 / 5

Rung 3: CYL 2 RET SEQ OUTPUT B12:4 / 3 — CYL 2 RET SOL O:1 / 7
CYL 2 RET PB I:0 / 4 — AUTO / MAN SEL SW I:0 / 20

Rung 4: CYL 3 EXT SEQ OUTPUT B12:4 / 4 — CYL 3 EXT SOL O:1 / 9

Rung 5: CYL 3 RET SEQ OUTPUT B12:4 / 5 — CYL 3 RET SOL O:1 / 11
CYL 3 RET PB I:0 / 6 — AUTO / MAN SEL SW I:0 / 20

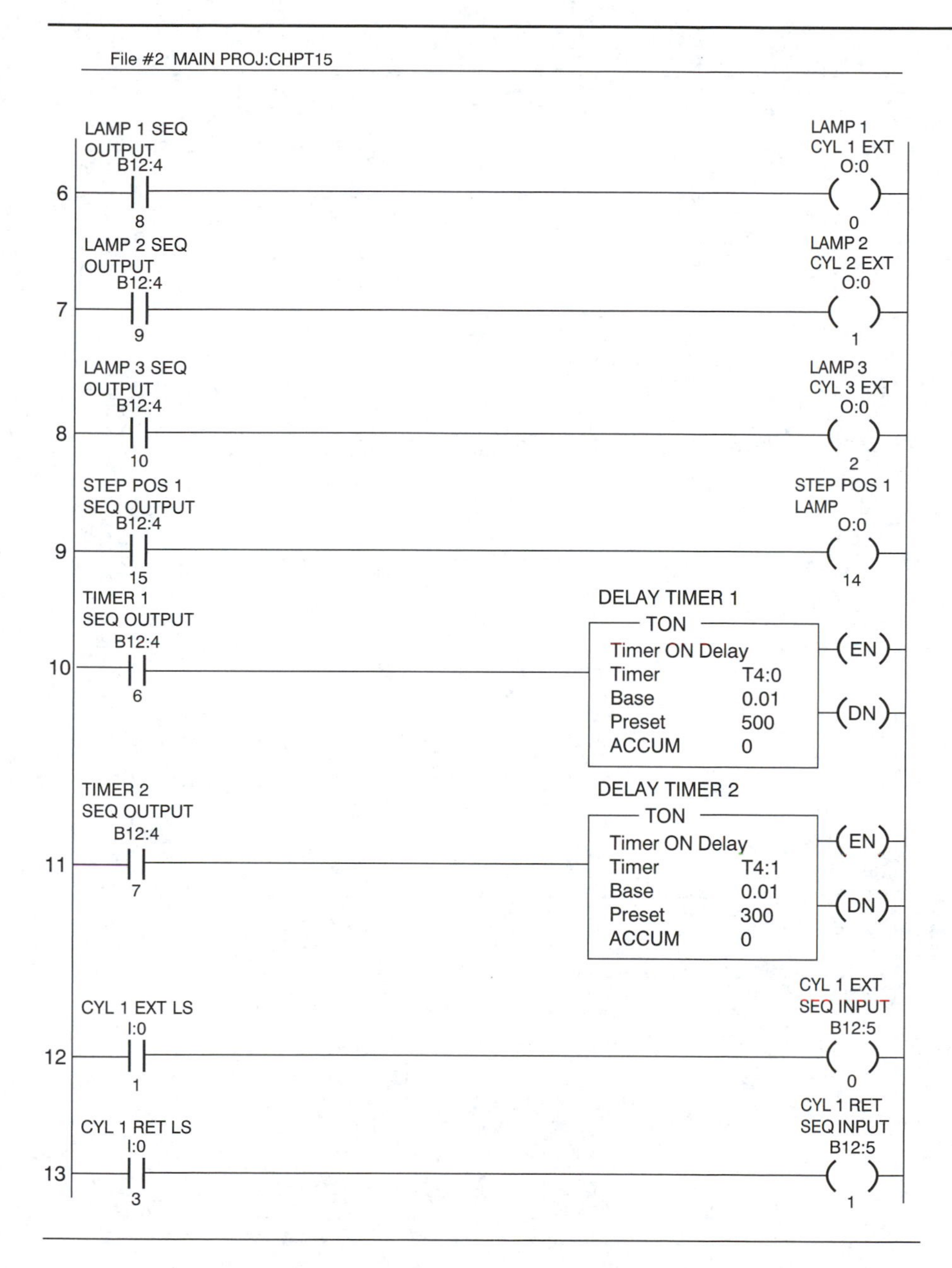

Figure 15–16 Sequencer program, *continued*

Rungs 0 through 9 read the sequencer output word bits and drive the real outputs for the solenoids and lamps.

Rungs 10 and 11 take the two timer bits from the sequencer output word and use them to enable timers T4:0 and T4:1.

Figure 15–17
Sequencer program, *continued*

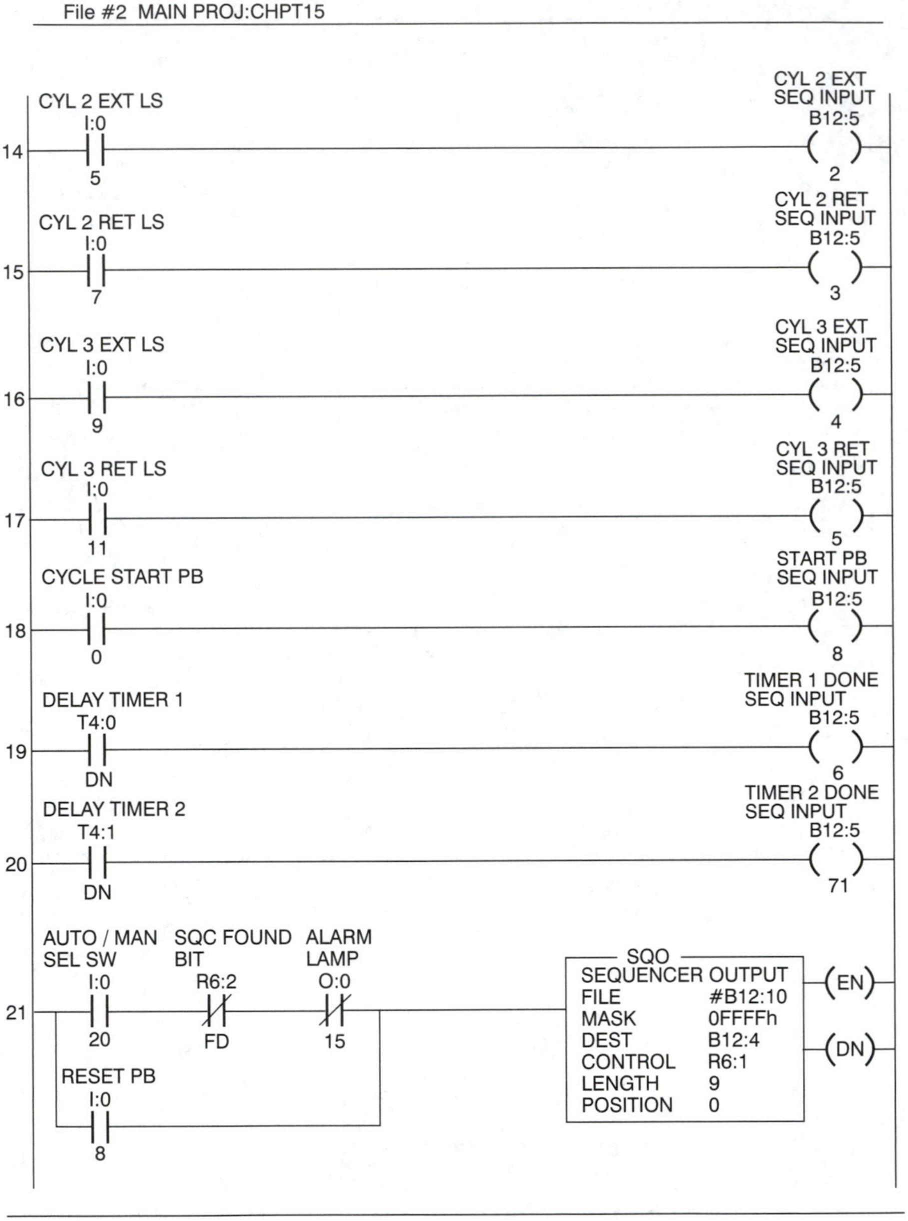

File #2 MAIN PROJ:CHPT15

14
CYL 2 EXT LS
I:0
5
CYL 2 EXT
SEQ INPUT
B12:5
2

15
CYL 2 RET LS
I:0
7
CYL 2 RET
SEQ INPUT
B12:5
3

16
CYL 3 EXT LS
I:0
9
CYL 3 EXT
SEQ INPUT
B12:5
4

17
CYL 3 RET LS
I:0
11
CYL 3 RET
SEQ INPUT
B12:5
5

18
CYCLE START PB
I:0
0
START PB
SEQ INPUT
B12:5
8

19
DELAY TIMER 1
T4:0
DN
TIMER 1 DONE
SEQ INPUT
B12:5
6

20
DELAY TIMER 2
T4:1
DN
TIMER 2 DONE
SEQ INPUT
B12:5
71

21
AUTO / MAN SEL SW
I:0
20
SQC FOUND BIT
R6:2
FD
ALARM LAMP
O:0
15
RESET PB
I:0
8

SQO
SEQUENCER OUTPUT
FILE #B12:10
MASK 0FFFFh
DEST B12:4
CONTROL R6:1
LENGTH 9
POSITION 0
(EN)
(DN)

Rungs 12 through 18 read the real inputs and set the bits in the sequencer input word.

Rungs 19 and 20 read the DN bits from the two TON timers and set the proper bits in the sequencer input word.

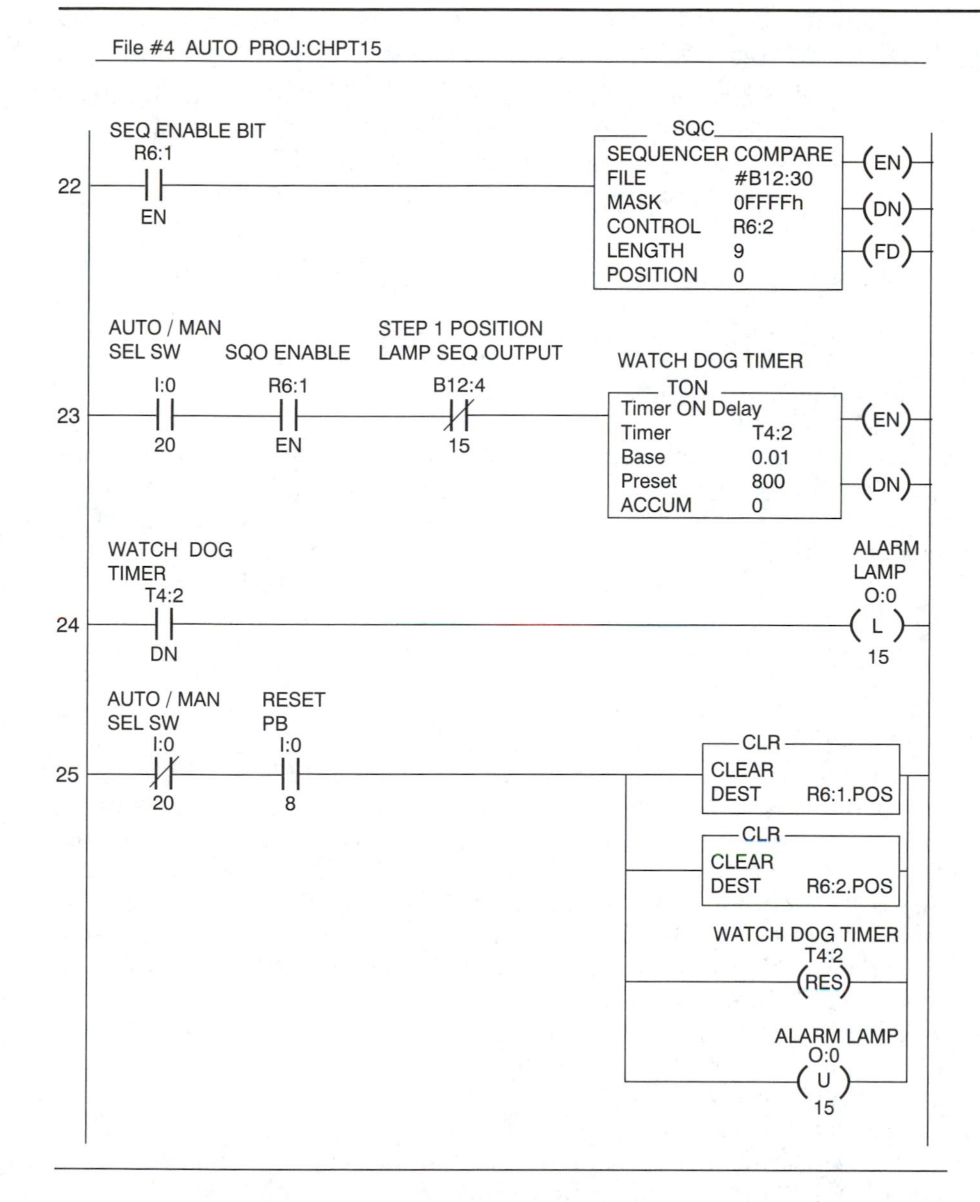

Figure 15–18 Sequencer program, *continued*

Rung 21 contains the SQO instruction. Note that it is active when we have selected auto mode, have no alarm bit set, and the SQC instruction's FD bit is not set. The bottom branch of the input rung contains the reset push button. This is necessary because the outputs of an SQO instruction are not updated when it is cleared to position 1 unless the instruction is enabled.

Rung 22 contains the SQC instruction. It is enabled when the SQO instruction finishes its step and turns on the EN bit.

Rung 23 contains a watch dog timer to kill the run mode if the mechanical system jams up. Note that we have set the time for about 8 seconds. If the sequencer sits on any step other than 1 for more than 8 seconds we will time out and shut down the SQO instruction.

Rung 24 latches the alarm bit if the watch dog timer ever times out.

Rung 25 resets the whole system if a jam-up occurs and the alarm bit was latched. Note that the two CLR instructions reset both sequencers to step 0.

At this point we should explain how the two sequencer instructions shake hands during the cycle. We start out with rung 21 being TRUE when the auto/manual switch is turned on. This causes the SQO instruction to step one address. When the new data is presented to the output word the DN bit is turned on. The DN bit then enables the SQC instruction in rung 22. The SQC instruction steps to the next address, and compares the contents of this new word to the input word. When the mechanisms complete their movements the two words match and the FD bit is set on. The FD bit in turn causes the input of rung 21 to change to FALSE and the SQO instruction clears its EN bit. This drops out the SQC instruction, which in turn clears the FD bit. The FD bit going FALSE restarts the SQO instruction, and the process continues stepping through the data files.

At this point you should run the program and observe how the various rungs interact. Stop cylinder 3 from moving on one cycle and ensure that the watch dog timer shuts down the sequence. Switch to manual mode, press the reset button, and return the cylinders to their home position. Switch back to run mode and verify that the cycle will restart. Try removing the branch rung with the reset push button from rung 21. Jam up the cylinder again and wait for the watch dog timer to time out. Switch to manual mode and try to reset the cylinders. What happens when you try to return the cylinders home?

Debugging a sequencer program can be a frustrating process. When debugging ladder logic you often look at the displayed ladder code to see what is missing for a true rung condition. When you're working on a sequencer instruction you need a way of viewing the file and the input or output word being acted upon. One way to do this is to right click on the sequencer instruction, and then click on the *Display Special* selection. This will bring up a pop-up box showing the sequencer's data file along with the input or output word for that instruction. You can do this for both instructions in our sample program and have both sequencers displayed as the program runs (see Figure 15–19).

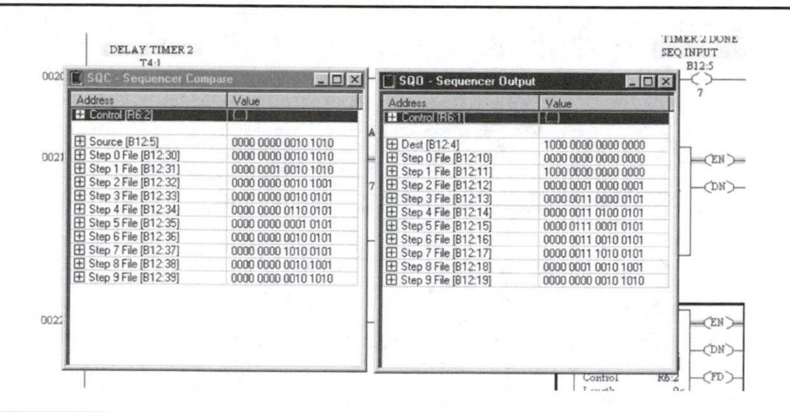

Figure 15–19
Sequencer data files displayed

Note that the source or destination word for the instruction is shown at the top of each box. When the program is running you would see the current step highlighted, and in the case of the SQC instruction, be able to look at the input word's contents to see if it matches the current file word's contents. If the sequencer cycle is stuck, you can then see which bit is holding up the process. You would then have to look at the actual ladder to see what real I/O is feeding that input bit.

SQL INSTRUCTION

Allen-Bradley's sequencer load instruction requires three words of memory in the control file #6 (see Figure 15–20). This instruction is only available in SLC-500/02 and higher processors.

EN The enable bit, which is on any time the rung conditions are TRUE.

DN The done bit, which is set on when the sequencer steps to the last position in the file. The next FALSE-to-TRUE transition of the rung clears this bit and resets the position counter to position 1. Note that position 0 is encountered only when the PLC is switched from program to run mode.

ER The error bit, which is set on if the processor sees a negative position value or no length value. This error condition would be encountered only if you were changing the sequencer's setup values as part of the ladder program.

Figure 15–20
Memory address layout for SQL instruction

15	14	13	12	11	10	9	8	0
EN		DN		ER				
SEQUENCER LENGTH								
STEP POSITION								

Instruction Elements

Figure 15–21 shows the instruction rung format for the SQL instruction. As in any instruction the input can have multiple elements.

File

The file address is the first word in the processor file where data will be stored by the sequencer instruction. It can be a bit or word type file, and you must use the (#) character in front of it to indicate that it is a file address. Using the preassigned B3 file can cause problems if the programmer is not careful when allocating new bit functions.

Source

The source address can be a word address or a file address. In both cases it is the input address for the SQL instruction. Each time the rung conditions go from FALSE to TRUE, the contents of this address are stored at the file word pointed to by the position counter. If the address is a file address, then the two files will be stepped in sequence as the load instruction runs.

Control

This is the position of the three status words allocated in the control file that are used to keep track of status bits and position counts for this instruction. The programmer must be careful to keep track of the control addresses used in his program. If you assign the same control address to more than one instruction, unpredictable operation will result.

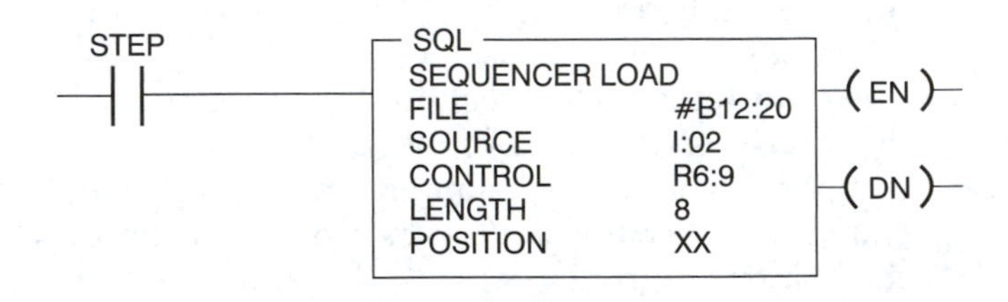

Figure 15–21
Sample rung format for SQL instruction

Length

The length value contains the number of steps assigned to the instruction beginning at step 1. All SQL instructions start at position 0 when the PLC is switched from program mode to run mode. Due to this extra state all sequencer files use the file length value +1 number of words.

Position

The actual position in our sequence file being loaded from the input word.

The SQL instruction can be used to load a sequence of data values into successive memory locations or to transfer a sequence of memory locations into a sequencer file. Each time the input rung goes from FALSE to TRUE the SQL instruction will copy the source word to the position in the file currently being pointed to by the position counter. In this way a programmer can load different prestored programs into a sequencer operation or record the actual input conditions that exist when an SOQ instruction runs.

CONCLUSIONS

Sequencer instructions can be a very powerful tool for controlling complex mechanisms if they are used properly. While SQO instructions are a good tool for manual operations they should never be used alone for continuously running cycles. For a continuously running operation you should always use an SQO-SQC combination. This will provide you with circuit feedback. Most equipment designs need position feedback to make the program safe to operate, eliminate most jam-up conditions, and allow the mechanisms to run at the fastest possible cycle times. While sequencer instructions can make the control design seem easier, you will find them somewhat hard to debug. When debugging ladder logic you often can look at the displayed ladder code and see what is missing for a true rung condition. When debugging an SQO-SQC indexer pair the process is not so simple. While you can look at each sequencer file in a window, you also need to look at the load and driver rungs for the operation. For this reason you may find debugging a sequencer program more difficult than debugging a regular ladder program.

REVIEW QUESTIONS

1. What is the purpose of the mask word in SQO instructions?
2. What are some of the advantages of using an SQO instruction for manual control?
3. How many words of the R6 file does each sequencer instruction use and what are they used for?
4. How do we enter the sequencer file data into the program?

5. What are the main problems with using an SQO instruction by itself to control the auto machine cycle?
6. Change the sample program to return the three cylinders in the reverse order.
7. Explain the function of the SQC instruction's FD bit when using an SQO-SQC combination.
8. List some reasons why you would want to store and then alternately load two different sequence programs for one mechanism.
9. Could you use a sequencer compare instruction to find the step that a mechanism was stopped in if we had an E-stop condition?
10. How do you display a sequencer file when you want to debug a running program?

LAB ASSIGNMENT

Objective

To gain an understanding of SQO instructions and how to implement them in machine control manual programs.

Create a sequencer program to perform the following sequence. Use an internal bit file as an output word. Use **PB 4** as the step input. Use **PB 5** as a reset function input and be sure it resets the sequencer to step 1. See the I/O list in Appendix A for address assignments.

Manual Program Sequence

1. Cycle the indexer one time. Be sure that PB 4 is disabled until the index is done.
2. Extend cylinder #1 and light lamp 1.
3. Extend cylinders #2 and #3 and turn on lamp 2.
4. Extend cylinder #4 and turn on lamp 3.
5. Extend cylinder #5 and turn on lamp 4.
6. Return cylinders #2 and #3 and turn off lamp 2.
7. Return cylinder #4 and turn off lamp 3.
8. Return cylinder #5 and turn off lamp 4.
9. Return cylinder #1 and turn off lamp 1.

Turn on lamp 8 when the sequencer is in the step 1 position.

When you have your program running correctly, print out a copy of the ladder program and demonstrate the program for the instructor.

CHAPTER

16

Diagnostic Programs

LEARNING OBJECTIVES

After completing this chapter the reader should understand:

1. Designing the proper level of diagnostic system for an application.
2. Methods for error monitoring typical mechanisms.
3. Methods for timing each station's cycle and generating warnings if cycle times increase.
4. Allocating error bit files.
5. Detecting a single error bit within a large error bit file.
6. Setting up the error message transfer protocol.
7. How to design meaningful HMI screens for error and warning messages.
8. Integrating the error detection program into the main control program.

All of the programming we have covered so far has been directed at controlling machines. We now come to the diagnostic programming section of this text, which covers how to monitor a control program and verify that the machine is really operating properly. The single biggest cause of downtime in manufacturing operations is the time spent troubleshooting machine problems. Most of these problems are a result of I/O failures and because most of our I/Os are inputs, these devices will be the most common failure points.

On average 70% of machine downtime is expended in figuring out what is wrong and only 30% is spent on actually fixing the problem. If we can reduce the 70% figure by half or better, we can justify the costs involved with writing a diagnostic program and adding the necessary display devices.

DIAGNOSTICS LEVELS

Our display and diagnostic requirements are divided into four main categories. Each step up in diagnostic capabilities significantly increases costs, but also greatly reduces troubleshooting time. The main purposes of a machine diagnostic system are to identify the machine malfunction, shut the machine down, and identify the actual defective part in a mechanism as accurately as possible.

An additional diagnostic function that can be added is an early warning of possible problems before they actually happen. This additional diagnostic function is only provided in the more advanced and therefore more costly systems.

Level 1

This is the lowest cost diagnostic system but also gives the least information to the maintenance personnel. In this level of diagnostic system, we are usually limited to a few simple lights that identify the malfunctioning part of the machine. Although this is better than no system at all, we can only give the maintenance person a general idea of where the problem is.

Level 1 is suited for use on small machines where the operation of the machine can be easily understood and most of the machine elements are in the open and visible. In its simplest form, we would assign an indicator light to each major machine function. If that function fails to complete its cycle, we stop the machine and latch on the proper light. The malfunctioning station may consist of several cylinders, valves, and limit switches. With our limited display ability, we can only point to the problem station and not to any particular mechanism contained in it.

Level 2

In level 2 we have the ability to display more descriptive messages as to where the problem mechanism is located. Many dumb display devices are available in the controls marketplace. Dumb displays are one- or two-line, 20- to 40-character, alpha-numeric display units. They can hold a number of messages, usually 256 or more, that are preprogrammed into memory by means of a plug-in keyboard. Most units have an 8-bit parallel input port that the programmer uses to select which message will be displayed. In this way, we can assign a specific message to each designated error bit that is generated by the ladder program.

We are unable to display any real-time data with this system, but we can be very descriptive as to where the problem area is. For example, we could display "STA 3 load cylinder failed to extend" or "STA 4 clamp failed to close." This type of detailed message can help a maintenance person find the problem area considerably faster than can a level 1 system. It is also more expensive than the level 1 system; most of these display devices cost about $400 to $800.

We also need a more detailed diagnostic program to feed the display device. In a level 1 program, we merely need to determine which station failed to complete its cycle. In level 2, we must be able to detect which mechanical element in a given station is at fault. We will be showing how to write this type of detailed error condition ladder programming later in this chapter.

Level 3

Level 3 allows us for the first time to display real-time data generated by the ladder program. In this level we incorporate a bidirectional display device also known as an HMI (Human Machine Interface). Many different manufacturers make these types of display interface devices, including Nematron, Screen Ware, TCP, and Allen-Bradley.

All of these display devices require us to learn an additional programming language, such as Opti-Basic or VB, which run on Nematron devices, or Panel View's Panel Builder software. These software packages allow us to communicate between the display device and the PLC controller. While these devices have considerable programming abilities they are not IBM PC-based machines and are about half the price of an IBM PC industrial computer system.

One important limitation of level 3 equipment is the lack of a long-term data storage device such as a hard or floppy disk drive. Due to this limitation, storing large amounts of data in level 3 systems is a problem. What these devices can provide are real-time windows into our ladder program.

In addition to the standard error display messages, we can show relative cycle times for each station on the line. One of the advantages of real-time data is that by monitoring each station's cycle time, we can warn maintenance personnel when a potential problem exists. Many mechanical failures are caused by the lack of lubrication or dirt buildup, which in turn can cause slowed movement of slides or cylinders. In most of these cases, cleaning or adding lubrication will fix the problem before real damage is done to the components. What the maintenance department needs is some way of knowing a problem exists before the mechanism completely fails. By displaying the normal cycle time of each station and the last cycle time, we can see if any station is slowing down. If any station slows below its limits we can turn the display red for that station's data and light a warning lamp on our stack light. In this way we can notify maintenance before physical damage is caused to our equipment.

Level 3 devices are able to send and receive data from the PLC and therefore they enable the machine operator to enter data through the keyboard or keypad attached to the display system. This numeric information can then be sent to data file addresses in the PLC program. This allows us to change program variables from the HMI display or select different display screens for additional information from our program. In the case of touch screens the display device can become our entire manual control panel. A system of this type is in the price range of $2000 to $5000 depending on HMI screen size, type of display (color or black and white), and keyboard/touch screen options.

Level 4

Level 4 systems usually contain an industrial IBM PC as the brains of the diagnostic system. In this level, we not only have all of the abilities of level 3 systems, but we also have mass data storage, a network communication hookup, the ability to export data into standard spreadsheet formats, and the unlimited graphic abilities of real computer systems.

Note that only an *industrial-hardened* PC should be used. The normal office desktop computer will not stand up to the environmental conditions found on the factory floor. They are not rated for high ambient temperatures or heavy dirt areas. They are also not designed to withstand the typical line voltage variations and spikes found in most factory floor power systems. Systems of this type start at $3,000 and go up depending on which options you install. While this may seem like a lot of money, the benefits gained with a real computer system are considerable. Having a hard disk on board allows us to accumulate product test and line downtime data for later report generation.

With the addition of a network card, we no longer have to retrieve data or read line status at the control panel. We can look at any line's test or performance information from our own PC through the network connection. This can be of great help if the plant floor is many miles from your office site. Collected data can be formatted and output in Excel spreadsheet formats, which will save considerable time in report generation. Using a true PC-based graphic display screen, we can store and display photographic-quality images to aid in the debugging process. All of these advantages are only available in the high-level diagnostic systems.

ERROR DETECTION PROGRAMMING

Writing programs to monitor machine operations requires a different approach than pure control programming. In control programming, we write a set of control rungs that cause the machine to move through a tightly controlled set of steps. In diagnostic programming, we must first determine all possible states in which a mechanism can exist. We then need to determine which of these states are error conditions and whether they will exist for a short time in normal operation as we move between good states. Once we have determined the error condition we must find a way to detect when these states occurred and be able to shut down the machine.

Let's look at how to generate a set of ladder rungs to detect mechanical error states. In this example, we use an air cylinder controlled by a double-acting valve and two limit switches (see Figure 16–1).

Figure 16–1 shows two prox switches that tell us if the air cylinder is at the returned or extended position. Note that neither switch will be made while the cylinder shaft is moving between positions. Table 16–1 assumes that we keep each solenoid energized until the cylinder reaches its proper position. It shows every possible state of the example mechanism and indicates if that state is an error immediately or only after some set time period.

Our programming task is to write a set of ladder rungs to detect all error conditions shown in Table 16–1. For this example we will use a normal cylinder movement time of 1.5 seconds. We will assign a separate error bit

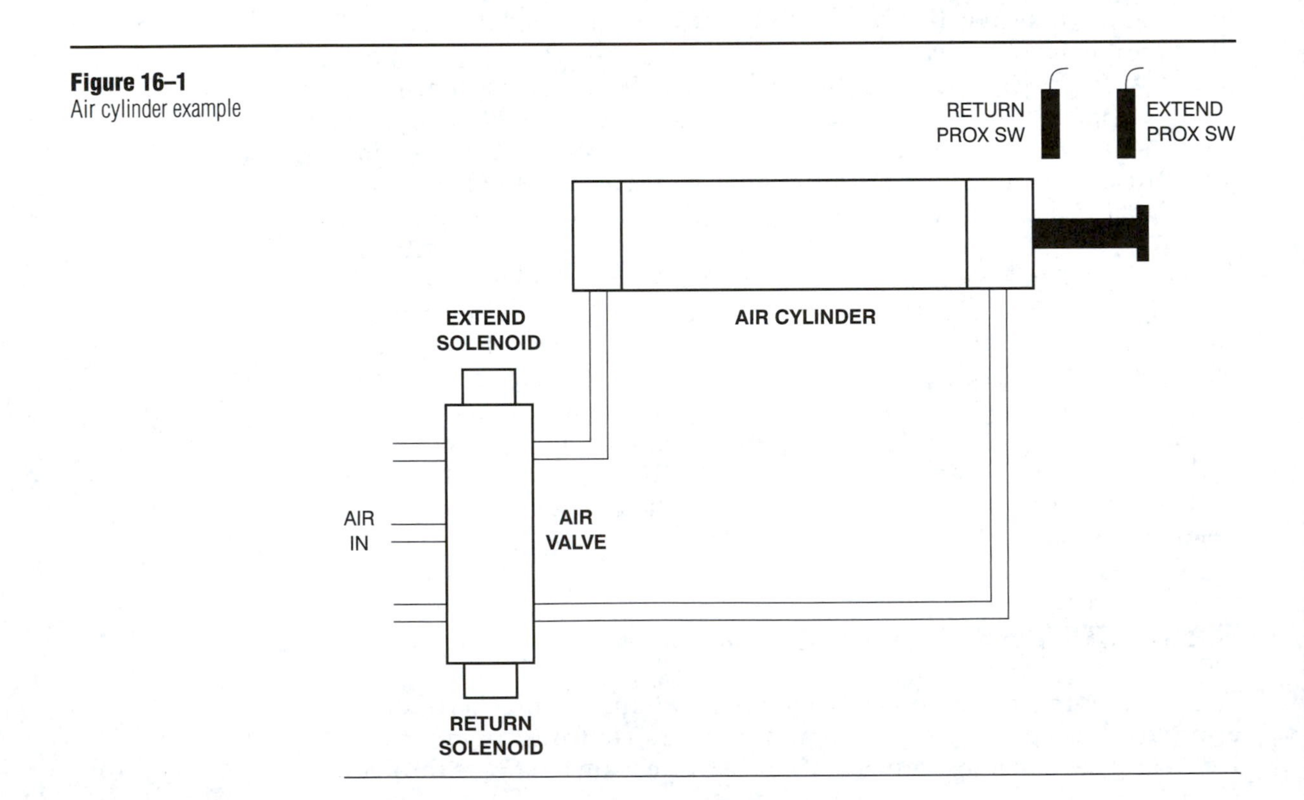

Figure 16–1
Air cylinder example

Table 16–1
State Chart for Typical Air Cylinder

RET SOL	EXT SOL	RET PROX	EXT PROX	Status	Condition
OFF	ON	ON	OFF	Error	Over Time
OFF	ON	OFF	OFF	Error	Over Time
OFF	ON	OFF	ON	OK	
ON	OFF	OFF	ON	Error	Over Time
ON	OFF	OFF	OFF	Error	Over Time
ON	OFF	ON	OFF	OK	
XX	XX	ON	ON	Error	Immediate

to each error condition. If we were using level 1 diagnostics, we would sum all error conditions to one bit since we are unable to display individual messages anyway.

Look at the top three states in Table 16–1. You can see the mechanism's I/O progression as the cylinder goes from its returned state to the extended position. Note that when the extend solenoid is first turned on, the two prox switches will be in the state shown in the first line of the chart. After a short time the mechanism will progress to the second line of the chart as the cylinder rod extends and both prox switches are off. If all goes well the mechanism should arrive at the final state shown in the third line of the chart about 1.5 seconds after the extend solenoid was turned on. The next three lines of the chart show the I/O progression as the cylinder is returned to its home position. The last line of the chart shows a possible state where both prox switches are on. This can happen if one switch is shorted or is loose and out of position.

We will also work under the conditions that all solenoids will be held on until the cylinder makes its complete move. Each solenoid needs a timer to time its normal completion time plus 50% to detect error conditions. As you can see from Table 16–1, all error conditions but one exist for a short time during the cylinder travel. Therefore, these error conditions must be timed because they are only errors if they exist longer than the normal air cylinder stroke completion time (see Figure 16–2). The last error condition on the chart, both prox switches on, should cause an immediate shutdown because at that point we have lost control of the mechanism and have no idea where the cylinder shaft is located.

Note in Figure 16–2 that all error bits are latched. This is necessary because the error condition may not be permanent. Air cylinders often slow down due to dirt buildup or they stick for a short time until the air pressure builds up and then they complete their stroke. If the error bits are not latched, the situation may arise in which the error bit is set on and the machine shuts down, but then no error message is displayed because the error condition cleared by itself. This is very frustrating for the machine operators because they have no idea why the equipment stopped.

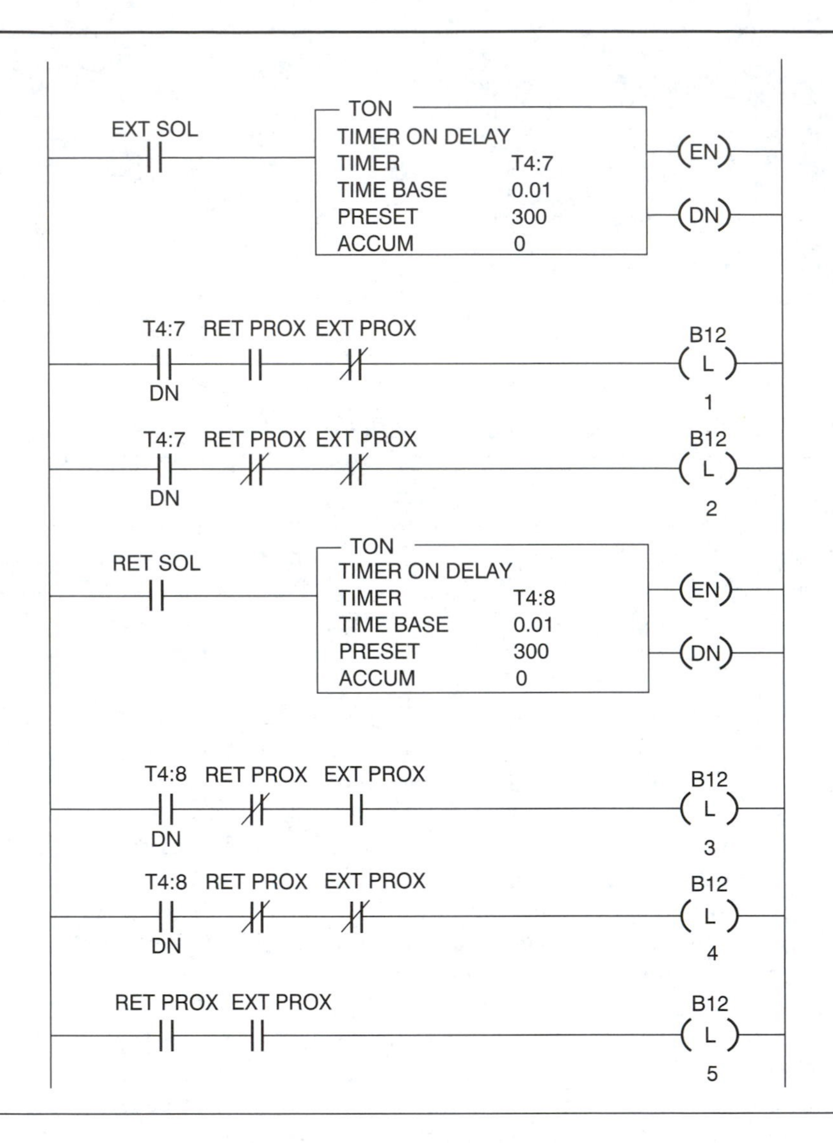

Figure 16–2
Program to detect all possible error conditions

Bits B12/1 and B12/2 at first glance look like they would have the same error message, but if we look closer we will find they detect different failures. In the case of B12/1 the cylinder is still in the returned position and therefore never moved. The most likely cause of this condition is a bad air valve or air supply line.

Error bit B12/2, on the other hand, tells us the cylinder did move away from the returned position but did not make it all the way to the extend position. This could mean a jam-up occurred or the extend prox switch is bad or poorly aligned. We will have five different error messages for the sample valve and air cylinder used to generate the error chart of Table 16–1.

Because we have to latch the error bits, it is important that they all be grouped into a defined section of memory. It is usually a good idea to use a separate bit file only for error bits. If you mix error bits with program latch bits, it will be extremely difficult to clear all of them when the problem is fixed. On large programs, it is not unusual to have 128 or even 256 error conditions. If they are all in one section of memory, we can clear 16-bit blocks of them quickly with the word CLR instruction. If not done in this way, you would have to use individual unlatch instructions on each error bit, which would require a large section of ladder just to clear error bits. This is a case where a little foresight and planning can go a long way toward making your programming easier.

An additional problem you will encounter in diagnostic programming is detecting if any one of our large number of error bits is set. Since we have to stop the machine and shut down the auto mode program if any one of the error bits is set, we must sum all of the error bits in some way. If all the error bits are grouped together, we can use the NEQ instruction to check if any 16-bit word in the error bit file has a value greater than 0000. If so, the master error bit can be set and the machine shut down. This is much easier than summing all error bits using branch ladder rungs. As mentioned earlier, a little foresight and planning makes a big difference in the size of your ladder program.

Another common mechanism for which we will need to write a diagnostic program is the machine's index motor and cam if one was used. If we look at the sample ladder in Figure 16–3, we will see two control problems. The top four rungs are the same as those introduced in Chapter 9 for controlling an index motor and cam assembly.

The first problem will occur if the index drive motor is off or jammed and does not turn at all. In this case the start timer T4:1 will time out, and because the cam is still in the off condition, the second timer T4:2 will generate an assembly start pulse just as if the index really did complete. If this mechanism places some object on the assembly line or into the product being assembled, it will try to place a second one on top of the one it just placed and cause a major jam-up.

The second problem condition occurs if the index starts and the line or table jams during the index. Since we keep the motor on until the cam switch opens, we will burn up the motor trying to move a jammed table. To detect these two error conditions we will have to add a few more rungs to our diagnostic program.

In Figure 16–3 we have programmed a set of rungs where B12/7 detects no movement of the cam switch when an index is called for. B12/8 detects the jam-up condition where the cam has started to move but did not complete its index cycle. B12/9 senses if the clutch overload plate has released due to an overload condition. Timer T4:1 is the index start timer and is controlled by the FALSE-to-TRUE transition of the master timer bit T4:0/TT. Each time the master timer TT bit goes high, the unlatch function

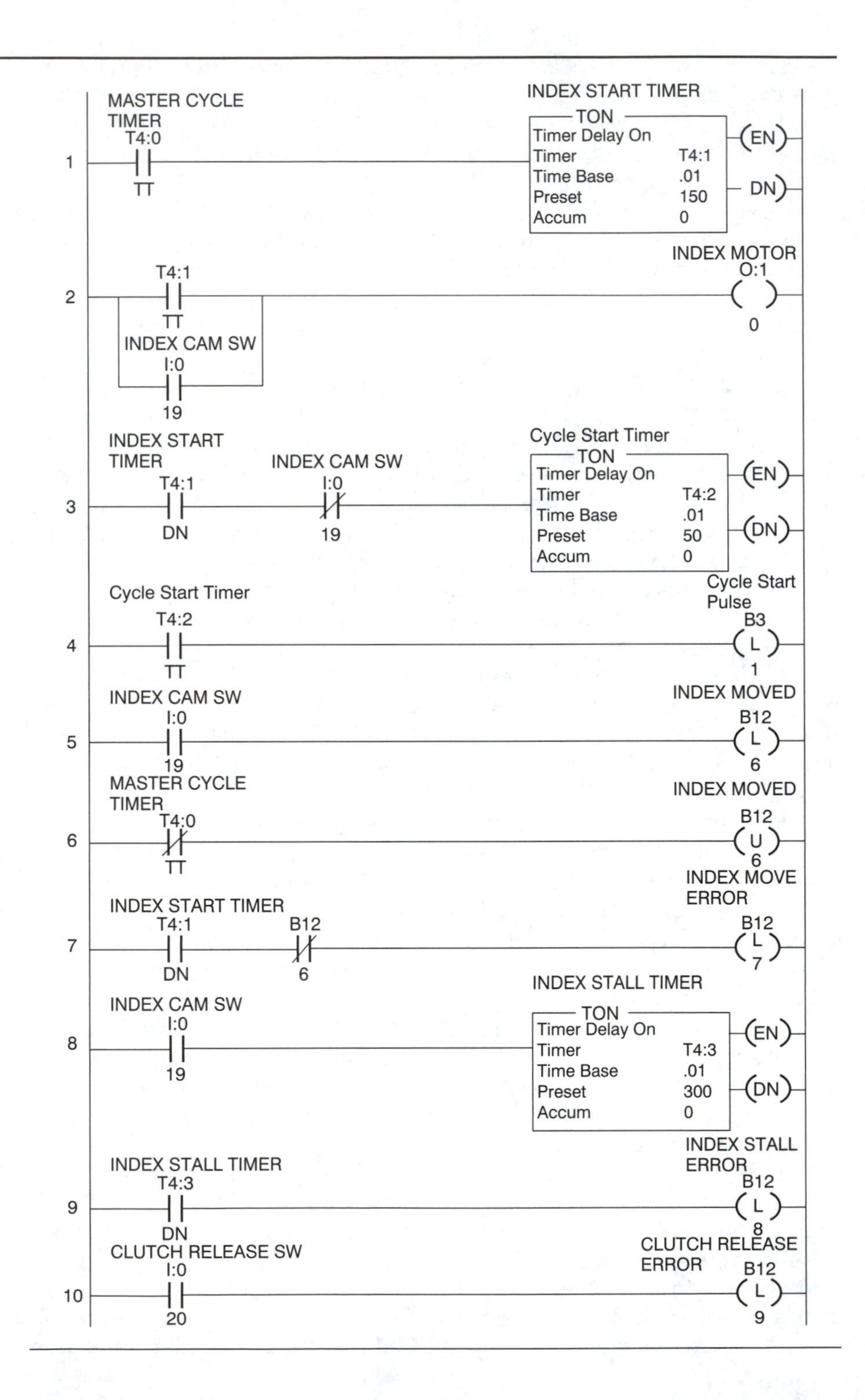

Figure 16–3
Control and diagnostic rungs for an indexer operation

is released on B12/6 (rung 6). If the index cam rotates normally, B12/6 will be latched on by the cam switch before the T4:1/DN bit is set on and B12/7 will never be latched. If the cam does not move B12/7 will be latched because rung 7 will be TRUE when T4:1 times out.

T4:3 looks for any condition that would cause the cam SW to be made for longer than the index time. This will detect jam-ups as well as a broken or poorly aligned prox switch. Rungs 8 and 9 of the sample program time the index cam switch input and latch the error bit if T4:3 times out. Rung 10 looks for the overload clutch release switch to make.

The next task in our diagnostic program is to generate an error message based on the error bit that was latched. How we accomplish this will depend on the type of diagnostic system we are using. For level 1 systems we need merely to group all error bits for each station together and turn on a single output to a lamp. For level 2 systems we need to generate a binary value from 0 to 255 and output this number to the I/O port connected to the parallel input port on the display device.

For level 3 and 4 systems the problem is one of cross program data identification, because in these systems we will be passing the whole error bit file as well as report data over to the HMI device. These HMI display devices will convert the transferred PLC file into individual bit and word variables. It is important that you create a cross-reference map that indicates how each PLC bit address or word address was assigned to a variable name (Tag Name) in the HMI display program.

In most of these higher level diagnostic display systems, we start with a control word as the first word in the file. This control word tells the display device what state the machine is in and what parts of the transferred file should be displayed. Figure 16–4 is a sample control word layout. This file is transferred at least once every second or faster between the PLC memory and the HMI display system. After each transfer, the HMI display system looks at the control word to determine the state of the machine. We will have previously designed and saved screens for each machine state. Each screen will contain static data that are displayed each time the screen is called and real-time data that will be pulled from the transferred data file. It is important that we know exactly how the data transfer file is laid out. A sample file is shown in Figure 16–5.

If the display program detects that the machine is in auto mode, the run screen is called up and the data to be filled in are read from the transfer file

05	04	03	02	01	00
DIAG ERROR	WARNING	AUTO MODE	MANUAL MODE	"E" STOP	CYCLE STOP

Figure 16–4
Sample control word layout

Figure 16–5
Sample PLC to HMI transfer data block

CONTROL WORD
ERROR BITS 1-16
ERROR BITS 17-32
ERROR BITS 33-48
ERROR BITS 49-64
WARNING BITS 1-16
WARNING BITS 17-32
STATIONS IN HOME POSITION BITS
STATION 1 LAST CYCLE TIME
STATION 2 LAST CYCLE TIME
STATION 3 LAST CYCLE TIME
STATION 4 LAST CYCLE TIME
STATION 5 LAST CYCLE TIME
MACHINE LAST CYCLE TIME
GOOD PARTS COUNT
BAD PARTS COUNT

and placed in the screen areas provided. A sample of the run mode screen is shown in Figure 16–6. Notice that the stations' normal cycle times are part of the display file and do not change. Only the last cycle time data are retrieved from the transfer file each time it is copied to the HMI display device.

If you look at STA 4's last cycle time in Figure 16–6 you will note that it is shown with a red background. (It will be in red on-screen.) This is due to the fact that it is larger than the normal time for that station. This indicates that a problem could be occurring in that station's mechanisms and should be checked by maintenance as soon as possible. Also note that the good/reject parts counts are pulled from the transfer data file each time it is sent. The total part run count and the efficiency ratio are calculated in the HMI display program from the other two numbers.

If the display program detects a cycle stop status bit, the display will change to a screen that is almost the same as that shown in Figure 16–6 but with the heading of "MACHINE STOPPED MODE" instead of "AUTO RUN MODE." The time and date of the change will also be displayed. If the operator then changes to manual mode, we again would change the screen to show the manual mode screen (Figure 16–7), which shows the

ASSEMBLY LINE 235
AUTO RUN MODE

STATION CYCLE TIMES

	NORMAL	LAST
STA 1	1.7	1.6
STA 2	3.4	3.1
STA 3	4.3	4.2
STA 4	3.7	4.6
STA 5	2.6	2.4
STA 6	5.5	5.5

SHIFT STATUS

TOTAL PARTS	12,485
GOOD PARTS	12,237
REJECT PARTS	248
EFFICIENCY	98.1%

Figure 16–6
Sample run mode screen

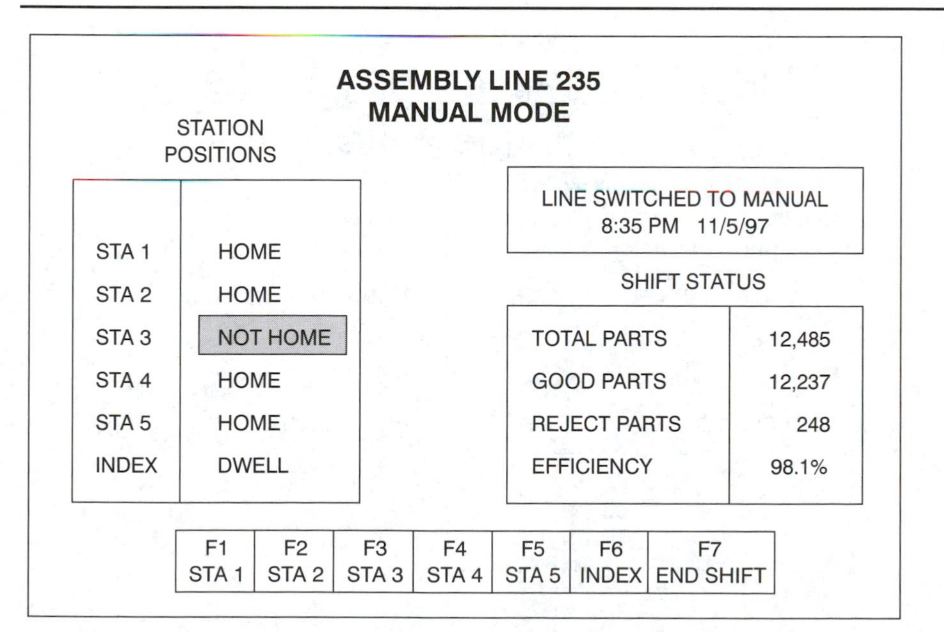

Figure 16–7
Manual mode

position status of each major station on the line. This allows the maintenance person to look at the manual screen and verify that all the stations are home. Real-time data for this screen originates in the station home status bits in the transfer file. Also note the [F] key descriptions shown at the bottom of Figure 16–7. It is often helpful in manual mode to be able to see the current status of each I/O address for a particular mechanical section. If the F1 key is pressed the screen showing all the I/O for station #1 is displayed (Figure 16–8).

From this sample you can see that the HMI display device is really a slave of the ladder program. It monitors the ladder transfer file and displays information based on the control word. Most high-level diagnostic programs also have some graphics abilities. These graphic files may be used in the display of diagnostic error problems. When an error is detected, not only do we display an error message but we can show a graphic picture of the machine section that is jammed. These graphic display files can be generated with a digital camera or by scanning actual photographic prints. Even in type 3 systems we can show a line drawing of the machine area that has problems. These graphic displays will help a maintenance person identify the problem area quickly and save on downtime. A sample layout of a level 3 diagnostic screen is shown in Figure 16–9.

Reviewing our discussion thus far, we can see that the decision as to which kind of HMI display system to use is not always an easy one.

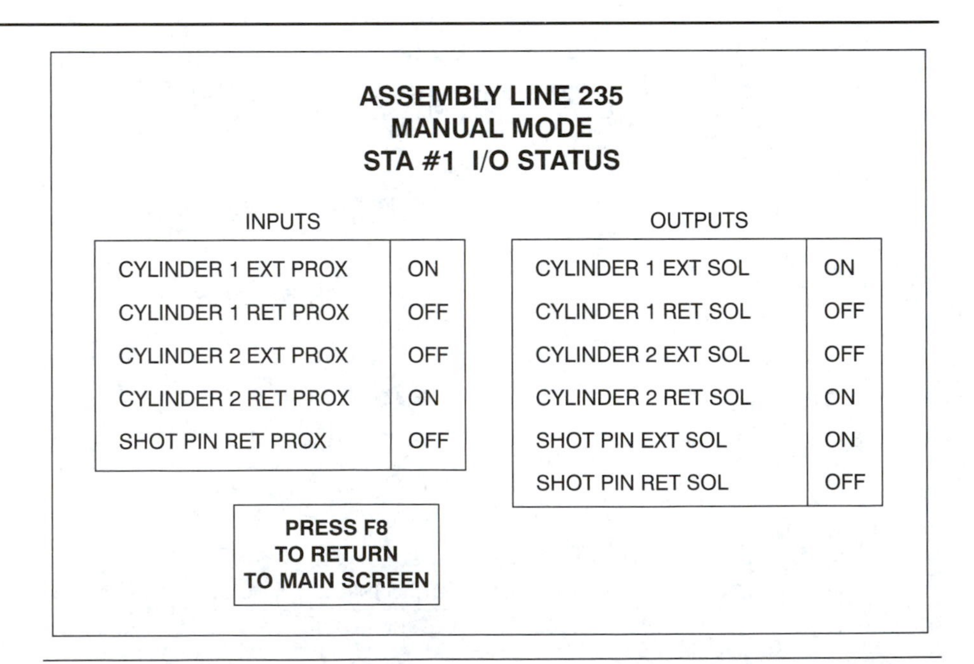

Figure 16–8
Manual mode showing I/O status of station #1

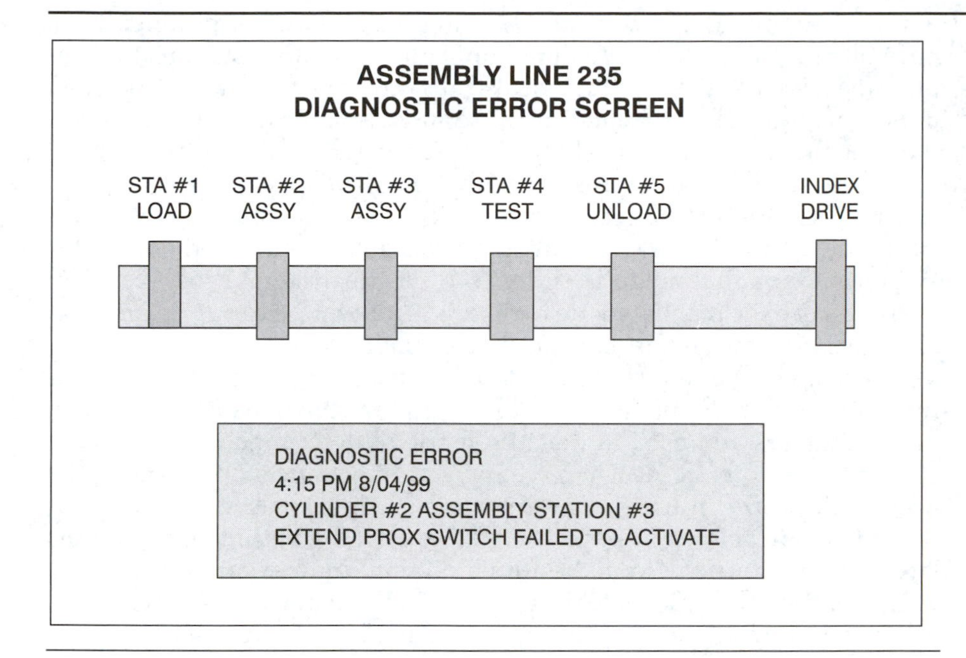

Figure 16–9
Sample layout of level 3 diagnostic screen

The main factors will be cost, machine size, and the level of maintenance expertise at the plant where the machine will be located.

To justify the costs of a diagnostic system, we must first gather information as to the normal amount of downtime on comparable equipment in similar situations. From this we can determine what the cost per minute is of downtime on this type of machine. If we can cut 30% to 40% from the downtime by having good diagnostics and an early warning of potential problems, we can now look at this as a cost-savings project. A project whose total cost is paid back in a year or less is usually quickly approved. Payback in 2 years or more is a harder sell.

Getting back to the ladder program, we need to be sure that we have maintained the three-state control system. We explained how to use the NEQ instruction to check each diagnostic error word to see if any bits were set. We need to latch an output bit on if any of these NEQ instruction rungs finds an error bit set. This will be our alarm error bit and it must be inserted in the unlatch rung for auto mode.

In the sample program shown in Chapter 13, this bit should be added in file 2, rung 6, as an additional branch. This will drop out the auto mode subroutine and kill the machine. You may want to delay this shutdown for a few seconds to give other parts of the machine time to finish their parts of the cycle. If you kill the auto mode immediately when an error bit is found, all machine mechanisms will halt and this may leave each station with a partially assembled unit. Delaying the shutdown for at least the normal machine cycle time will not only allow all other stations to complete their assembly cycles but will also allow other error bits to be latched if there are multiple error conditions. Be sure also to insert this error bit in the rung that controls the flashing alarm lamp and in the control word in the diagnostic file that is transmitted to the CRT display device.

We also must provide several word CLR instruction rungs in the manual mode subroutine to clear out any error latch bits when the operator switches to manual mode in order to reset the system. Most program designers use the manual mode to do this error word cleanup rather than have a separate push button to clear error messages.

Last of all, you must unlatch the alarm error bit to allow the machine to be started in auto mode again. You must carefully look at all the latch bits and timers you used in the auto mode to determine if any of these elements has to be reset when we have a diagnostic stop. Because the diagnostic stop can happen at any time during a cycle, we may exit the auto mode before our sequence is complete. If you don't clean up the latch bits and timers used in the auto mode subroutine, you could have a problem when you reactivate the auto mode. When reentering auto mode, if the auto mode program tries to start someplace in the assembly cycle other than at the beginning, it is a good bet that you have left some latch bits set from the previous cycle.

In conclusion, the designer must carefully check each error condition and verify that the correct message appears on the display device. Although this is a time-consuming project, it is the only way to verify that all the diagnostic rungs work. During the break-in of the machine you will almost certainly find a few jam-up conditions that you did not foresee. These new conditions will have to be added to your ladder rungs and message file.

During this break-in period you must also check your error timers to verify that you do not have nuisance error conditions occurring due to delay times being set too close to the normal cycle times. Although maintenance department personnel will benefit greatly from a good machine diagnostic system, they can become quickly disenchanted if they have to answer repeated calls due to constant false error messages generated by bad timing adjustments.

SAMPLE PROGRAM

We now modify the program we wrote in Chapter 13 to include full diagnostics even though we cannot yet display the error messages. In the next chapter, we look at how to implement an Allen-Bradley Panel View terminal for use as an operator control center and diagnostic display system. We then again modify this same program in the next chapter to interface with this new display terminal. For now we will only have one light over each station on our assembly line to show that a diagnostics error condition has occurred; in short, a level 1 diagnostic system.

Main Program File 2

The main program ladder file is shown in Figures 16–10 through 16–16. The first change to the program is in the main section, rung 6. Timer output T4:10/DN has been added to shut down the auto run mode and auto subroutine. This timer is driven by alarm bit O:0.0/3 and gives a 5-second

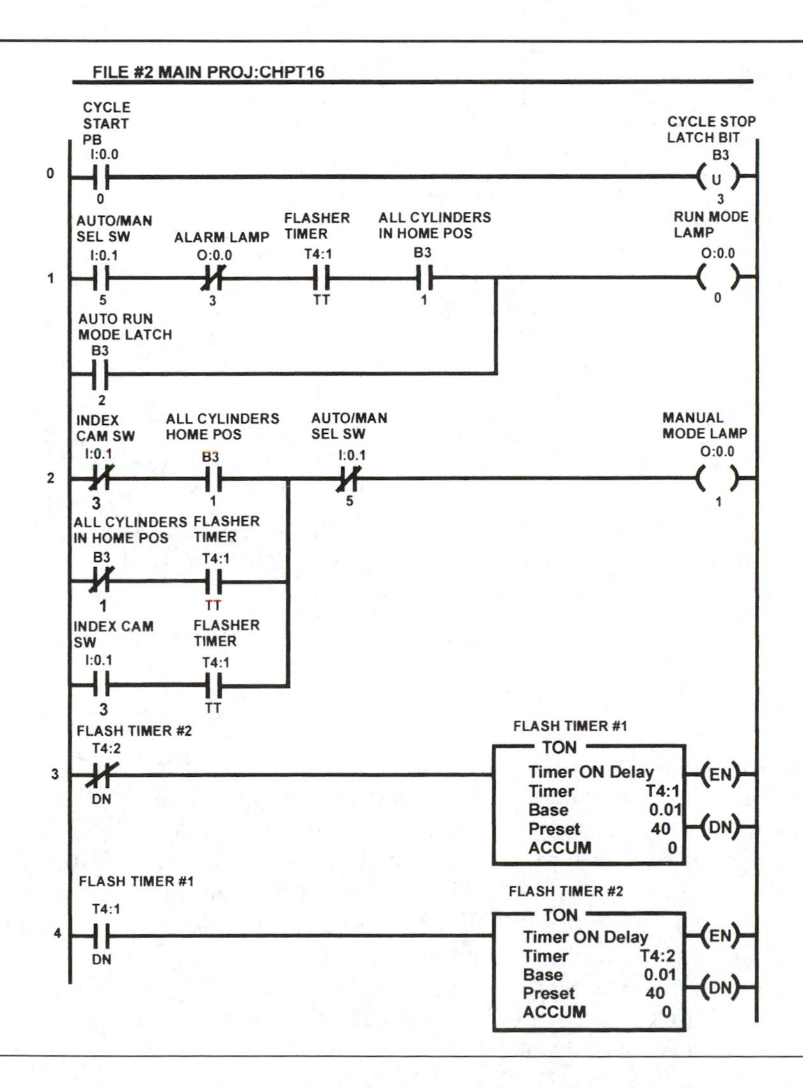

Figure 16–10
Main program ladder file 2

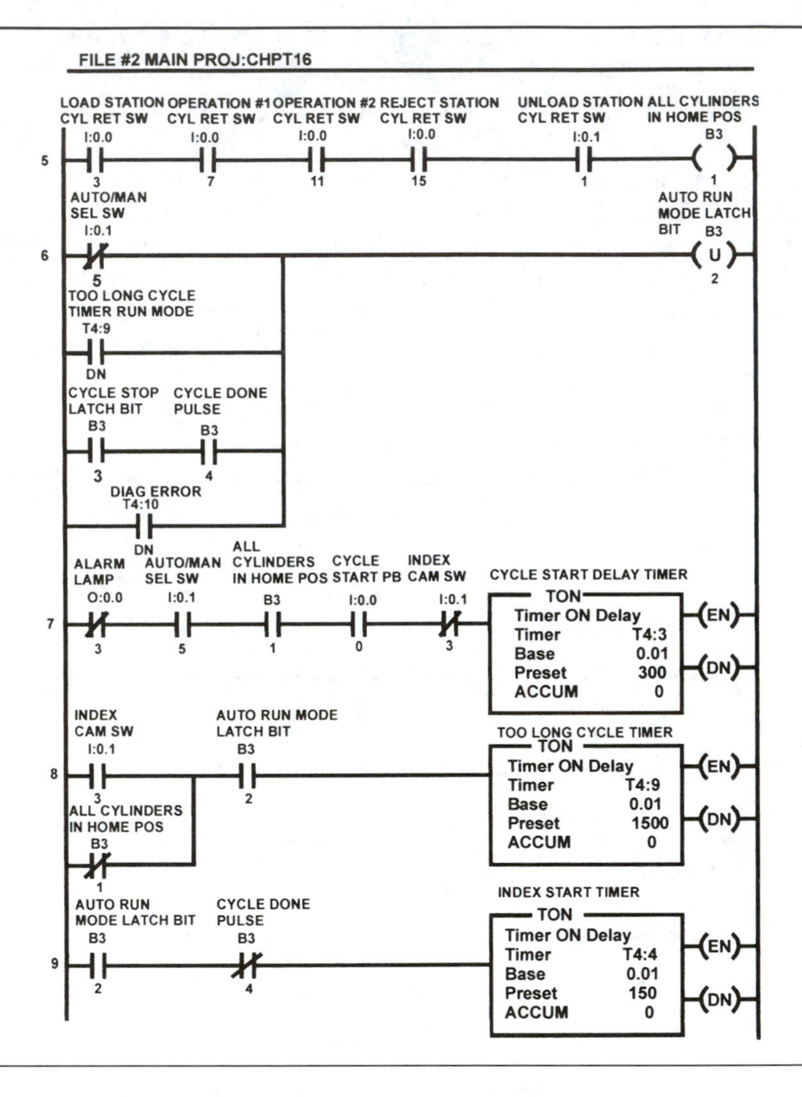

Figure 16–11
Main program ladder file 2, *continued*

delay from the time an error is found until the auto mode shuts down. This delay gives the other stations time to complete their cycles.

Notice that the "too long" cycle timer has been retained just in case we have an error condition that our diagnostic program does not find. This is our fail-safe timer and should always be included in any program. Note also that the alarm error bit O:0.0/3 is used in the indexer control rung 11 to stop the indexer motor as soon as an error is found. There is no reason to delay if an index error occurs because there are no operations going on during the index and any delay in shutting down the indexer drive motor may cause additional damage.

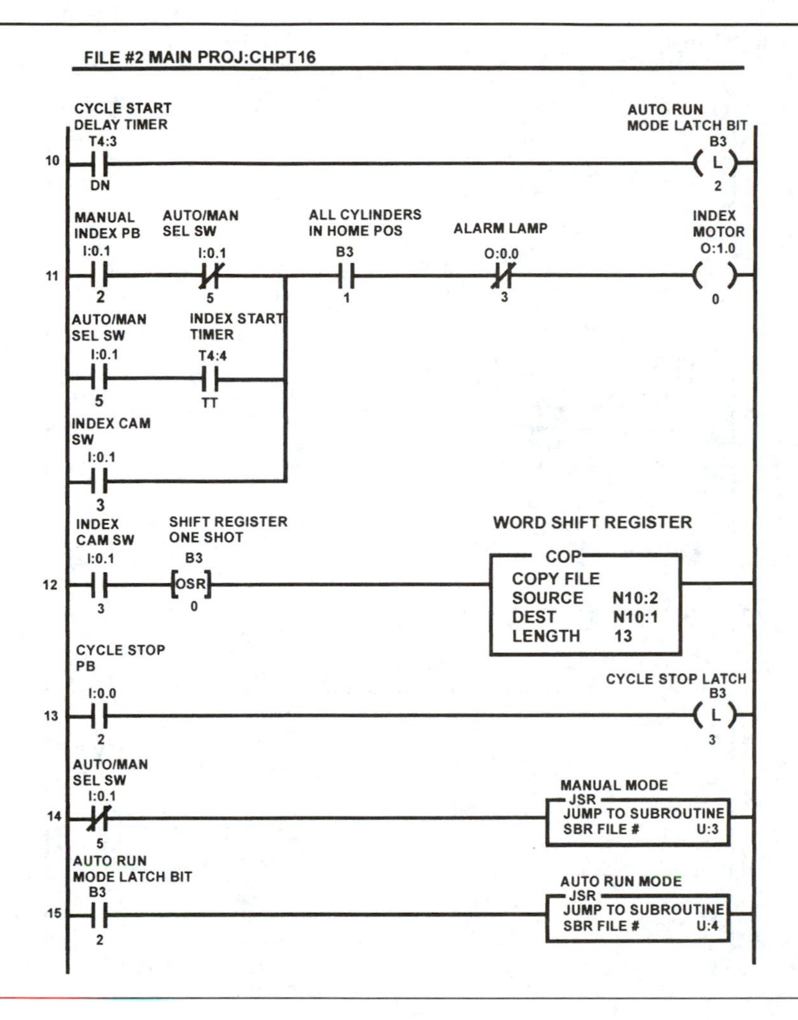

Figure 16–12 Main program ladder file 2, *continued*

The B12 bit file is used for all error latch bits. In this program we need only use the first two words of the file. We need to set the alarm bit on if any of the error bits are set. The easiest way to do this is to use the NEQ instruction. If you look at rung 17, you will see that we have added two NEQ instructions to the alarm rung. Each instruction checks one word of the error bit file and latches the alarm bit on if the word's value is not zero.

Rung 19 was added to drive the diagnostic delay timer if the alarm bit is set on. Rung 20 scans the diagnostic program if the auto run mode is active. We do not want to scan the diagnostic program if the program is in the manual mode or not running. Rungs 21 through 26 flash one of the error lamps depending on what error bit was found. Notice that multiple error bits can be set and more than one lamp will flash.

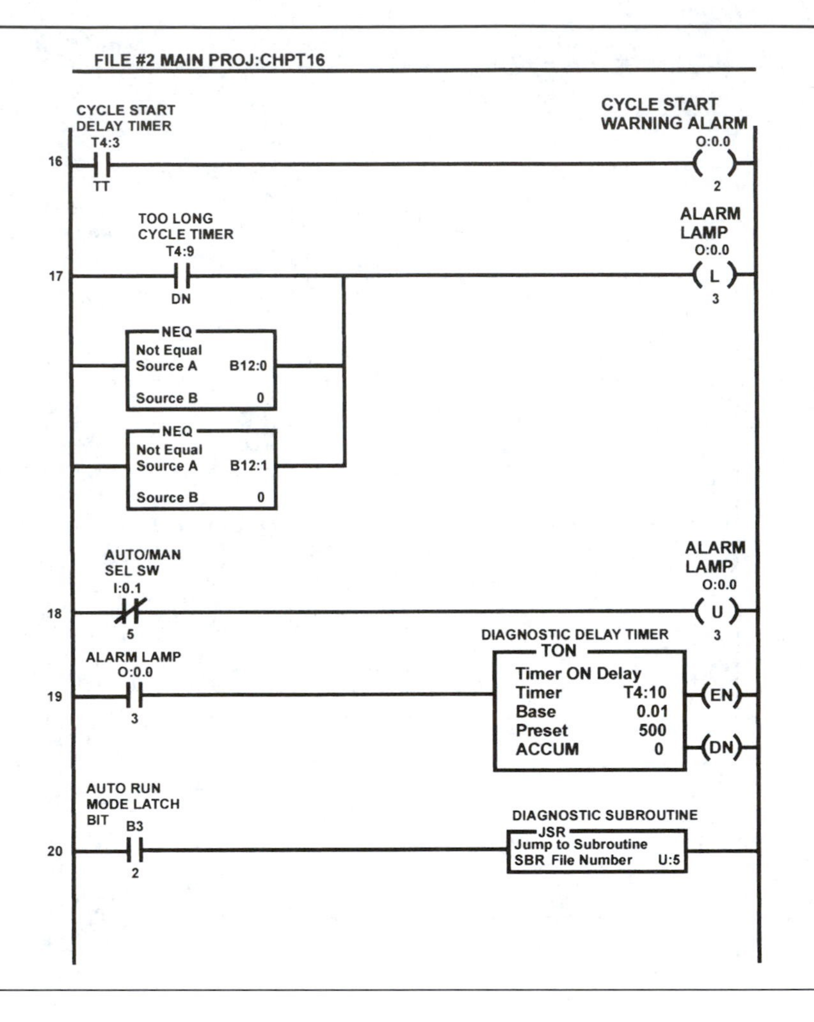

Figure 16–13
Main program ladder file 2, *continued*

Manual Mode File 3

In the manual mode file (Figures 16–17, 16–18, and 16–19), we must be sure to clean up any latch bits left on if we shut down from an error condition. The manual mode should also be used to clear all error bits and timers. If it is necessary to keep the error bits latched in manual mode, you can have a separate push button to clear all error bits and messages. Because in most cases you switch to the manual mode to clear error problems, it is convenient to clear all error bits automatically when the operator switches to manual mode.

The first change in the manual mode program is on rung 9. The reset to timers 21 and 22 was added. Rung 11 was added to clear both words of the error bit file and one word of the B3 latch bits is used to detect the return cylinder errors. Rungs 12 and 13 clear all the rest of the error timers.

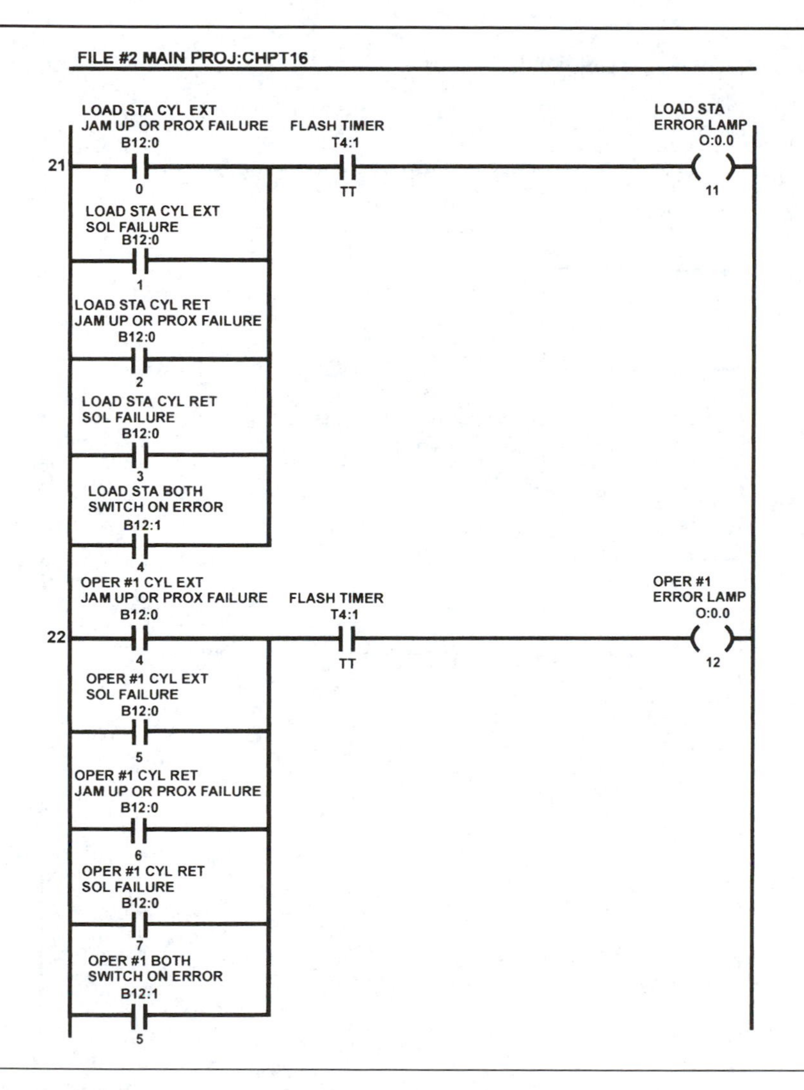

Figure 16–14
Main program ladder file 2, *continued*

Auto Mode File 4

No changes were made to this file, but the file is shown in Figures 16–20 through 16–23.

Diagnostics File 5

The rungs in the diagnostic file (Figures 16–24 through 16–29) are designed to time each air cylinder and the index cam and latch an error bit if the mechanism does not complete its operation within the allotted time window. As stated earlier, it is important always to latch the error bits to prevent intermittent contacts or slow-moving cylinders from causing

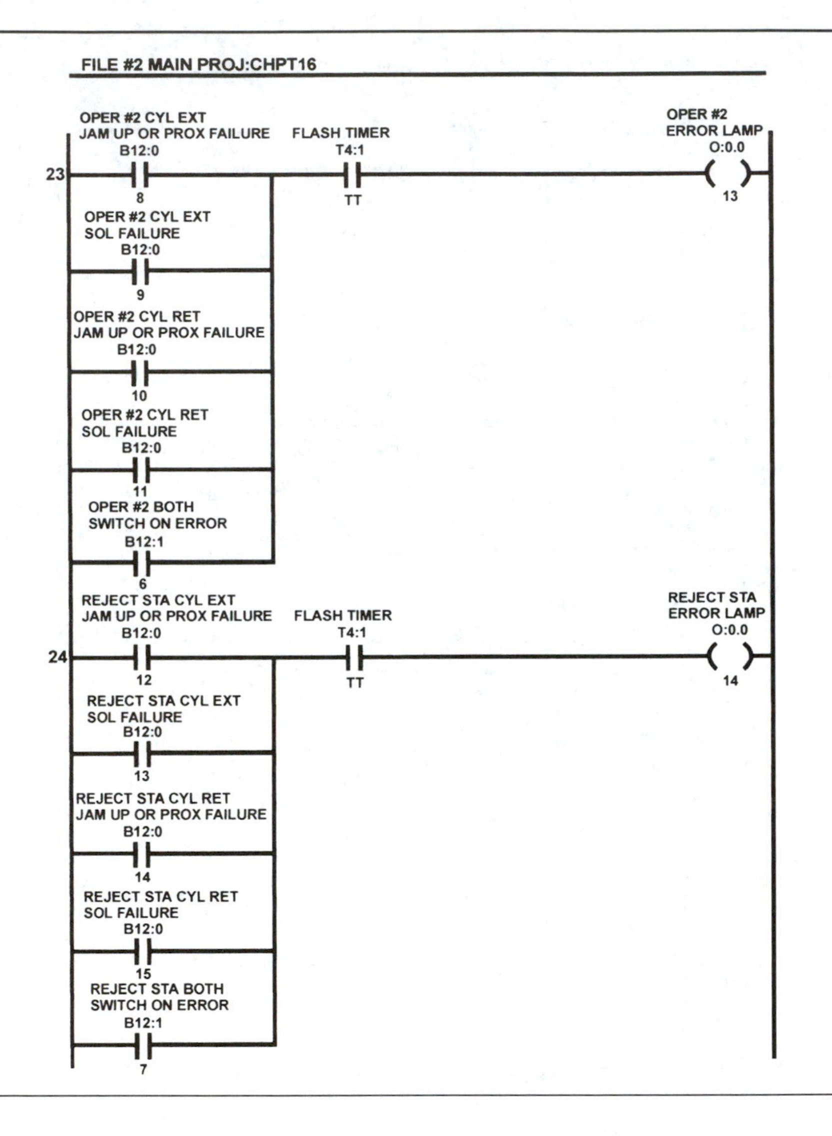

Figure 16–15
Main program ladder file 2, *continued*

a shutdown and then having no error message displayed or lamp lit because the error condition cleared on its own.

If we look at the auto mode rungs, we will see that most of the extend air solenoids are latched on and only turn off if the extend switch is made. The return solenoids are not controlled in the same way. They are only held on for a short time while the extend limit switch is made. As soon as the extend limit switch opens, they are turned off. This presents a problem for our diagnostic program in that we can time the extend solenoids but have a problem with the return solenoids.

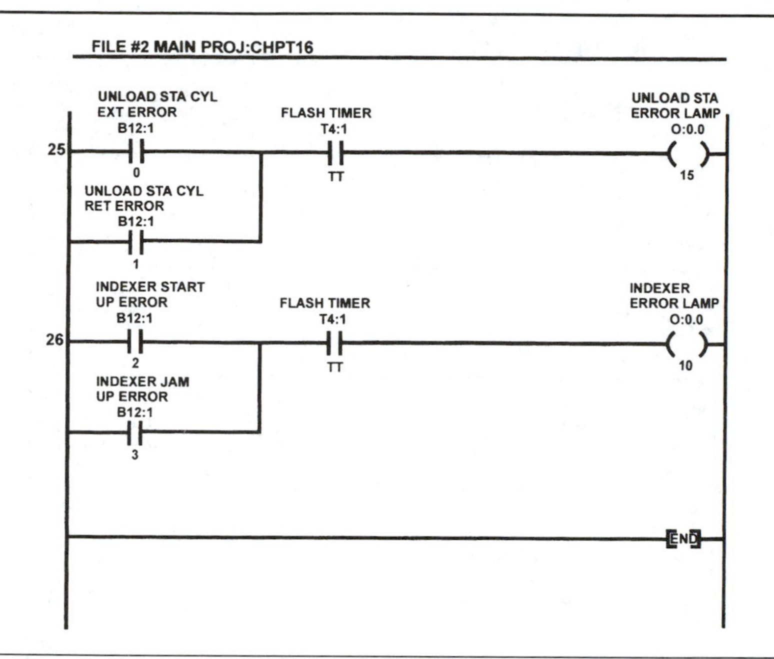

Figure 16–16
Main program ladder file 2, *continued*

Starting with rung 0, the extend solenoid is used to drive the cylinder extend error timer. The time setting is based on the average cylinder move time on the teaching kit of about 1.5 seconds. The two rungs that follow the timer instruction latch one of two error bits if the timer times out.

The first rung detects the fact that the cylinder did move from the home position but never made it to the extend limit switch. This indicates that the air supply and valve were probably all right, but something jammed the cylinder in midstroke or the extend limit switch has failed or moved out of position.

The second rung checks for the condition when the air cylinder does not move at all. This can indicate that the air supply is low, the air lines are damaged, the air valve is defective, or our electrical connection to it is bad. These are quite different error conditions and it will be a great benefit to the maintenance person if we can tell her what the actual error condition is.

As stated earlier, the return solenoid is not held on for the total move time. To get around this problem, we latch an internal B3 bit when the return solenoid turns on and unlatch it when the return limits switch is made. We can now use this bit in rung 5 to drive the return error timer. Following it are the same two rungs used after the first timer but looking for errors in the return direction. The last error bit for this mechanism is located in rung 39 and this looks for both limit switches to be on at the same time. No time delay is required for this error detection bit.

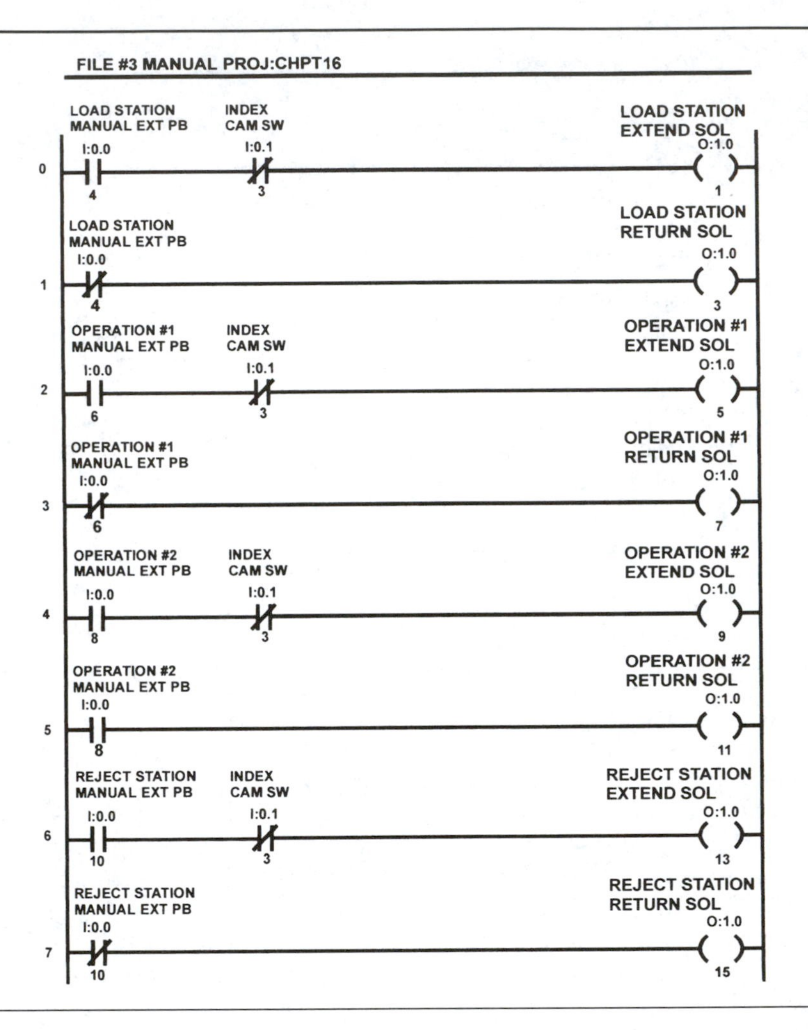

Figure 16–17
Manual mode ladder file 3

The next three mechanisms are written in the exact same manner but with their different addresses and timer numbers. We then come to the unload station. This has a different set of programming problems because it only has a single acting air valve, one solenoid spring return, and only one limit switch for the home position. Due to these limitations in I/O sensors, we can only indicate the cylinder moved off home position and returned to it.

In rung 32 the extend solenoid is used to drive the extend error timer. The next rung looks to see if the cylinder has moved off the home limit switch. If the timer times out and the home switch is still made, the error bit is latched. We then use the fact that the extend solenoid is turned off

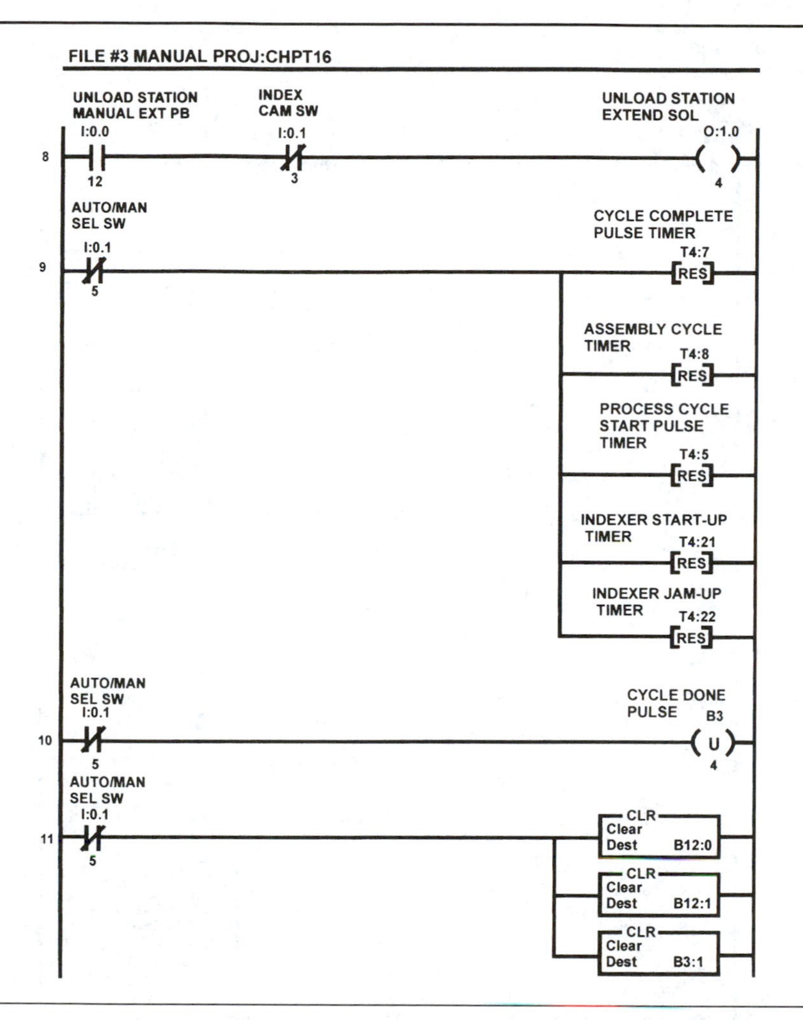

Figure 16–18
Manual mode ladder file 3, *continued*

to drive the return error timer. When this timer times out and the home switch is not made the other error bit is latched.

The next three rungs, 36, 37, and 38, detect error conditions in the indexer mechanism. In the first rung, the first timer is used to check that the cam switch is made within a reasonable time. If we look at main program file 2, rung 9, you will see that the index start timer is set for 1.5 seconds. The kit's indexer normally takes about 0.8 second to come up on the cam from the dwell position. It is important to make our timer shorter than this start time if we intend to stop the next assembly start pulse.

Note that in auto mode file 4, rung 3, the alarm lamp is used to block the next assembly cycle if an error is found. Because one of the other elements in this rung is the index start timer done bit, we must set the

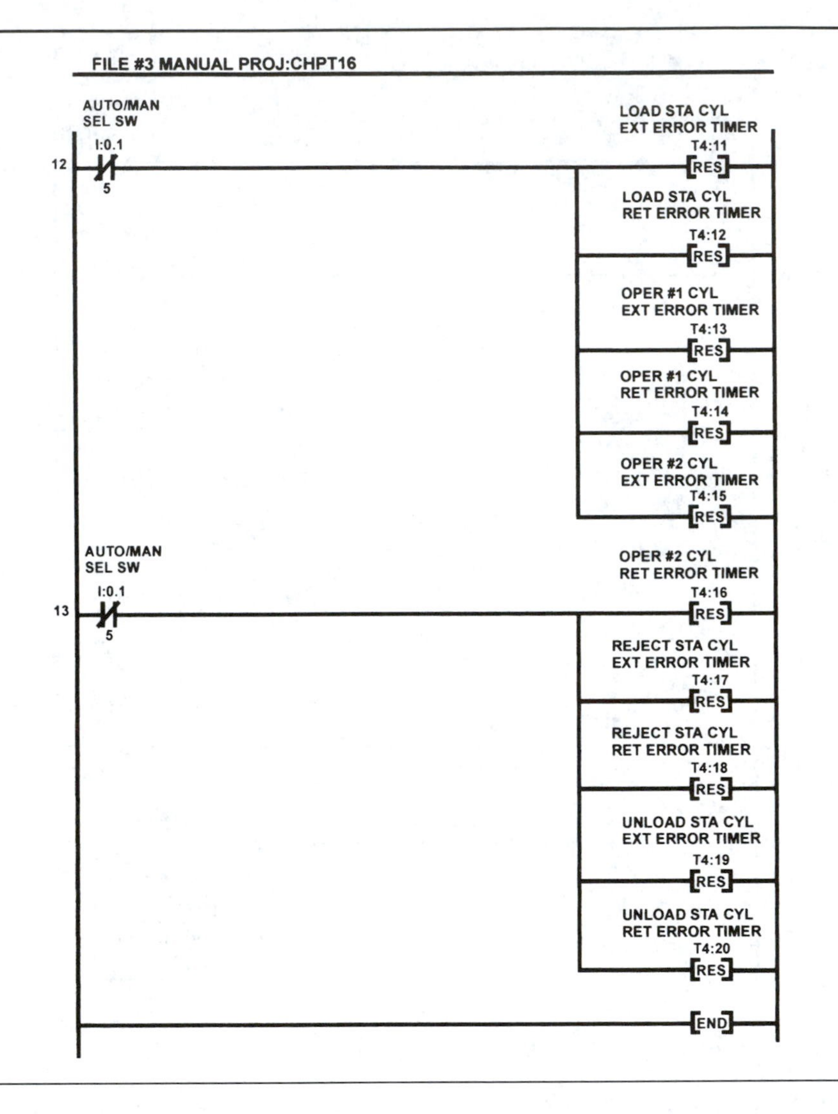

Figure 16–19
Manual mode ladder file 3, *continued*

alarm before this timer times out or another assembly cycle will start even though the index did not happen.

The second timer is used to detect if the indexer cam switch is made but the indexer is jammed or the torque clutch dropped out. On our kit we do not have a separate switch to indicate the clutch dropped out; normally you would have one on a real indexer assembly.

The last rungs, 39 through 42, detect if both limit switches are made on any of the mechanisms that have both extend and return limit switches.

This concludes the diagnostic program explanation.

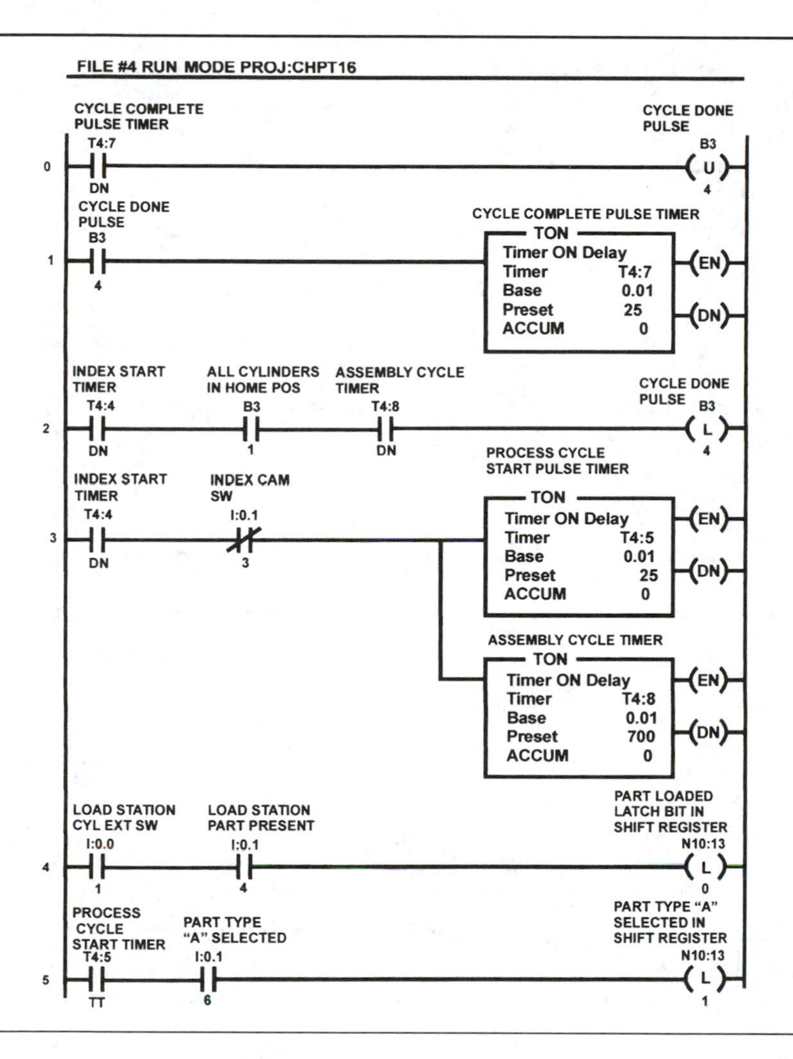

Figure 16–20
Run mode ladder file 4

CONCLUSIONS

In conclusion, a well-thought-out diagnostic system can enhance any electrical control system by providing accurate real-time feedback of the main system parameters. The focus of any diagnostic system is to provide a user-friendly man/machine interface for presenting system information. The program should provide warnings of potential problems, be able to pinpoint actual failures, and if possible record events for later analysis.

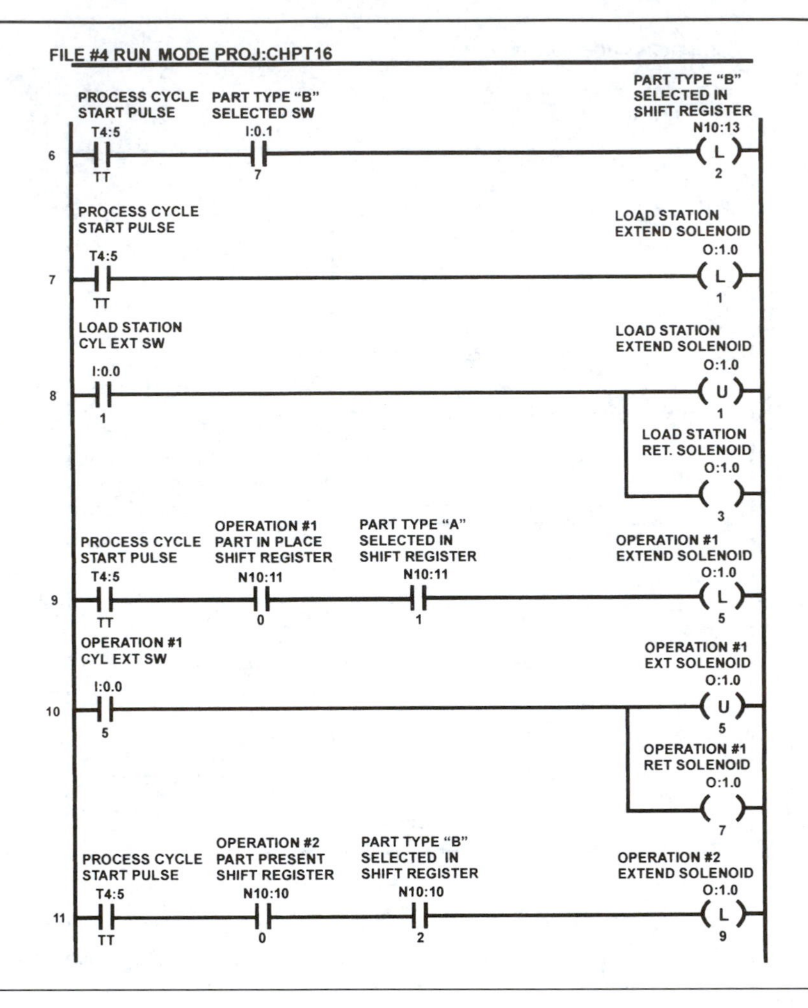

Figure 16–21
Run mode ladder file 4, *continued*

When a new design project is begun, the diagnostic system should be discussed in the first meetings with the mechanical groups. Most experienced PLC programmers have written a fair number of these diagnostic programs for machines after management became disenchanted with the equipment because it was difficult to maintain. This often happens because the original designers did not feel a diagnostic program was necessary. As a control systems designer you have an in-depth understanding of the equipment and, when something fails, will know right where to look for the problem. Any designer is making a big mistake if he or she believes the plant maintenance personnel can ever achieve this level of understanding. From a plant maintenance person's point of view, the more the machine can tell them about what is wrong the better. They may only work on the machine

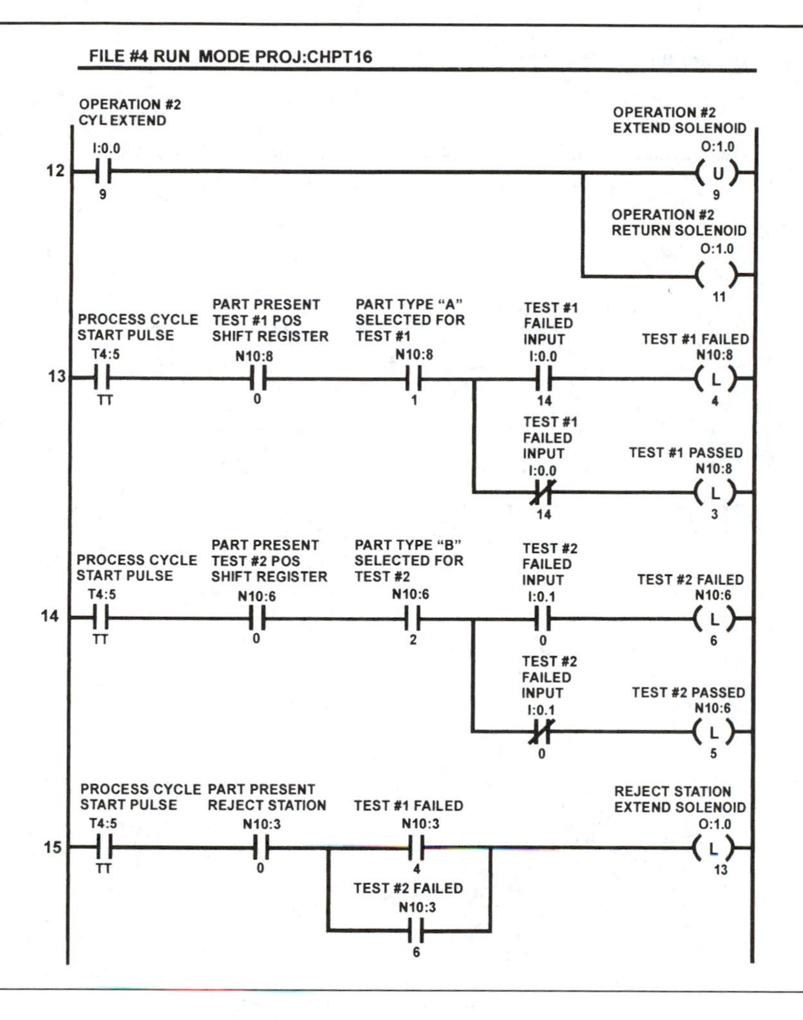

Figure 16–22
Run mode ladder file 4, *continued*

once or twice a year and therefore never develop an in-depth knowledge of the equipment. With the high level of personnel turnover at some plants, equipment is often repaired by someone who was not even on the staff when the machine was installed and maintenance people were trained. For a design to be successful, the machine must be very reliable, easy to set up and adjust, have a user-friendly interface that is convenient for the operators, and, in the event of a major problem, provide information to the maintenance department so they can repair it in the shortest possible time. A well-thought-out and designed diagnostic system will go a long way toward making any machine meet the above requirements.

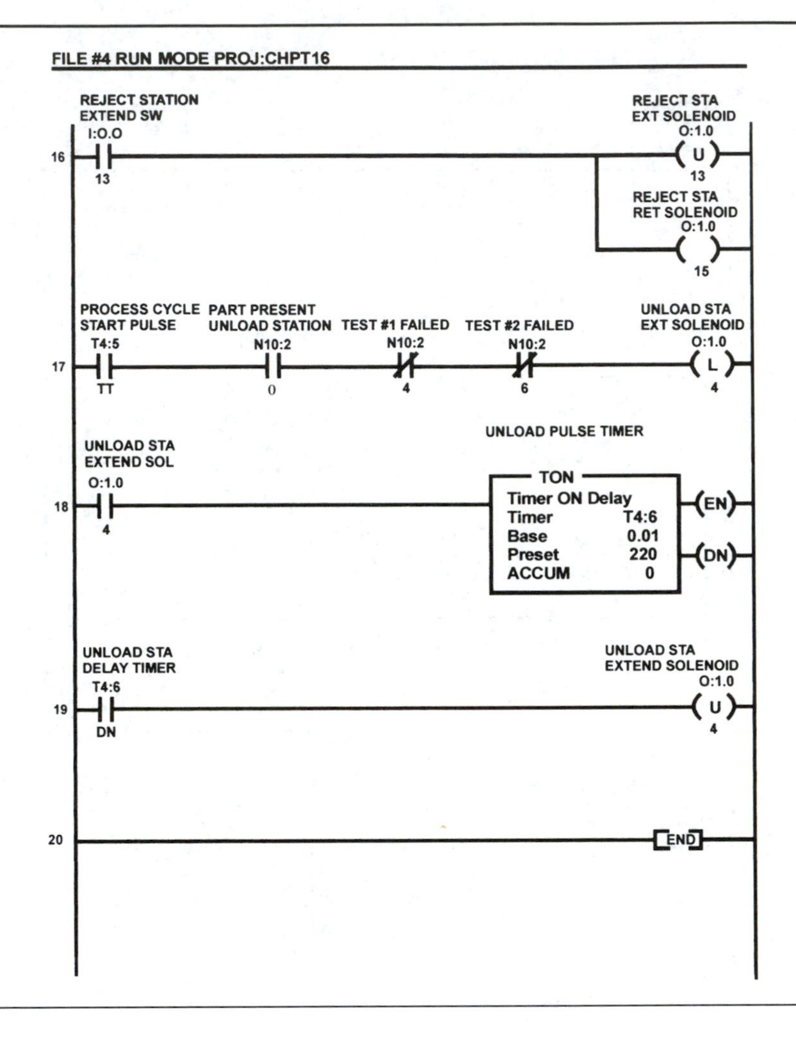

Figure 16–23
Run mode ladder file 4, *continued*

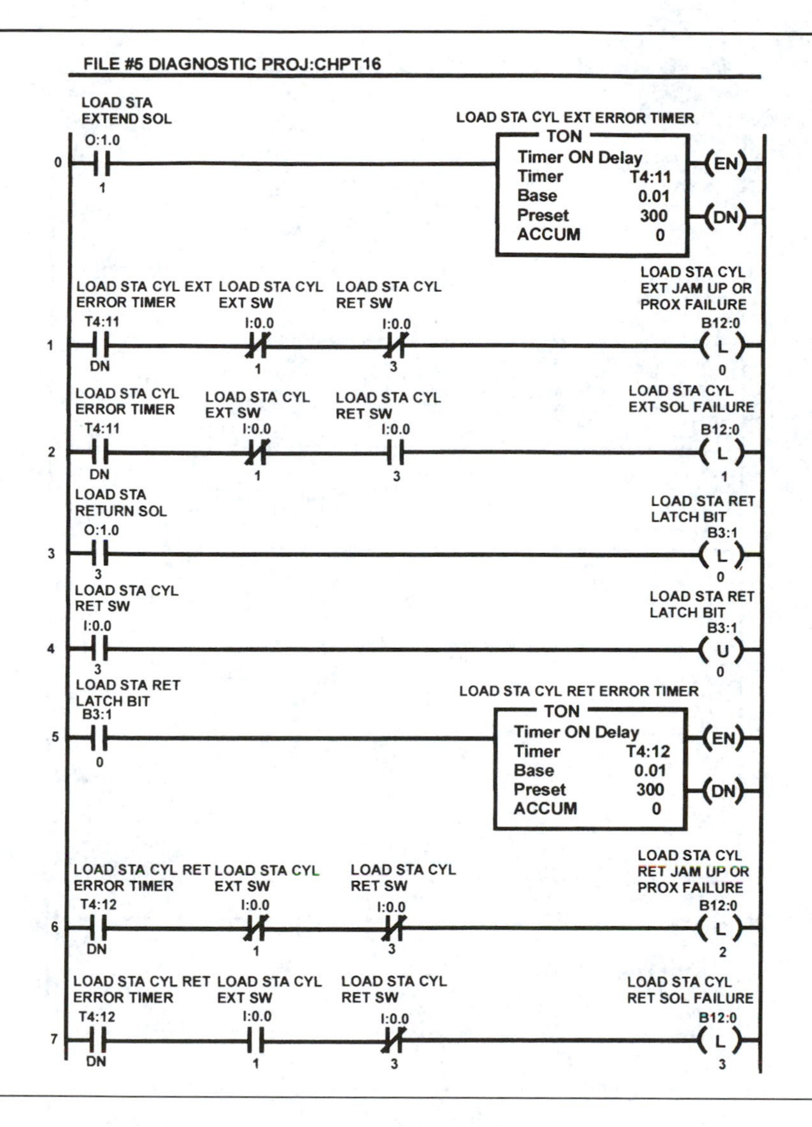

Figure 16–24
Diagnostics file 5

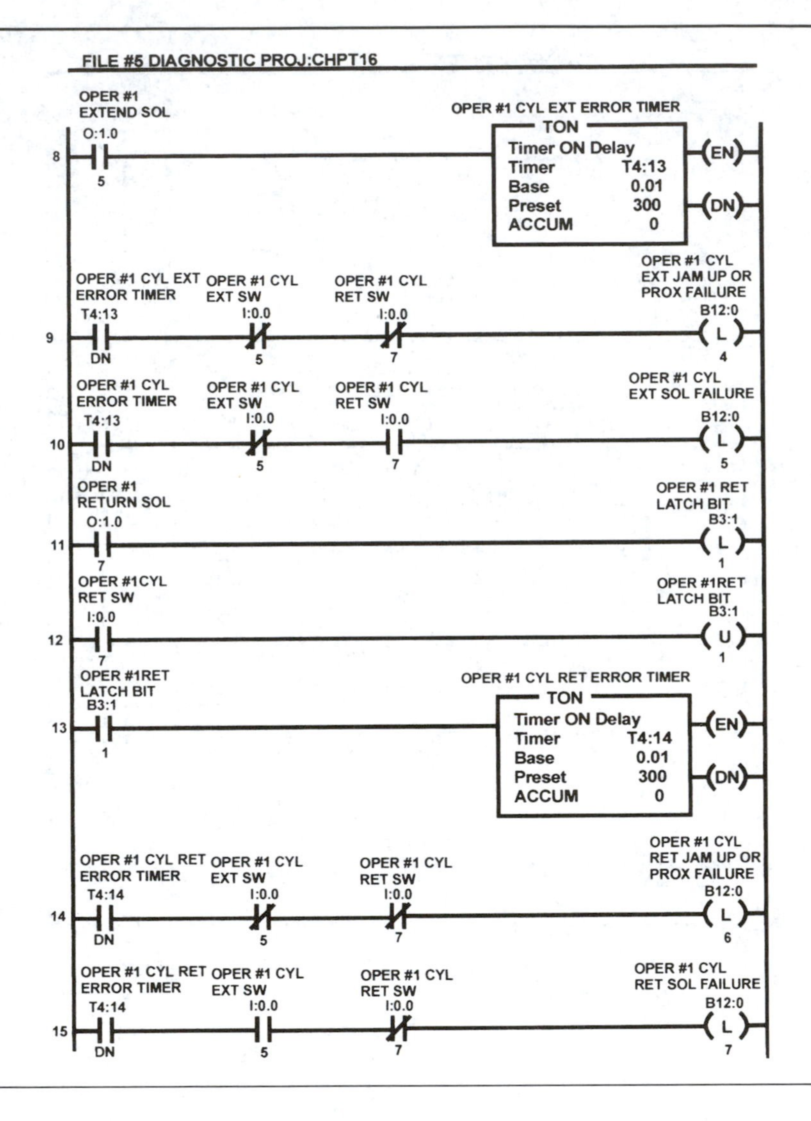

Figure 16–25 Diagnostics file 5, *continued*

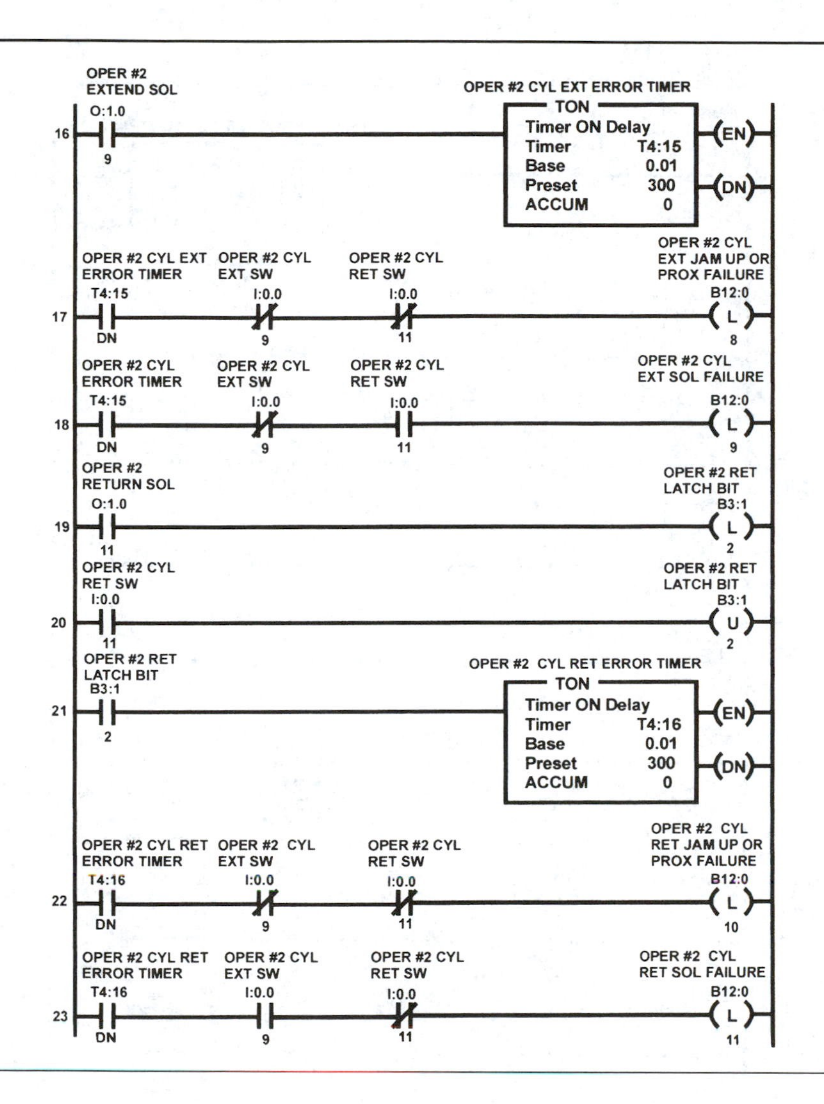

Figure 16–26
Diagnostics file 5, *continued*

Figure 16–27
Diagnostics file 5, *continued*

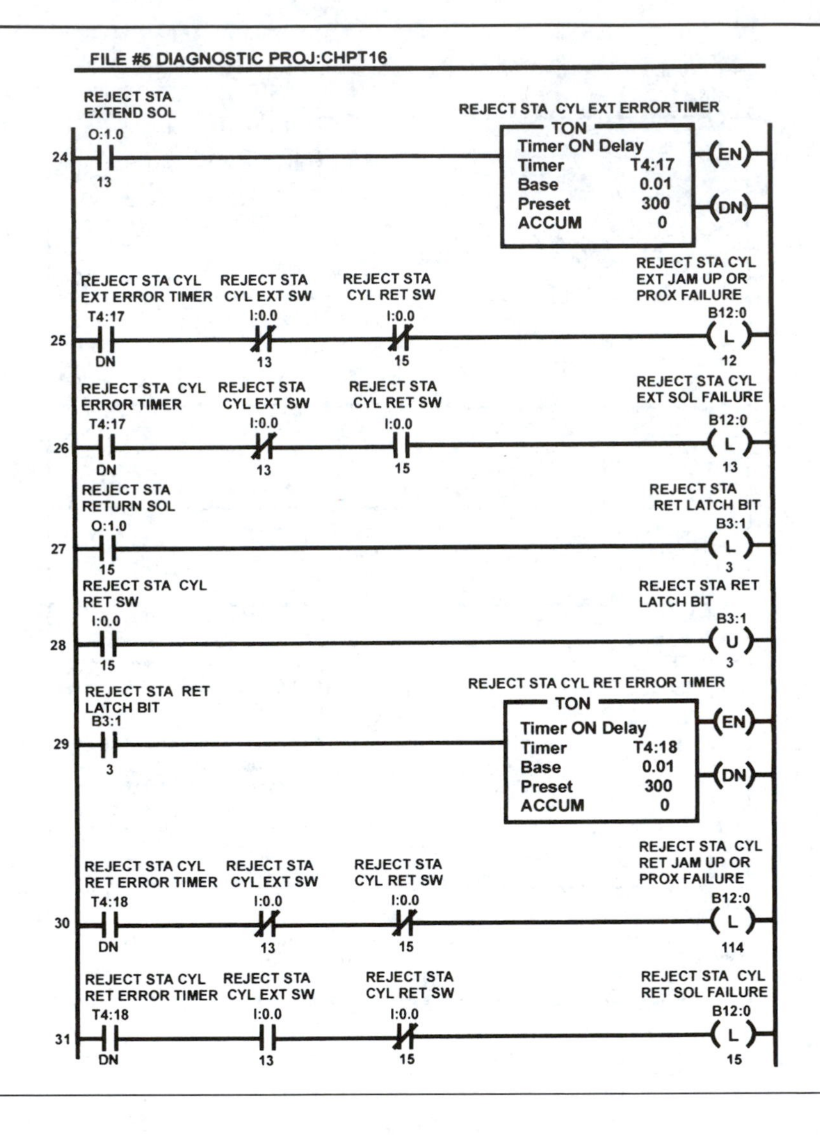

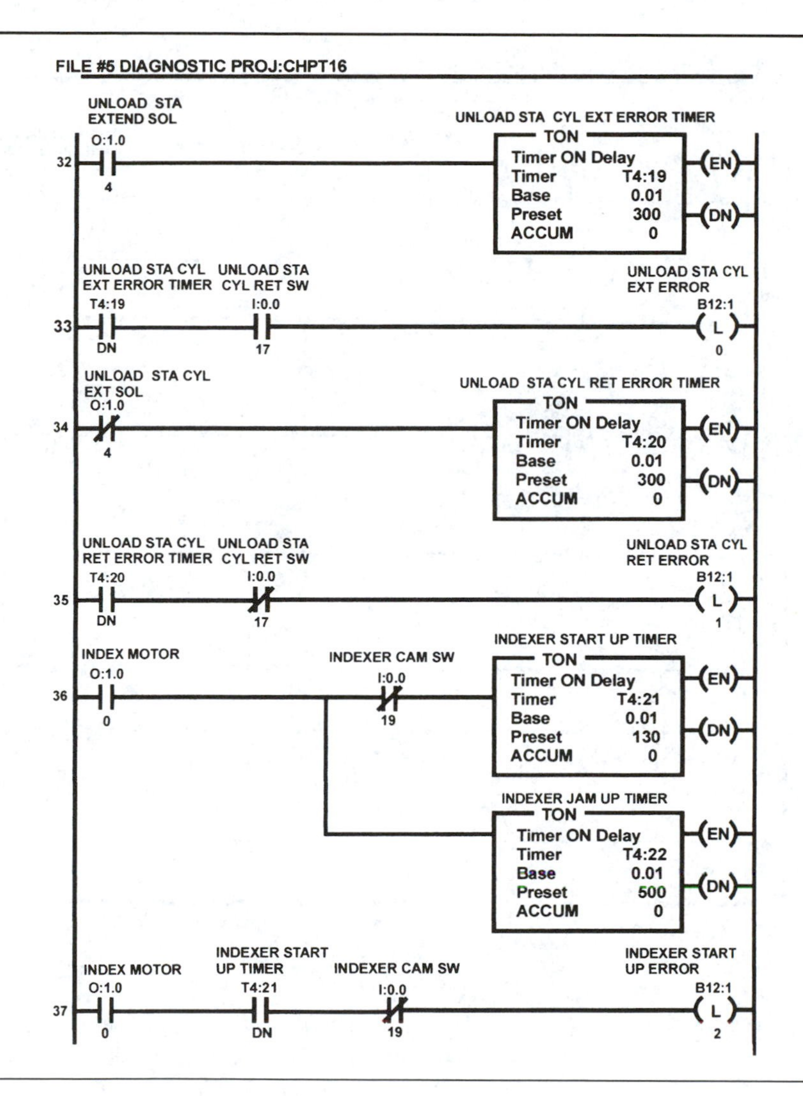

Figure 16–28 Diagnostics file 5, *continued*

Figure 16–29
Diagnostics file 5, *continued*

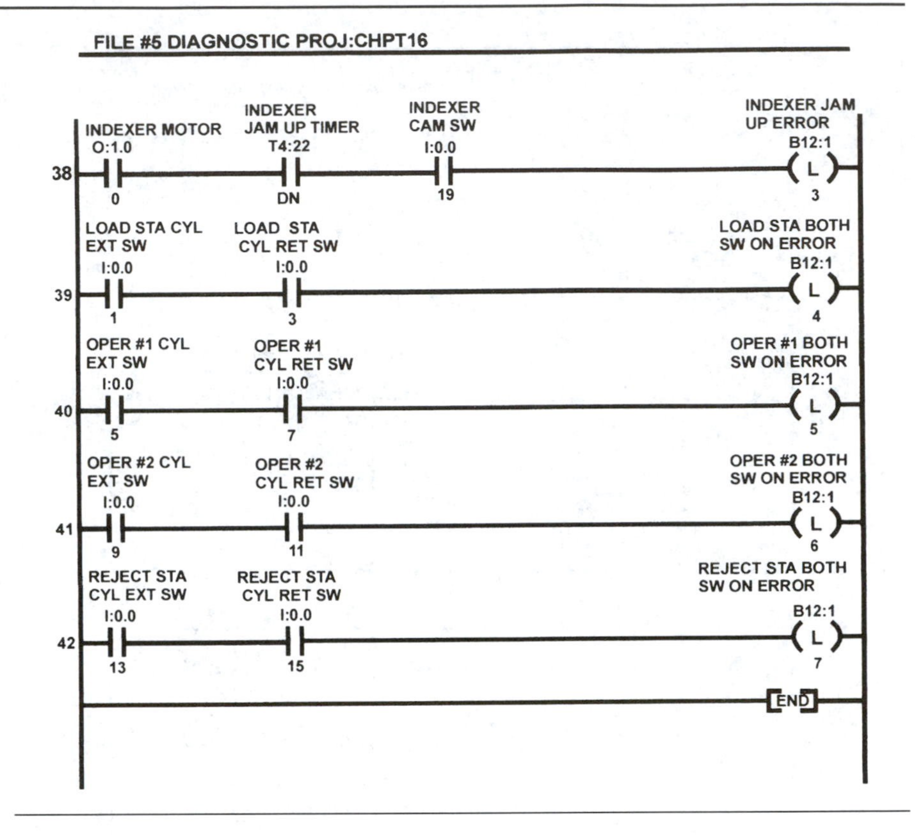

REVIEW QUESTIONS

1. Define the four main levels of diagnostic systems and the advantages of each.
2. Why must most error conditions be timed in our ladder programs?
3. What is the one switch condition that requires an immediate error notification and shutdown?
4. If an air cylinder has a stroke time of 2 seconds, to what delay time should the error timers be set?
5. What two error conditions must be checked for on indexer equipment?
6. When setting up an error data transfer file, what is the purpose of the control word?
7. From the maintenance department's point of view, what is the advantage of having a warning when a mechanism is slowing down?
8. On a manual run screen, what is the advantage of showing the position status of each mechanical element of the selected mechanism?
9. What is the typical percentage of mechanical downtime required to find the problem and how much time is required to actually fix it?
10. Why is it a good idea to print all error messages along with the time and date on which they occurred?

11. Why is it an advantage to be able to display a graphical image of the machine and the area of it in which the malfunction occurred?

12. How can we justify the additional costs incurred by diagnostic systems?

13. In the sample program, why was the indexer start-up error timer set to 0.2 second less than the index start pulse timer's value?

14. In the sample program, why did we delay in shutting down the auto mode for a period of time after the first error was detected?

LAB ASSIGNMENT

Write a program to monitor the application you wrote in Chapter 13. Monitor all five cylinders and the index motor. You must light the proper alarm lamp for each failure detected. Any error detected must shut down the auto run mode. Switching to manual mode should reset all the alarm lamps.

Note the additional error lamps that were added to the I/O list below. Lamps 0:0.0/14 and 0:0.0/15 are to be used to indicate if any air cylinder has both extend and return proximity switched on at the same time. Note that because cylinder #5 has only one limit switch you can only detect a return error.

I/O List

Inputs:

I:0.0/00	CYCLE START PB
I:0.0/01	CYLINDER #1 DOWN LS
I:0.0/02	CYCLE STOP PB
I:0.0/03	CYLINDER #1 UP LS
I:0.0/04	CYLINDER #1 DOWN MANUAL PB
I:0.0/05	CYLINDER #2 DOWN LS
I:0.0/06	CYLINDER #1 UP MANUAL PB
I:0.0/07	CYLINDER #2 UP LS
I:0.0/08	CYLINDER #2 DOWN MANUAL PB
I:0.0/09	CYLINDER #3 DOWN LS
I:0.0/10	CYLINDER #2 UP MANUAL PB
I:0.0/11	CYLINDER #3 UP LS
I:0.0/12	CYLINDER #3 DOWN MANUAL PB
I:0.0/13	CYLINDER #4 DOWN LS
I:0.0/14	CYLINDER #3 UP MANUAL PB
I:0.0/15	CYLINDER #4 UP LS
I:0.1/00	MANUAL INDEX PB
I:0.1/01	CYLINDER #5 UP LS
I:0.1/02	CYLINDER #5 DOWN MANUAL PB
I:0.1/03	INDEX CAM LS
I:0.1/04	AUTO / MANUAL SELECTOR SW
I:0.1/05	CYLINDER #4 DOWN / UP MANUAL SELECTOR SW
I:0.1/06	NOT USED
I:0.0/07	NOT USED

Outputs:

0:0.0/00	AUTO MODE LAMP
0:0.0/01	MANUAL MODE LAMP
0:0.0/02	NOT USED
0:0.0/03	MAIN ALARM LAMP
0:0.0/04	INDEX IN DWELL POSITION LAMP
0:0.0/05	CYLINDER #1 EXTEND ERROR
0:0.0/06	CYLINDER #1 RETURN ERROR
0:0.0/07	CYLINDER #2 EXTEND ERROR
0:0.0/08	CYLINDER #2 RETURN ERROR
0:0.0/09	CYLINDER #3 EXTEND ERROR
0:0.0/10	CYLINDER #3 RETURN ERROR
0:0.0/11	CYLINDER #4 EXTEND ERROR
0:0.0/12	CYLINDER #4 RETURN ERROR
0:0.0/13	CYLINDER #5 ERROR
0:0.0/14	BOTH PROX SWITCHES ON ERROR CYLINDER #1 OR #2
0:0.0/15	BOTH PROX SWITCHES ON ERROR CYLINDER #3 OR #4

0:1.0/00	INDEX MOTOR
0:1.0/01	CYLINDER #1 DOWN SOL
0:1.0/02	NOT USED
0:1.0/03	CYLINDER #1 UP SOL
0:1.0/04	CYLINDER #5 DOWN SOL
0:1.0/05	CYLINDER #2 DOWN SOL
0:1.0/06	NOT USED
0:1.0/07	CYLINDER #2 UP SOL
0:1.0/08	NOT USED
0:1.0/09	CYLINDER #3 DOWN SOL
0:1.0/10	NOT USED
0:1.0/11	CYLINDER #3 UP SOL
0:1.0/12	NOT USED
0:1.0/13	CYLINDER #4 DOWN SOL
0:1.0/14	NOT USED
0:1.0/15	CYLINDER #4 UP SOL

Program Checkout

Once you have the program written, verify the following conditions:

1. Check each cylinder for extend and return failures.
2. Verify the proper error lamp lights for each error condition.
3. Verify that changing to the setup mode clears the error.
4. Verify that you can restart the auto run cycle after resetting all the cylinder positions.
5. Hold the cam switch closed during an index and verify that the watch dog timer shuts down the auto run program and lights the proper error lamp.

When you have the program operating correctly print out a copy of the ladder program and demonstrate the program to the instructor.

CHAPTER

17

HMI Display Systems

In our previous chapter on diagnostic systems we discussed being able to display production information and diagnostic error messages. In this chapter we design a complete operator interface and diagnostic display system. As mentioned before, numerous HMI display systems are available from the controls marketplace. They cover a wide range from 40-character single-line alpha-numeric readouts to top-of-the-line IBM PC-driven systems with hard drives, network interfaces, and high-resolution graphic displays. The type of display you incorporate into your control system will depend on the cost restraints and construction budget for the project.

In this chapter we design a level 3 diagnostic system, but we will use an HMI display device that is at the low end of the range for this level. Allen-Bradley's Panel View display device makes a good low-end HMI operator display. Its main drawback is the lack of an internal programming language. Without the internal software we cannot perform complex graphic functions or use look-up tables for selecting data. HMI systems that have an internal Basic or VB type programming language are more suited for these types of needs.

Panel View terminals make a good button replacement panel and allow the operator a window into the PLC system. Panel View terminals can read or write to any I/O or memory address inside the PLC system but they have a problem with floating-point data. Messages can be displayed and data can be entered on the keypad and then saved into the PLC's memory. The cost of these terminals is quite low considering the good functionality they provide.

Remember that for this example we are going to modify the program written in Chapter 13 and modified again in Chapter 16 to add full diagnostics.

Designing an HMI operator display system requires some upfront planning before we generate our first screen. As system designers, we need to make an effort to understand how the operator will run the machine. It is important to ask the following types of questions and be comfortable with the answers before you start your project:

1. In what position relative to the control cabinet will the operator spend most of his time while working on this machine?
2. What production data should be displayed and in what format?
3. Is the operator panel located close enough to the mechanisms to be used as a manual control panel thus replacing numerous push-button panels?
4. Would mounting the display on a swing arm be advantageous over mounting it on the front of the control cabinet where it might be hard to see?

If the machine is already in existence or a similar machine is currently being used, ask the operators what they do and do not like about the current setup. Getting operator feedback can save you a lot of time and help you avoid mistakes at the design stage. Study the operator for a short while as she operates the equipment and try to determine the best place to mount the display for ease of operation and convenience to the people that work around the machine. Failure to consider the operator's habits and working conditions can often result in hundreds of hours of engineering work sitting unused and collecting dust because the operator finds the HMI display too difficult or inconvenient to work with.

After the mounting position has been decided and the operators interviewed, the next step is to lay out paper sketches of the screens needed to operate the equipment. Based on your input from the future operators, or your own understanding of how the operator will interface to the equipment, you should lay out screens for both the manual mode and the run mode of operation.

The manual mode should enable the setup person to manipulate each mechanism during setup and adjusting as well as to reset all mechanisms from a jam-up condition. You should incorporate feedback so that for every action initiated from the panel, the operator can see a response on the HMI when the action is completed. A good feedback example would be when any push button is pressed to extend a cylinder, the push button changes color when that cylinder reaches the end of its stroke. You should also provide some type of indicator to tell the operator that all stations are in their home positions.

The auto mode should provide the operator a means to start and stop the machine as well as process control information if required. As with the manual mode, we should have some type of indicator here to show that all stations are in their home positions and the system is ready to start. Generally speaking, it is a good idea to have a separate set of push buttons to start and stop the machine that are not part of the HMI display. Because the HMI display is easily damaged by operators using sharp objects to press the buttons, such as fingernails, screwdrivers, or pencils, you should have a duplicate set of buttons available to start and stop the machine's cycle. This will allow you to run the machine and build product even if the HMI is damaged. It could take several days to get a new display and reload the software. If your HMI is the only way to start and stop auto mode, you will be down for the entire time it takes to get a replacement. For these reasons we highly recommend you have an alternate way to run the machine without the HMI. You will not receive all the production data and diagnostic messages in the meantime, but you can still build product while you wait for a replacement screen.

ASSIGNING ADDRESSES

The next step in our design process is to identify all of the PLC addresses that will be required to interface the PLC to the HMI display screens. We then assign each PLC address a tag name on the HMI side. In addition to the tag name we also need to define the data type of each tag name variable. The two main data types we will use are bit addresses and word addresses. In the Panel Builder software a tag name can be up to 32 characters long, must start with an alpha character, and cannot contain any spaces. If you need to use separate words to make a tag name, use the underline character between them; Start_Push_Button is a good example. Try to create tag names that closely identify the real function being performed by that address and not a name like Button_1.

We start this step of the design process by looking over our paper sketches of the proposed screens and making a list of all the push buttons that connect to the PLC side of the system. Each HMI button device will need an address on the PLC side. Next, look at all the indicators and numeric displays you have showing. Each of these must have a corresponding bit or word address on the PLC side. Most designs, at the very least, will require one word address to select screens based on the machine's operating mode and a second word address to transfer error messages to the HMI.

Each display manufacturer has a different way of assigning PLC addresses to variables in the display system. Be sure to read the manual for your particular system before working on a tag list. Once you have designed a few systems you can jump right in and begin creating screens

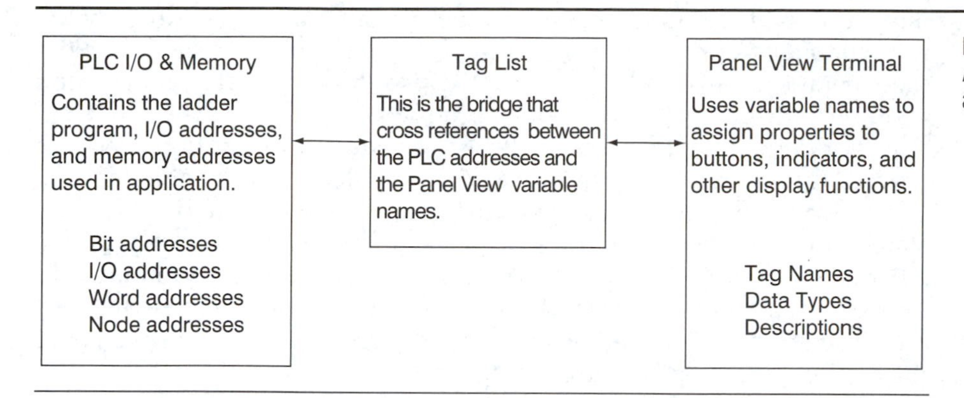
PLC I/O & Memory
Contains the ladder program, I/O addresses, and memory addresses used in application.
Bit addresses
I/O addresses
Word addresses
Node addresses
Tag List
This is the bridge that cross references between the PLC addresses and the Panel View variable names.
Panel View Terminal
Uses variable names to assign properties to buttons, indicators, and other display functions.
Tag Names
Data Types
Descriptions

Figure 17–1
A tag list functions as a cross-reference map

and editing your tag list as you go. Because this example is our first design effort, we will make a list up ahead of time to help collect thoughts about what needs to be displayed and what addresses are needed to perform this application. The tag list acts as a cross-reference between the PLC systems and the HMI display (see Figure 17–1).

SCREEN DESIGN

Our program will have to have three main screens: auto mode, manual mode, and system data. Note that there is no screen for diagnostics. In the Panel View system, diagnostic messages are handled as alarm messages and these will pop up on top of whatever screen is currently showing.

Auto Mode

In the auto mode screen, we need to be able to start and stop the machine from the panel and be able to see if all the stations are in their home positions before the machine is started. We also want to be able to switch to the system data screen to get production and maintenance information.

Manual Mode

This screen will appear when we switch to the manual mode setting. We want to be able to extend or return any of the five cylinders from this screen and have feedback as to the prox switch status on all five cylinders. We also want to be able to index the line from the manual screen and be able to see that the index cam is working.

System Data

This screen is entered from the auto mode screen and will show production and maintenance information. The top half of the screen is intended for troubleshooting a slow system or to warn of possible problems by indicating which station is slow. On the bottom half we will display the number of good parts and bad parts the line has run and be able to clear these counters at the end of the shift. We also want to display information about what is being run on the line: product A, B, or C.

Working from the screen design specifications outlined in the preceding paragraphs, we know we will need bit tags for all cylinder prox switches and the cam switch because we know we need to display their status. We also need 10 bit tags, which will represent the manual push buttons on our manual screen that activate the five air cylinders and indexer motor in manual mode. We need two more to cover the start and stop push buttons on the auto mode screen and one more to cover the clear counters push button.

We will require three word tags to cover the two counters for good and bad parts and one for the type of product being run. In addition, we will need one word tag for the alarm menu and one to select between the auto and manual screens and five for the station timer values. The tags are shown in the following list:

Tag Name	Type	Description
Load_ext_sw	Bit	Load station extend prox sw
Load_ret_sw	Bit	Load station return prox sw
Oper1_ext_sw	Bit	Operation #1 extend prox sw
Oper1_ret_sw	Bit	Operation #1 return prox sw
Oper2_ext_sw	Bit	Operation #2 extend prox sw
Oper2_ret_sw	Bit	Operation #2 return prox sw
Reject_ext_sw	Bit	Reject station extend prox sw
Reject_ret_sw	Bit	Reject station return prox sw
Unload_ret_sw	Bit	Unload station return prox sw
Index_cam_sw	Bit	Indexer cam prox sw
Run_mode	Bit	Indicates the machine is in run mode
Stop_latch	Bit	Indicates the stop latch bit is set
Load_ext_pb	Bit	Load station manual extend panel push button
Oper1_ext_pb	Bit	Operation #1 manual extend panel push button
Oper2_ext_pb	Bit	Operation #2 manual extend panel push button
Reject_ext_pb	Bit	Reject station manual extend panel push button
Unload_ext_pb	Bit	Unload station manual extend panel push button
Index_pb	Bit	Indexer manual panel push button
Clear_cntr_pb	Bit	Clear the good/bad parts counters

Start_cycle_pb	Bit	Start the auto cycle panel push button
Stop_cycle_pb	Bit	Stop the auto cycle panel push button
Alarm	Word	Alarm word
Screen	Word	Selects which screen to display
Good_cnt	Word	Good parts counter
Bad_cnt	Word	Bad parts counter
Type	Word	Type of parts being run
Load_sta_time	Word	Cycle time for load station
oper1_sta_time	Word	Cycle time for operation #1 station
oper2_sta_time	Word	Cycle time for operation #2 station
Reject_sta_time	Word	Cycle time for reject station
Unload_sta_time	Word	Cycle time for unload station

At this point we have the major tags identified. Note that while the prox switch inputs already have assigned addresses, we have to assign addresses to the new panel push buttons and the 10 new word addresses.

A number of different types of Panel View terminals are available. Some have touch screens and some have keypads. There are several different interfaces also, RS232 and DH485 to name a couple. The type you select will depend on the type of PLC you are using and the system requirements of your design project. The environment in which your system must operate has a large impact on the type of screen you pick. If your equipment is going to run in a plant where the operators work with a material that is gummy or sticky you probably don't want to use a touch screen. You must also be sure that the interface you pick is compatible with the interface on your PLC system. A DH485 interface and a keypad front panel were chosen for our teaching kit.

Before we begin entering data to the Panel Builder software, a few words of caution are necessary. A whole chapter could be written on serial communication types and problems because this will be the most common problem you will face in trying to get your system to work. The designer should have a good understanding of the basic serial communication types and their advantages and disadvantages before you start. Many good books dedicated to this subject are available. This chapter is not intended to be a software manual for the Panel Builder program. Often, walking a first-time user through a design project is the fastest way to gain experience and a foothold on using the entire package.

One problem you may encounter if you decide to use a DH485 unit is in downloading from the 1747PIC module. None of the Allen-Bradley documentation warns you that the DH485 interface needs a 12V DC supply and neither the Panel View terminal nor the PIC module contains one. The PIC module interfaces between your computer's RS232 COM port and the Panel

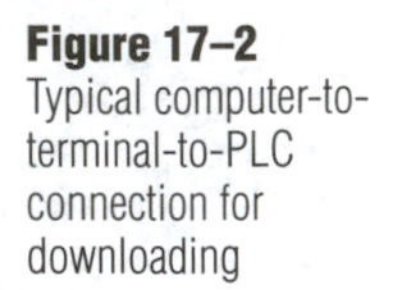

Figure 17–2
Typical computer-to-terminal-to-PLC connection for downloading

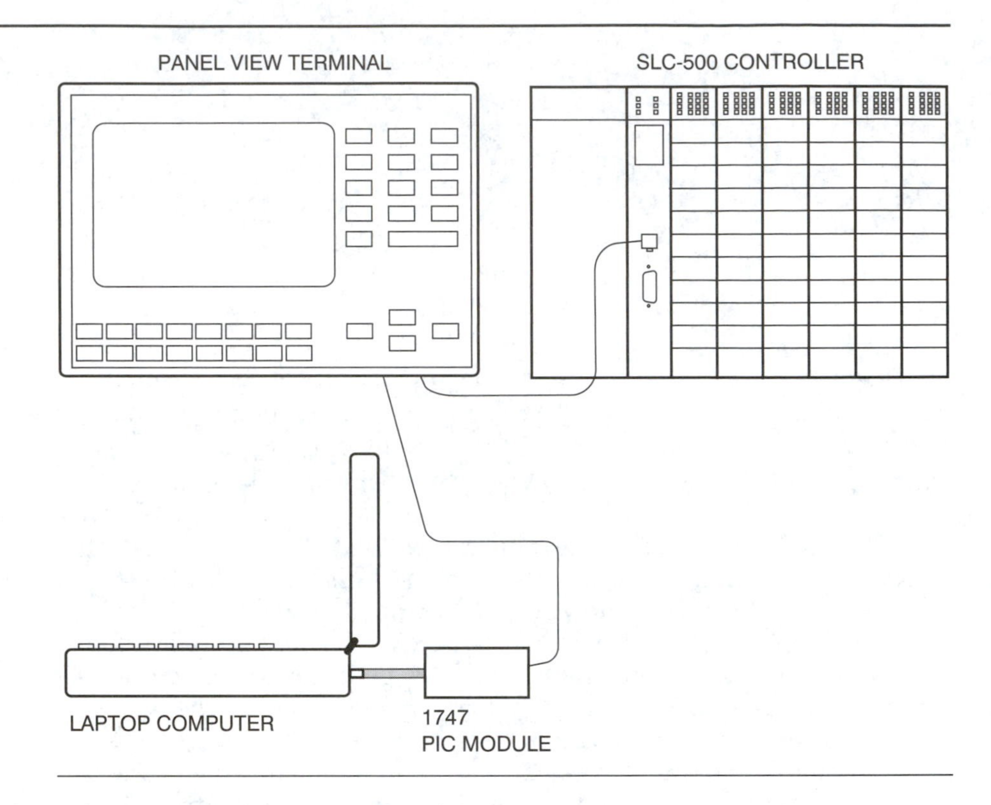

View terminal or PLC controller. This is not a problem when interfacing to a PLC controller because the controller has the required power supply inside. The Panel View terminal does not have a supply and will not accept a download without one present. The easiest solution is to also connect the PLC to the Panel View terminal and in this way the PLC power supply will power both the Panel View terminal and the PIC module (see Figure 17–2).

An additional advantage to this setup is that you can download both the PLC ladder files and the Panel View files without changing any connections. You can also switch back and forth between the two programs during the debugging process as needed.

The PIC module is currently being replaced with the USB to DH485 interface converter as most newer laptop computers now come equipped with only a USB port.

BEGINNING A PANEL BUILDER PROJECT

Once the Panel Builder software opens, click on the New tab. This will bring up a menu for new applications (Figure 17–3). Fill in the application name first. The first eight digits of this name will be used as a file

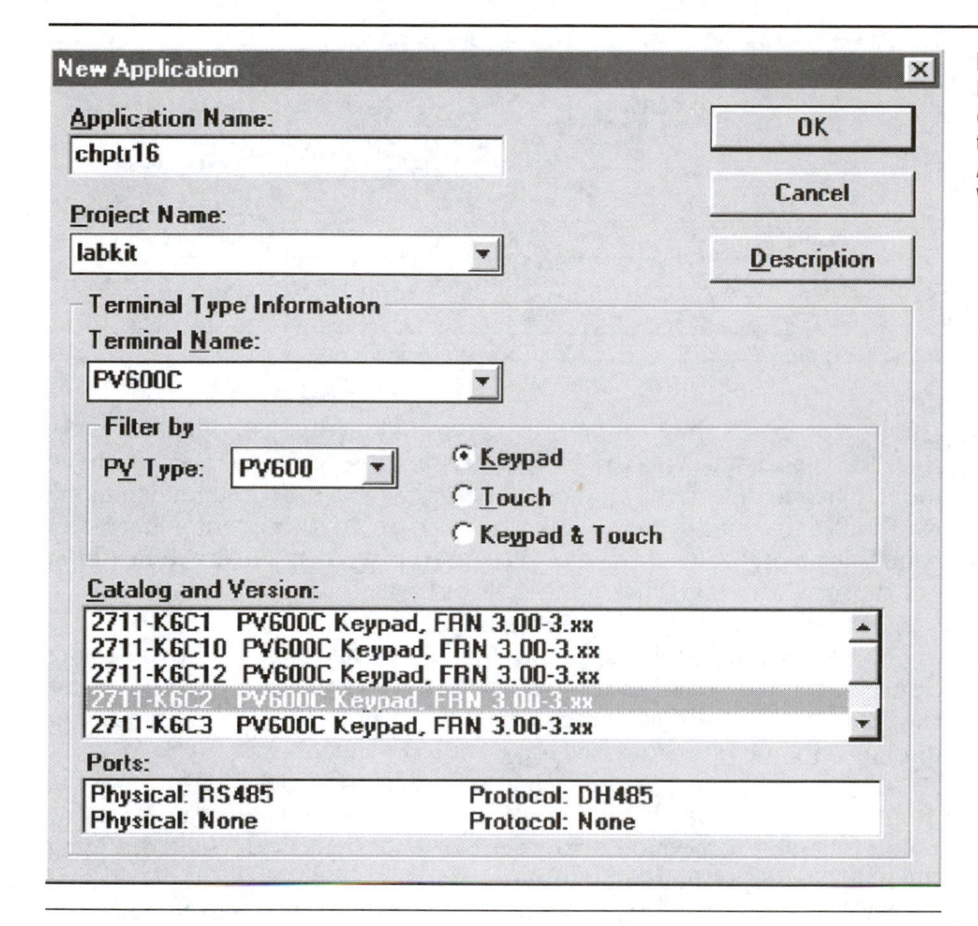

Figure 17–3
New Application screen. (All screen captures in this chapter courtesy of Allen-Bradley, West Allis, Wisconsin.)

name for the project. Next input the project name. This name can be the same as the application name or, if you use the same tag lists for multiple projects, a default project file.

Next we must select the actual Panel View terminal we will be using. In our teaching kit that is the PV600C–2711–K6C2 Rev 3.00. If your model number is not shown, it may have a touch screen. Clicking on the touch screen selection will bring up a different set of model numbers. When you have done this, click on the OK tab and the display now shows the main programming screen.

Note that no screens are showing yet because we have not generated any screen names yet. Before we begin the screen layout process, let's set up our terminal communications first. We do this by clicking on the Application tab at the top of the screen and then clicking on Terminal Setup. We need to specify the information as shown on the screen print of Figure 17–4.

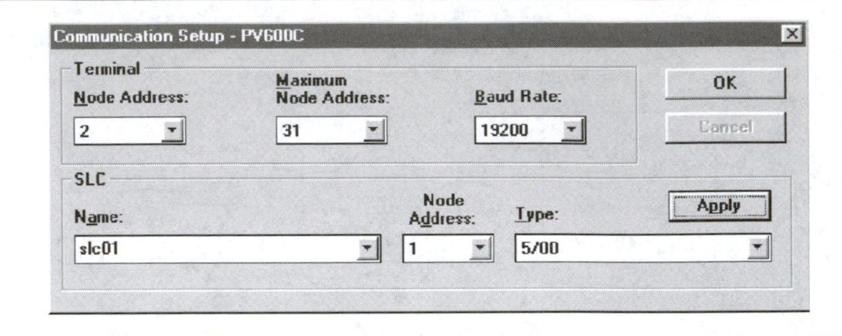

Figure 17–4
Communication Setup screen

In most cases the programming terminal is always node address 0, the PLC is node address 01, and the terminal is address 02. If you have more than one PLC in the system this setup may be different. Enter a name for the PLC in your system. This name can be anything, but is normally something like PLC01. Be sure to click on Apply when you are done filling out the form and then click on the OK button.

The next step in the process is to fill in our tag list. We do this by clicking on the Tools tag and then on the Tag Editor tab. This will open the Tag Editor screen with no tag addresses showing. To enter tag addresses, press the Insert key on the keyboard several times until you get a good number of blank tag lines showing (see Figure 17–5).

You can expand the name field or description field by moving the mouse pointer to the division mark between the fields, waiting until the pointer changes to a double arrow symbol, and then clicking and dragging the field border to the right.

We will start by entering all the PLC switch inputs since these already have addresses assigned. When you click on the Node Name box a selector arrow will pop up. You need to pick the node name of the PLC we entered in the Communications Setup box. If you have more than one PLC

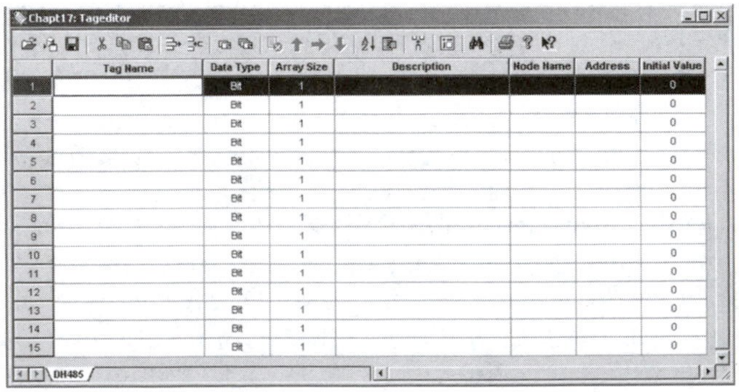

Figure 17–5
Tag Editor screen

on this network you must select the PLC name where this address resides. Be sure to click on the Data Type box for each name even if the data type you're entering is Bit as shown in Figure 17–5. The bit that is shown in the blank line is just the first selection in the data type list table. If you do not click on the Data Type box and then select the Bit type, nothing is selected. When you try to save the tag list you will get errors saying that no Data Type was selected.

We now assign the rest of the tag names and assign addresses to them as we go. We assign B3 addresses to all the push buttons that will act on a PLC address and eight new word addresses are added along with two counter addresses. Not all of the new tags are visible on the screens shown in Figure 17–6.

None of the addresses we have added currently exists in our PLC program. We will have to do some extensive editing of our ladder program after we have completed our display layout. It is a good idea to print out your tag file to help in editing the ladder program later.

Be sure to save the tag file before you click the exit form box. If you do not save the file before exiting the form, everything is lost. There is no warning message when you click on the exit form box, your file is just gone!

When you save the form the system will check for errors. If errors are found you must fix them before the file can be saved. You can save the file with errors as a Draft File, in case you want to work on fixing the errors at a later date. Be sure you have deleted any blank lines before you try to save the tag list. Any blank lines will be detected as errors.

Our next step is to create the three screens with which we will be working. Click on the Screen tab at the top of the screen and click on New. A pop-up menu will appear showing the screen number and requesting a name (Figure 17–7). Enter our first screen name, Automode. The default

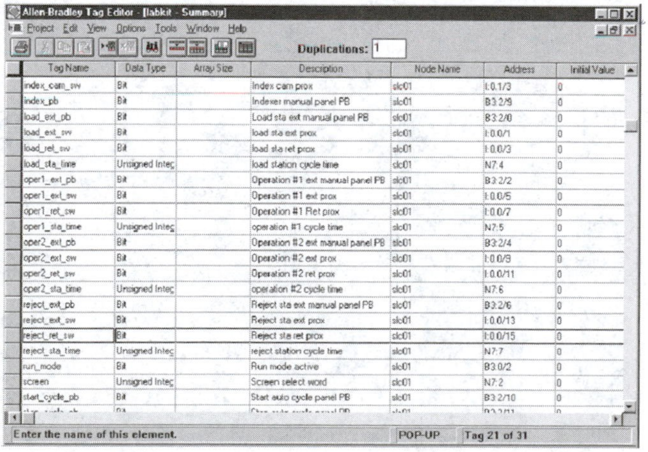

Allen-Bradley Tag Editor - [labkit - Summary]
Project Edit View Options Tools Window Help
Duplications: 1

Tag Name	Data Type	Array Size	Description	Node Name	Address	Initial Value
index_cam_sw	Bit		Index cam prox	slc01	I:0.1/3	0
index_pb	Bit		Indexer manual panel PB	slc01	B3:2/9	0
load_ext_pb	Bit		Load sta ext manual panel PB	slc01	B3:2/0	0
load_ext_sw	Bit		load sta ext prox	slc01	I:0.0/1	0
load_ret_sw	Bit		load sta ret prox	slc01	I:0.0/3	0
load_sta_time	Unsigned Integ		load station cycle time	slc01	N7:4	0
oper1_ext_pb	Bit		Operation #1 ext manual panel PB	slc01	B3:2/2	0
oper1_ext_sw	Bit		Operation #1 ext prox	slc01	I:0.0/5	0
oper1_ret_sw	Bit		Operation #1 Ret prox	slc01	I:0.0/7	0
oper1_sta_time	Unsigned Integ		operation #1 cycle time	slc01	N7:5	0
oper2_ext_pb	Bit		Operation #2 ext manual panel PB	slc01	B3:2/4	0
oper2_ext_sw	Bit		Operation #2 ext prox	slc01	I:0.0/9	0
oper2_ret_sw	Bit		Operation #2 ret prox	slc01	I:0.0/11	0
oper2_sta_time	Unsigned Integ		operation #2 cycle time	slc01	N7:6	0
reject_ext_pb	Bit		Reject sta ext manual panel PB	slc01	B3:2/6	0
reject_ext_sw	Bit		Reject sta ext prox	slc01	I:0.0/13	0
reject_ret_sw	Bit		Reject sta ret prox	slc01	I:0.0/15	0
reject_sta_time	Unsigned Integ		reject station cycle time	slc01	N7:7	0
run_mode	Bit		Run mode active	slc01	B3:0/2	0
screen	Unsigned Integ		Screen select word	slc01	N7:2	0
start_cycle_pb	Bit		Start auto cycle panel PB	slc01	B3:2/10	0

Enter the name of this element. POP-UP Tag 21 of 31

Figure 17–6
Tag Editor screen with data entered

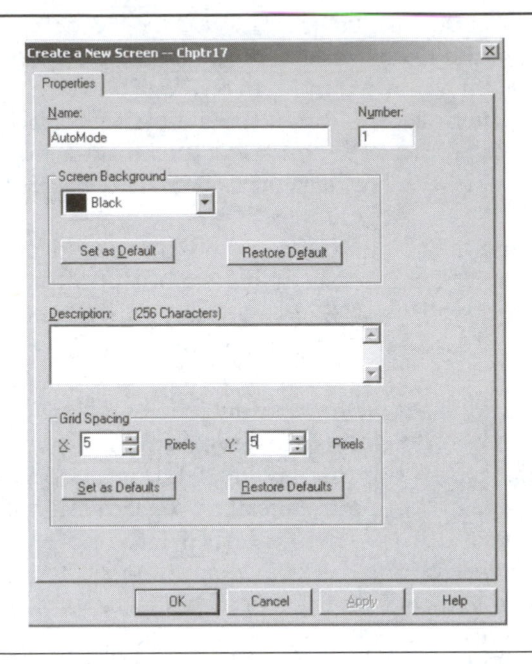

Figure 17–7
Create a New Screen

grid spacing is much too large. Change both the X and Y grid spacing to 5, as shown in Figure 17–7; then click the OK button. The screen will change to a background showing the front of the terminal and its keypad (Figure 17–8). Because we have several more screens to add, click on the X at the top right corner to clear this working screen and repeat the process to create the manual mode and system data screens. Be sure to change the X and Y grid spacing in each new screen to 5.

After creating the other two screens our screen selector box should look like that shown in Figure 17–9.

Figure 17–8
Blank auto mode screen

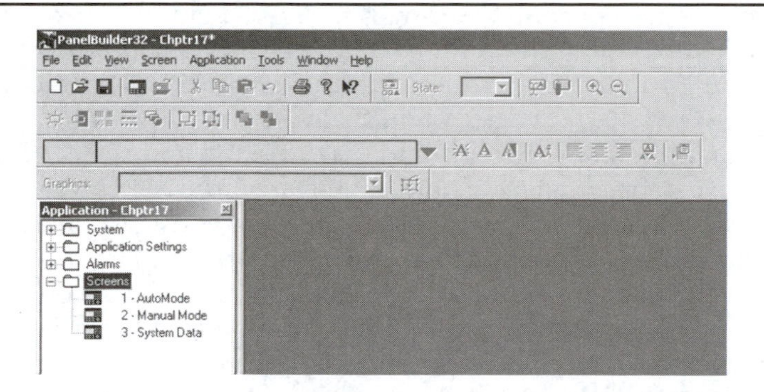

Figure 17–9
Main Panel View screen showing existing screen files

Auto Mode Screen

We can now begin laying out and editing our first screen. Just double click on the auto mode screen name and the working screen will open again. You can switch between screens any time by closing one and clicking on a different screen name. The screen printout shown in Figure 17–10 is how the screen should look after we enter the main elements.

We start by placing the AUTO RUN SCREEN text at the top of the screen. This is accomplished by clicking on the Objects tab, and then the Text selection. The arrow pointer changes to a crosshair cursor. Place it about five grid points down from the top left side of the screen and then drag it to the right and up. This opens a text box into which the text will go. On the top of the menu screen there is an Edit Text box into which you type your heading.

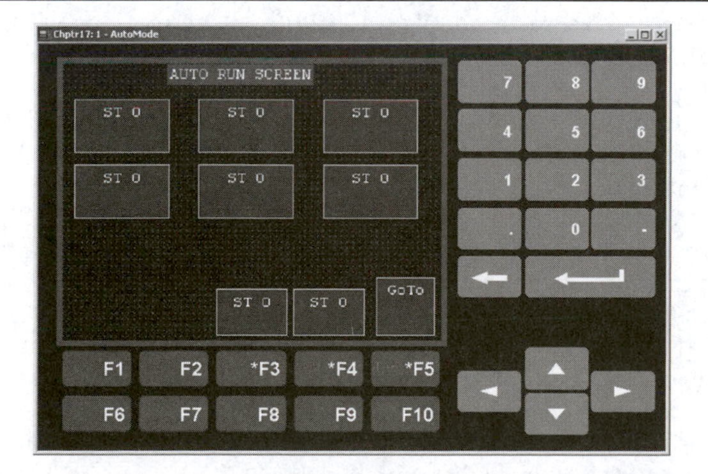

Figure 17–10
Auto mode screen with basic elements showing

To release the text entry command, click outside the text box. Click again on the text to highlight the text box. With the text box highlighted click on the Format tab and then the Background Color selection and change it to Dark Red. The text should already be shown in white. If it is not, you can click on the Foreground Color option and select White.

The next step is to place six multiple-state indicators on the screen to show the status of the five air cylinders and the index cam. We do this by clicking on the Object tab, then the Indicator tab, and last the Multistate tab. The cursor changes again to a crosshair pointer. Place it on the screen at about five grid points down and three grid points to the right of the left edge and drag it to open a 7 × 15 grid point box. Repeat the process six times, placing three boxes across the screen and then one row of three below. Do not be too concerned if you do not position the boxes correctly the first time, because you can always move them or resize them later.

We now need two push buttons for the Cycle Start and Stop functions. We begin creation of these buttons by clicking the Object tab, then the Push Button tab, and finally the Momentary tab. Again the crosshair cursor appears; we open two boxes about 8 × 12 grid points in size, directly above the [F3] and [F4] keypads.

The last object to be added is the push button that will open the system data screen. This is not a normal push-button selection because the button will not be assigned to any PLC address and is only seen within the Panel View system. We create it by clicking on the Object tab, Screen Selector tab, and then the Goto tab. Then open a 8 × 12 box directly above the [F5] keypad. At this point we have our screen setup but have not assigned the indicators or buttons to any PLC addresses.

Before we start defining the push buttons and indicators on our screen, we need to discuss how Panel View assigns colors to its objects. Figure 17–11 shows the assignment variables and indicates the section of the object that each variable controls. You can have many different colors on your screen, but most people find it difficult to work on screens with lots of different colors. For this reason keep your color combinations on the conservative side and consider the operator's needs. Always put dark text on a light background or light text on a dark background. An indicator that the operator can't read will only create frustration. Pick your color combinations carefully and stay consistent across all the screens.

We start our object editing by double clicking on the first indicator box on the left side of the screen. This will be our Load Station in returned position indicator. When we double click on the first indicator box we get the Multistate Indicator object edit screen shown in Figure 17–12.

The default setting for multistate indicators is four states plus an error state. We only need two states for our program, so our first step is to click on the States tab to edit the number of states (Figure 17–13). Place the pointer on the second state and hit delete twice. We can now enter the indicator button messages for each state. State 0 will be labeled "LOAD STA NOT HOME"

Panel View Push Button
Color Assignments

TEXT
Text Background
Text Foreground
Object Background
Object Foreground

Push button with text foreground and object foreground set to the same color and object background and text background set the same but to a different color.

Push button with the text foreground set to one color and the text background, object background, and object foreground all set the same.

Figure 17–11
Button color assignments

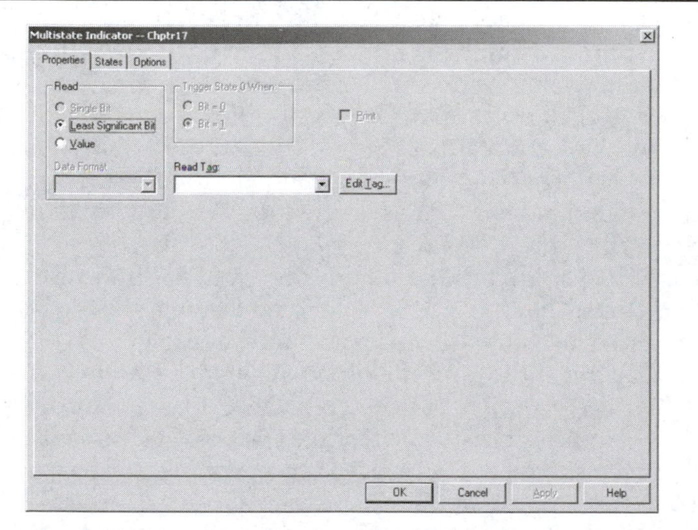

Figure 17–12
Multistate Indicator screen

and state 1 will be labeled "LOAD STA HOME". In Figure 17–13 note the /*R*/ between the two halves of the message, indicating that a [ENTER] key was pressed. This means the text "LOAD STA" will appear on the top line and the text "NOT HOME" will appear below it. We set the object foreground and background color along with the text background color to light

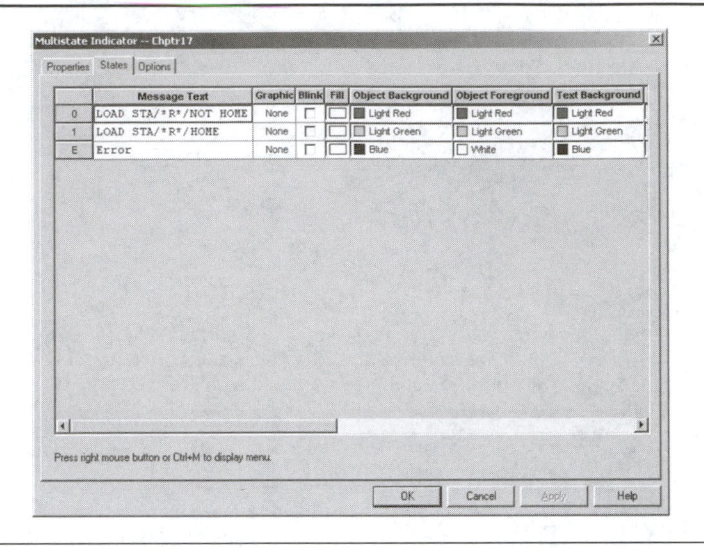

Figure 17–13
Multistate Indicator edit screen showing two states and text messages

red for the indicator state of 0. We then set the same three variables to light green for the 1 state. This will cause the indicator to turn red when the load station is not home and green when the load station is home. Both the text foreground colors were set to black but are not visible in Figure 17–13.

We must now assign a tag address to the indicator, select the single bit read property, and decide which state is active when the bit goes high. To do this, we click on Properties to go to the Properties tab selection, where we set the Read status to Single Bit and the Trigger State 0 When status to Bit = 0. Last we set the tag name in the Read Tag box for this indicator to "load_ret_sw". The screen should now look like the screen print shown in Figure 17–14.

We now repeat the process on all the rest of the indicator objects. Be sure to remember that the indexer is in dwell state if the tag bit is set to a 0 rather than a 1, as were the rest of the indicator tags.

Now that this indicator edit process is complete we can set up the cycle start and stop push buttons. Setting up push buttons is a little different than indicator displays. Double click on the first button box and the setup screen shown in Figure 17–15 appears. We need to set a number of things on this screen. The first selection is to set the push button type to Momentary. We then select to write a single bit to the PLC side when the push button is pressed. We next assign the [F3] key to this push button function and the hold time to 500 ms. We want the contacts to be Normally Open and this is selected next. We have set the write tag to Start_Cycle_PB and the indicator tag to Run_mode.

We can change the background color of a push button based on the state of some bit in the PLC. For operator feedback in this case we will be

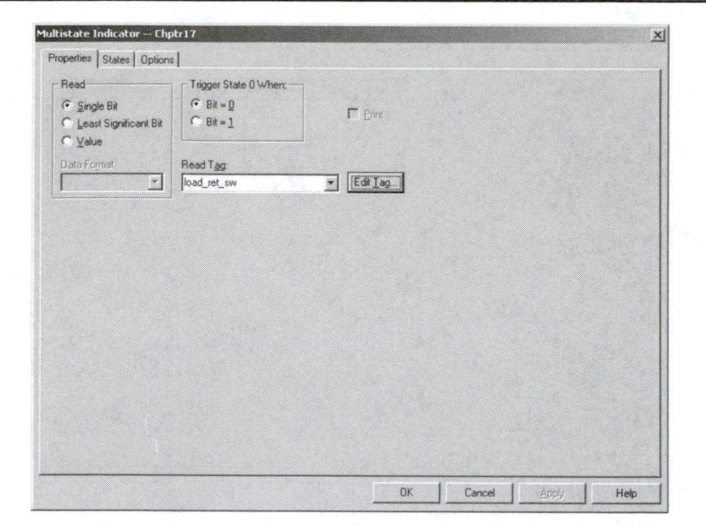

Figure 17–14
Screen showing Properties edits for tag "load_ret_sw"

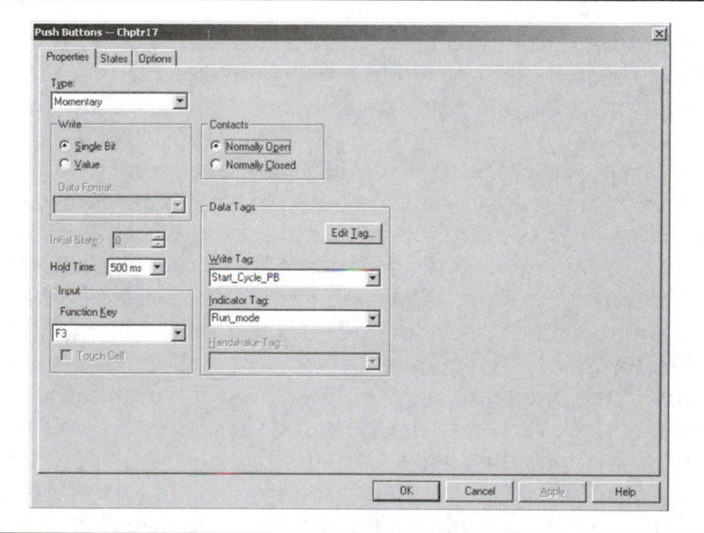

Figure 17–15
Push Buttons Properties edit screen

using the tag Run_mode to show that the machine is in the run mode. To set the color and text on this push button we click on the States tab. The text is the same in both states so we enter the same text in both: CYCLE START. The return is entered between each word because we want the word Cycle to be above the word Start. To set the button colors we set both background colors and the object foreground to yellow and the text foreground color to dark red for the state 0. In state 1 we will change both background colors

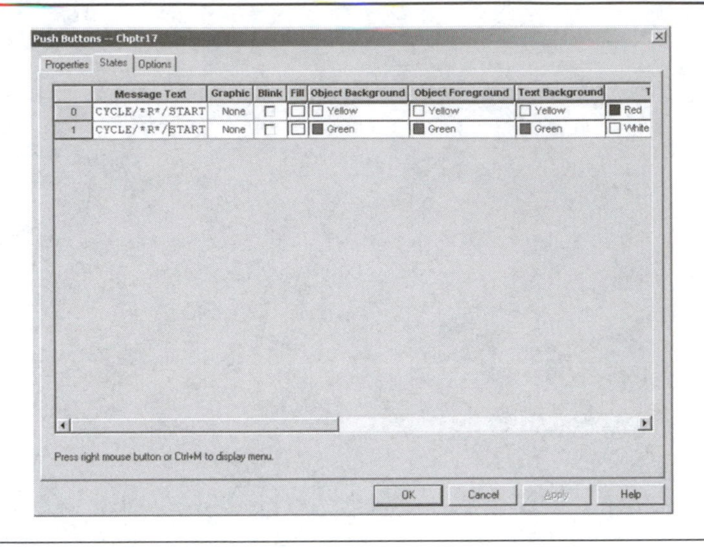

Figure 17–16
Push Buttons States edit screen for the start cycle push-button function

and the object foreground color to green and the text foreground color to white. In this way the button will be yellow when the machine is in the stopped state and turn green when the machine switches into the run state (Figure 17–16).

We now repeat the same process on the cycle stop push button, with its Write Tag being stop_cycle_pb and its Indicator Tag being stop_latch. We set the button colors to have a yellow background if the latch bit is not set and red if it is set on. When the button is pressed and the cycle stop latch bit in the PLC turns on, the button will turn red to show that the PLC has received the button input. When the cycle start button is pressed and the stop latch is reset, the stop button will return to its yellow background color.

We now have one more push button to set up and that is the button that will change the screen to the system data screen. By double clicking on the button we open the Goto menu. In this case we want to set the type to Goto Specific Screen, the screen to system data, and the key to [F5] (Figure 17–17). This screen selector button does not have the states field like our previous two push buttons did.

We need to change the text on the Goto button. With the button selected, click on the Format tab and then the Inner Text tab. This opens the Text Edit box on the top of the screen. Type in "SYSTEM DATA F5" with an enter between each word. We should also change the background color to yellow and the color of the text to dark red for better appearance. Click again on the Format tab and then the foreground color tab. Note that this is now the text color since we have the Text Edit box open; set it to dark red. Click again on the Format tab and change the background color to yellow.

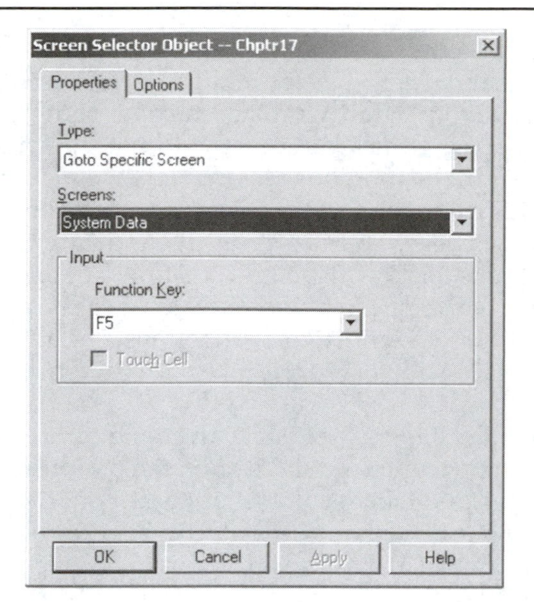

Figure 17–17 Screen Selector Object screen

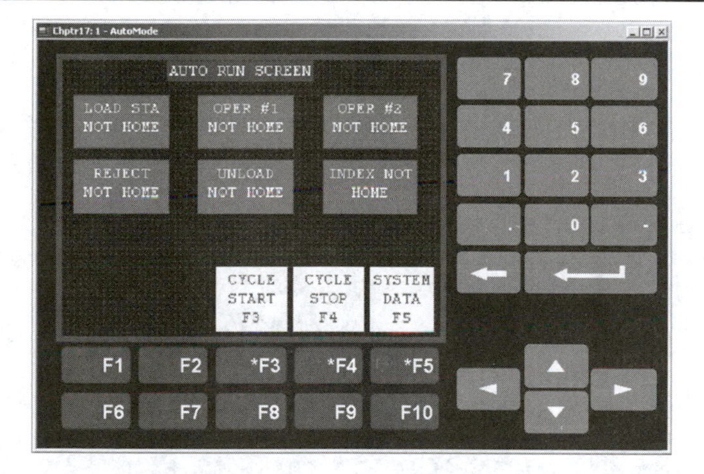

Figure 17–18 The completed auto mode screen

Click anywhere in the open space area of the display window to close the Text Edit box. Click on the button again to highlight it and then on the Format tab again. This time we want to set the foreground and background to both yellow. This will change the button to a yellow background with red text like the first two buttons. We have now completed this screen and it should appear like the screen print shown in Figure 17–18.

Manual Screen

The next screen we will create is the manual screen. Open it by double clicking on the manual screen selection in the Screen Selector box. The completed screen layout we are working toward is shown in Figure 17–19.

We start by placing the header on the top of the screen in the same manner as we did in the previous screen. Next we place the five station extend push buttons on the screen and arrange them to line up with the top row of [F] keys below. Next we put five indicator boxes under the extend push buttons to show the home position. We then place a second graphic text box between the rows of push buttons and the four indicators placed below to identify their functions. Next the indexer cycle push button is placed above the [F10] key and given its own text identifier above it.

For the four air cylinders with dual valves and prox switches, we want the feedback to change the color and text on the buttons and indicators to show the position of the cylinder. We set up the top row of [F] buttons to be the cylinder extend functions and set state 0 to be on a yellow background. When the extend switch makes, we set state 1 to be green so the color will change to green. Note that each push button has its corresponding [F] key shown. See the screen printouts of Figures 17–20 and 17–21. Next we set up the lower indicator to be the return cylinder function (home position) and again set state 0 to red and NOT HOME for text. The state 1 condition is set to green and the text set to HOME. The same process is repeated on the next four positions being careful to select the correct write tag and indicator tag names for each function.

For the unload function we have a problem in that we only have one air valve and one prox switch to tell us we are in the home position.

Figure 17–19
Manual mode screen

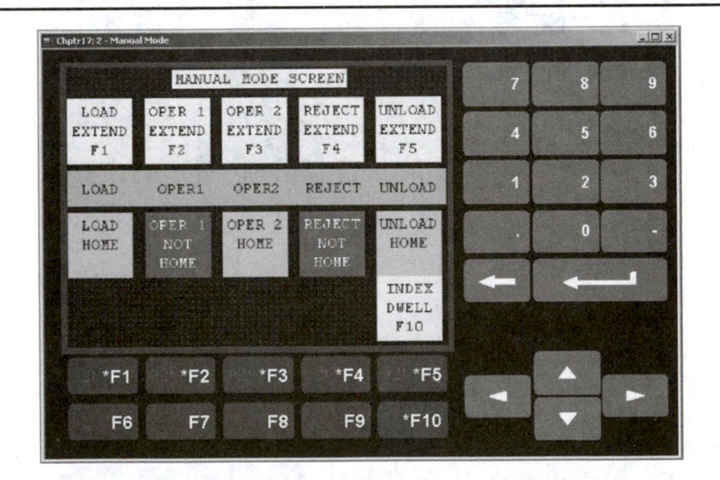

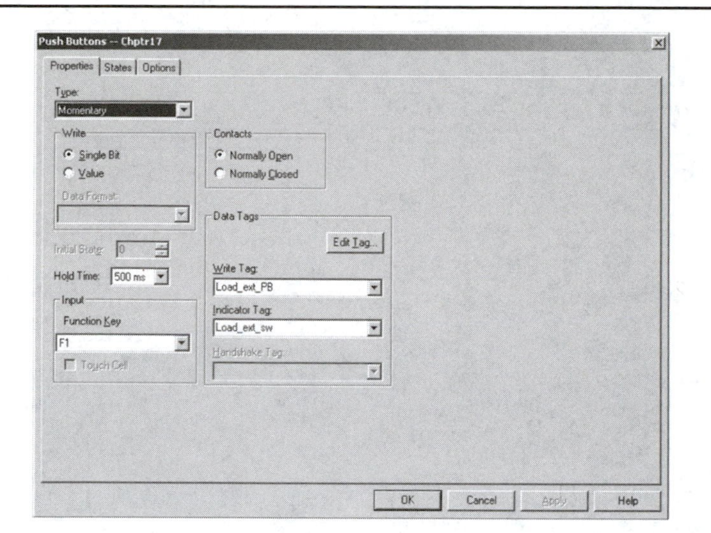

Figure 17–20
Setup properties edit screen for first push button

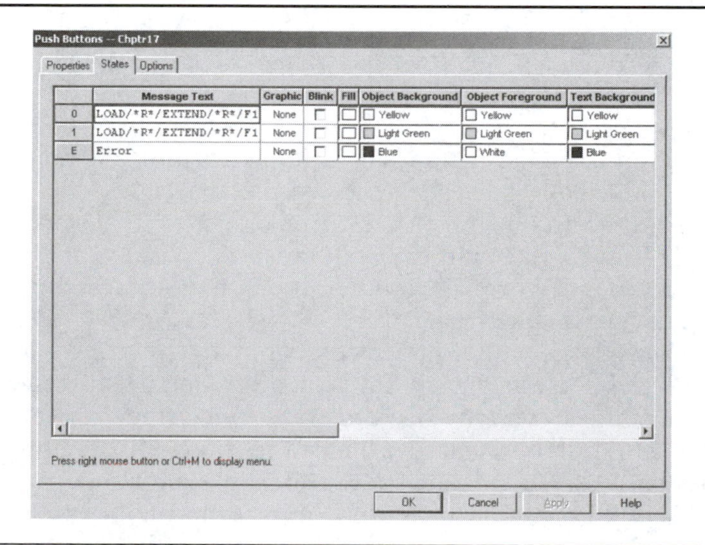

	Message Text	Graphic	Blink	Fill	Object Background	Object Foreground	Text Background
0	LOAD/*R*/EXTEND/*R*/F1	None			Yellow	Yellow	Yellow
1	LOAD/*R*/EXTEND/*R*/F1	None			Light Green	Light Green	Light Green
E	Error	None			Blue	White	Blue

Figure 17–21
Text and color setup screen for first push button

For this function we set the push button to be the cylinder extend function with state 0 set for a green background. We then set state 1 to a yellow background.

The index push button is set up in much the same manner. We set state 0 to a yellow background with a text setting of DWELL since the indexer is in the dwell position if the switch is not made. We set state 1 to a red background and set the text to read ON CAM. In this way, when the

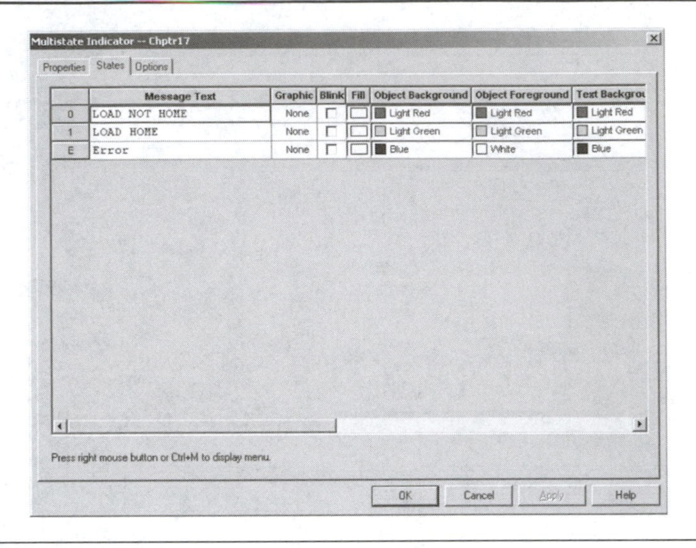

Figure 17–22
Setup screen for first indicator display

operator presses the index button, the indexer will start moving and the button will turn red when the cam switch makes and back to yellow when the index is complete.

We now set up the five indicator displays, one under each push button, to indicate when the station is in its home position. We do this by setting the 0 state to be a red background and text that states "NOT HOME". The 1 state is then set for a green background and text that says "HOME". Figure 17–19 shows three indicators in the home position and two in the not home position. Figure 17–22 shows the color and text setup screen for the first indicator display.

Note here that even with this very simple program we are almost out of [F] keys for use as push buttons on this manual control screen. For complex machines you may need to create one screen for each station or mechanism on the machine. This can become a little more complicated because you will need to first create a main manual mode screen and it will contain only a number of Goto buttons on it. You would then use these Goto buttons to get the operator to the individual manual control screens for the various stations on the machine. In any case the example here is a good starting point for understanding how to create manual control screens.

System Data Screen

We have one more screen to create at this point, the system data screen. We need to display two different types of data on this screen, station timing data and production data. Therefore, we split the screen into two

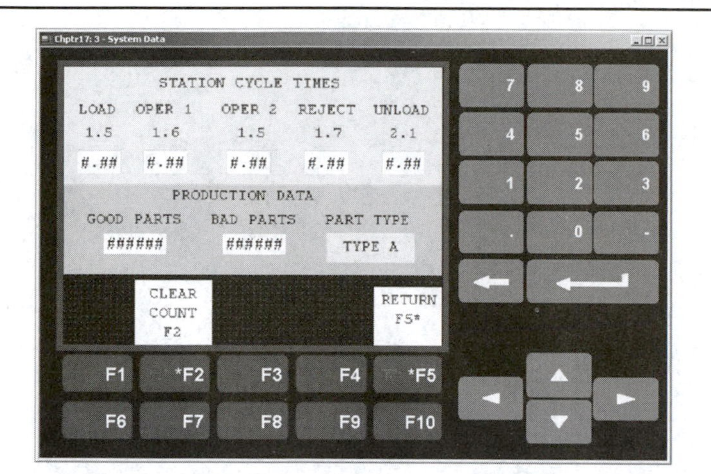

Figure 17–23
The completed System Data screen

halves and insert a graphic rectangle covering about half the screen for each data type. We want the background color to be different for each section.

When you create overlapping objects, as we will do on this screen, you can always change the order in which they are shown on the screen. You do this by clicking on the Edit tab and using the Bring to Front or Send to Back functions to get the right object on top. The completed screen is shown in Figure 17–23 for reference. Please refer to it as we begin to create the system data screen.

Timing Display

We start the top half by creating two text fields: one with the heading "STATION CYCLE TIMES" and a second that contains the station names and below them the normal cycle times for each station. Below this we place five numeric data fields, which are created by clicking on Objects, then Numeric Data Display (Figure 17–24). The cursor changes into a crosshair, which is used to create five boxes, one under each station heading. If you make a mistake, keep going because you can always move or resize them later. To set up each display, double click on the box to open the Numeric Data Display menu.

Once this menu is open, you will need to enter the Read Tag name for the word you want to be displayed. Since this will be a timer value and we have our timers set for 0.01 second, we need to set the decimal point two positions from the right. We should also set the maximum number of digits to 4 from the default setting of 6. Because the timer value will be a 4-digit unsigned integer, we need to scale it for 2 decimal points. We do this by clicking on the Edit Tag box and then set the scale factor on the pop-up menu to .01 instead of the default value of 1.

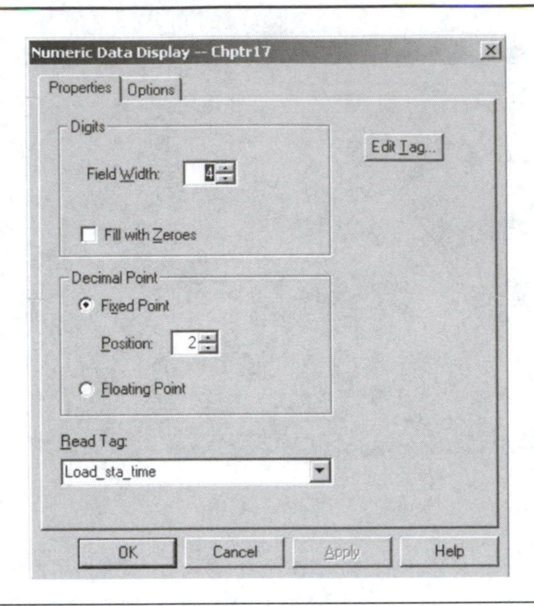

Figure 17–24 Numeric Data Display screen

Let's look at an example of one of the weaknesses of this low-end HMI display system. If we had an internal basic language program, we could compare the normal time reading with the latest reading and change the background color to red if the new value exceeded the normal value. The Panel View terminal does not have this ability. We could still get this function by performing the math in the PLC itself and use a bit to change a multistate indicator to show the error condition. For this example, however, we will settle for just displaying the actual numbers and let the operator be the judge of when a problem is present. The other four stations are set up in the same manner.

Production Data

For this section we begin as we just did for the timing display by placing two text boxes on the bottom rectangle to define the display header and function headers. We then create two more numeric indicator boxes under the GOOD PARTS and BAD PARTS headings. These should be set up in the same manner as the top ones with the exception that the decimal point is left at the default setting and we also leave the default setting of 6 places.

If the shift production volume was greater than 32,000 pieces we would have to use two counters and combine their values to display the whole value. Below the counter fields we place the counter reset push button and set it for the [F2] key. This is a standard push button with the tag CLEAR_CNT assigned to address B3:2/13.

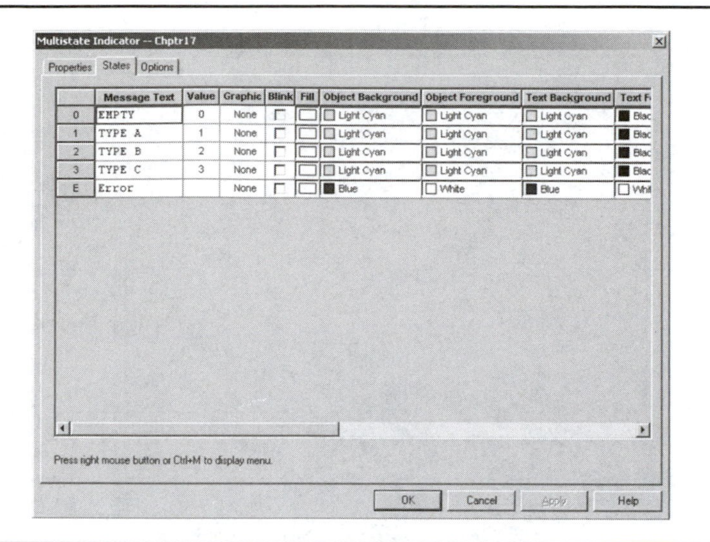

Figure 17–25
Multistate Indicator setup box for type display

The type display is a normal multistate indicator box. The only difference is that we have four states and we use the word tag "TYPE" to control it. We set the state 0 text to EMPTY, state 1 to TYPE A, state 2 to TYPE B, and state 3 to TYPE C. See Figure 17–25 for the correct setup parameters.

We have one more element to add to the screen and that is the return push button. Remember that we got to this screen by pressing the Goto button on the auto mode screen and therefore we need a way to get back to the auto mode screen. We enter this button by clicking on the Objects tab, Screen Selector, and the return box. Open the box on the lower right corner of the screen above the [F5] key. If you double click on the button now you can open the Setup menu. Note that the type "Previous Screen" has been selected and the [F5] key has been assigned to this button function (Figure 17–26).

At this point in our application design, we need to take care of one more detail. We want the auto and manual screens to appear automatically depending on the state of the PLC system. A word tag named "screen_call" was assigned to be used to select which screen the HMI should be showing. To set this function, we need to click on the Application tab, Terminal Setup tab, and then the Control Tags tab. This will open a menu as shown in Figure 17–27. All we need do is set the Screen Number Tag to "screen_call" using the drop-down box.

This ends the screen creation process but we still have one very important function to define and that is the alarm banner function. The alarm banner is used to display our diagnostic messages and is discussed in the next section.

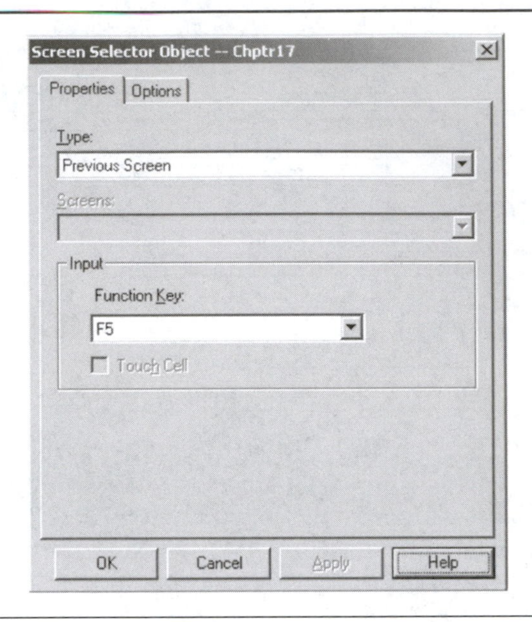

Figure 17–26
Screen Selector Object screen

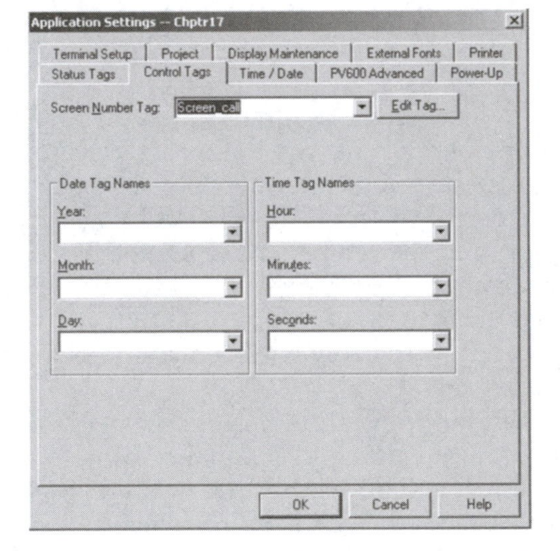

Figure 17–27
Application Settings screen

ALARM BANNER

We start this process by clicking on the Screen tab, then on the Create Alarm Banner tab. This brings up the default alarm banner screen as shown in Figure 17–28. We should modify this default screen a little by enlarging the clear button and the text field. We should also expand the

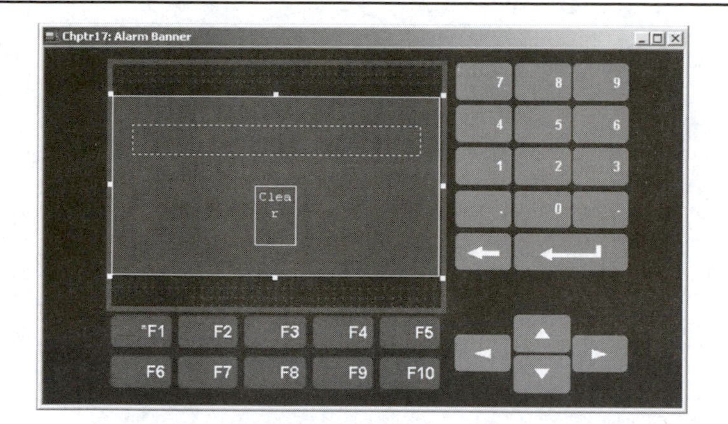

Figure 17–28
Default alarm banner screen

alarm banner to make it show up better; and change the Clear button to a yellow background with the [F3] key assigned (see Figure 17–29).

Save the modified screen by clicking on the X in the top right corner. You will notice that you now have a new screen showing on the top of the screen list but it has no screen number. That is because the alarm banner is controlled by the alarm tag word. Our next task is to set up the alarm tags and messages. We do this by clicking on the Applications tab and then the Alarm Setup tab. This opens the first of three menus for alarm message setup (Figure 17–30). On the first menu we will change the history size to 5 and set it to clear on power up.

Saving more than a few alarm messages is a waste of time and memory in most cases. If you are getting that many alarms, something

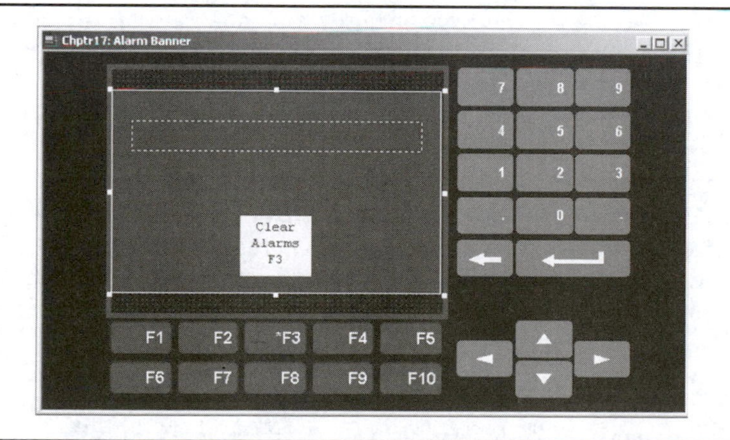

Figure 17–29
Modified alarm banner screen

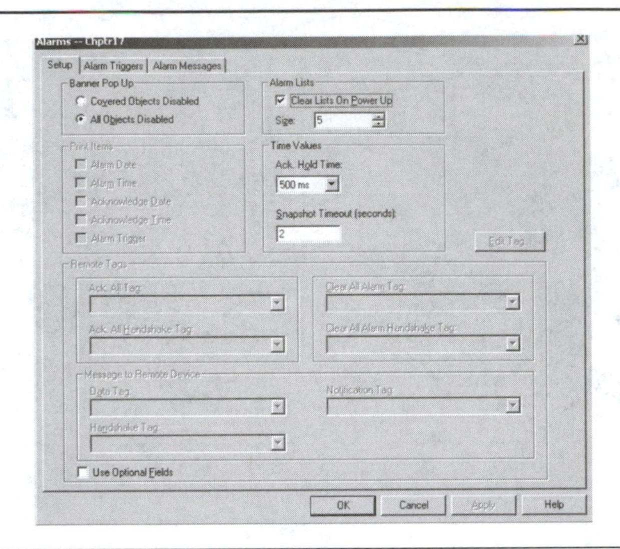

Figure 17–30
Alarm Setup screen

is probably wrong with your system in the first place. You can always change this setting later if you need a longer alarm record. You can also select to have all covered buttons disabled or the whole screen. Because our alarm banner covers most of the screen, we will leave it set to the default of All Objects Disabled. See Figure 17–30 for complete setup details.

We will not be using any acknowledge buttons so we can ignore this setting. The next tab sets up the word or words that will be used to signal that an alarm condition is present. Each word is 16 bits wide and therefore can represent 32,000 messages. In larger systems we could have more than one PLC connected to our display terminal and we would then have a different alarm word assigned to each PLC system.

We need to set up the trigger tag to our alarm_word tag name and indicate that it is a value type tag (Figure 17–31). This means that if the value of the word alarm_msg is anything but 0, an alarm condition is present. The next tab will allow us to define what word value represents each alarm message.

Before we start filling in this menu, we need to go back and look at our Chapter 16 program's B12 file and make a list of which bits are assigned to which error messages. It works best if we can use the B12 value + 1 as our alarm trigger value because it makes things easier to keep track of. We need to add 1 because a word value of 0 means no alarm present. A list follows of the 20 alarm messages we are required to display, the alarm word value assigned to them, and the PLC bit that will generate the error message. We will have to modify our PLC ladder program to move the assigned value into N7:1, which is the address assigned to the tag alarm_word in our tag file. After you have created a few operator terminal programs you can skip

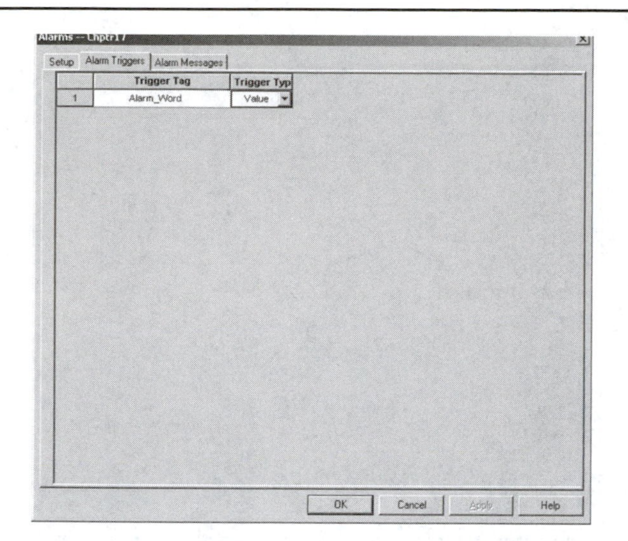

Figure 17–31
Setting the trigger tag

the following chart, but it is handy to have in the documentation package if someone else has to make changes to your program at a later date.

Alarm Message Chart

Bit	Value	Message
B12:0/0	1	Load station extend jam up or prox failure
B12:0/1	2	Load station extend solenoid failure
B12:0/2	3	Load station return jam up or prox failure
B12:0/3	4	Load station return solenoid failure
B12:0/4	5	Operation #1 extend jam up or prox failure
B12:0/5	6	Operation #1 extend solenoid failure
B12:0/6	7	Operation #1 return jam up or prox failure
B12:0/7	8	Operation #1 return solenoid failure
B12:0/8	9	Operation #2 extend jam up or prox failure
B12:0/9	10	Operation #2 extend solenoid failure
B12:0/10	11	Operation #2 return jam up or prox failure
B12:0/11	12	Operation #2 return solenoid failure
B12:0/12	13	Reject station extend jam up or prox failure
B12:0/13	14	Reject station extend solenoid failure
B12:0/14	15	Reject station return jam up or prox failure
B12:0/15	16	Reject station return solenoid failure
B12:1/0	17	Unload station extend failure
B12:1/1	18	Unload station return failure
B12:1/2	19	Indexer start up error
B12:1/3	20	Indexer jam up error

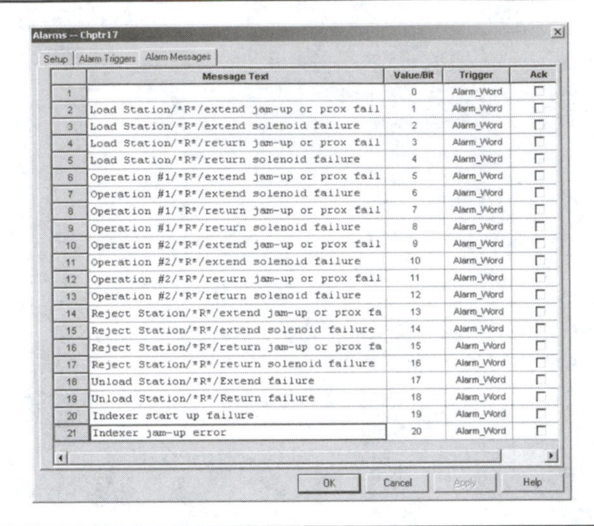

	Message Text	Value/Bit	Trigger	Ack
1		0	Alarm_Word	
2	Load Station/*R*/extend jam-up or prox fail	1	Alarm_Word	
3	Load Station/*R*/extend solenoid failure	2	Alarm_Word	
4	Load Station/*R*/return jam-up or prox fail	3	Alarm_Word	
5	Load Station/*R*/return solenoid failure	4	Alarm_Word	
6	Operation #1/*R*/extend jam-up or prox fail	5	Alarm_Word	
7	Operation #1/*R*/extend solenoid failure	6	Alarm_Word	
8	Operation #1/*R*/return jam-up or prox fail	7	Alarm_Word	
9	Operation #1/*R*/return solenoid failure	8	Alarm_Word	
10	Operation #2/*R*/extend jam-up or prox fail	9	Alarm_Word	
11	Operation #2/*R*/extend solenoid failure	10	Alarm_Word	
12	Operation #2/*R*/return jam-up or prox fail	11	Alarm_Word	
13	Operation #2/*R*/return solenoid failure	12	Alarm_Word	
14	Reject Station/*R*/extend jam-up or prox fa	13	Alarm_Word	
15	Reject Station/*R*/extend solenoid failure	14	Alarm_Word	
16	Reject Station/*R*/return jam-up or prox fa	15	Alarm_Word	
17	Reject Station/*R*/return solenoid failure	16	Alarm_Word	
18	Unload Station/*R*/Extend failure	17	Alarm_Word	
19	Unload Station/*R*/Return failure	18	Alarm_Word	
20	Indexer start up failure	19	Alarm_Word	
21	Indexer jam-up error	20	Alarm_Word	

Figure 17–32
Partial alarm messages list

We now return to the alarms menu and begin filling in the messages from our list. If you look at the screen printout of Figure 17–32 you will see that 21 messages have been entered. A return has been placed between the station location and the error message. This will cause a two-line display with the station location on top and the error message below.

To enter the next message, right click anywhere on the screen and click on Append Alarm in the pop-up menu that appears. This opens the next message number. Note also that the trigger word is shown in each line. This is required because we could have multiple trigger words if more than one PLC is connected to this same display terminal.

At this point we are finished with our first revision of this application. Before we download or close the application, we should verify that the application is good. You do this by clicking on the Application tab and then clicking Validate All Selections. The program will check all the screens and data files to verify that everything is correct. If it finds any errors, it will list them for you. If any errors are found you must clean them up before you can download to the real terminal.

CHANGES TO SAMPLE PROGRAM

With our display terminal application complete we now have to rework the PLC ladder program to generate the additional bits and words we have assigned in our HMI display application. We will start with the main program file because we need to make several additions to it. Instead of showing the whole program we show just the rungs that were changed or

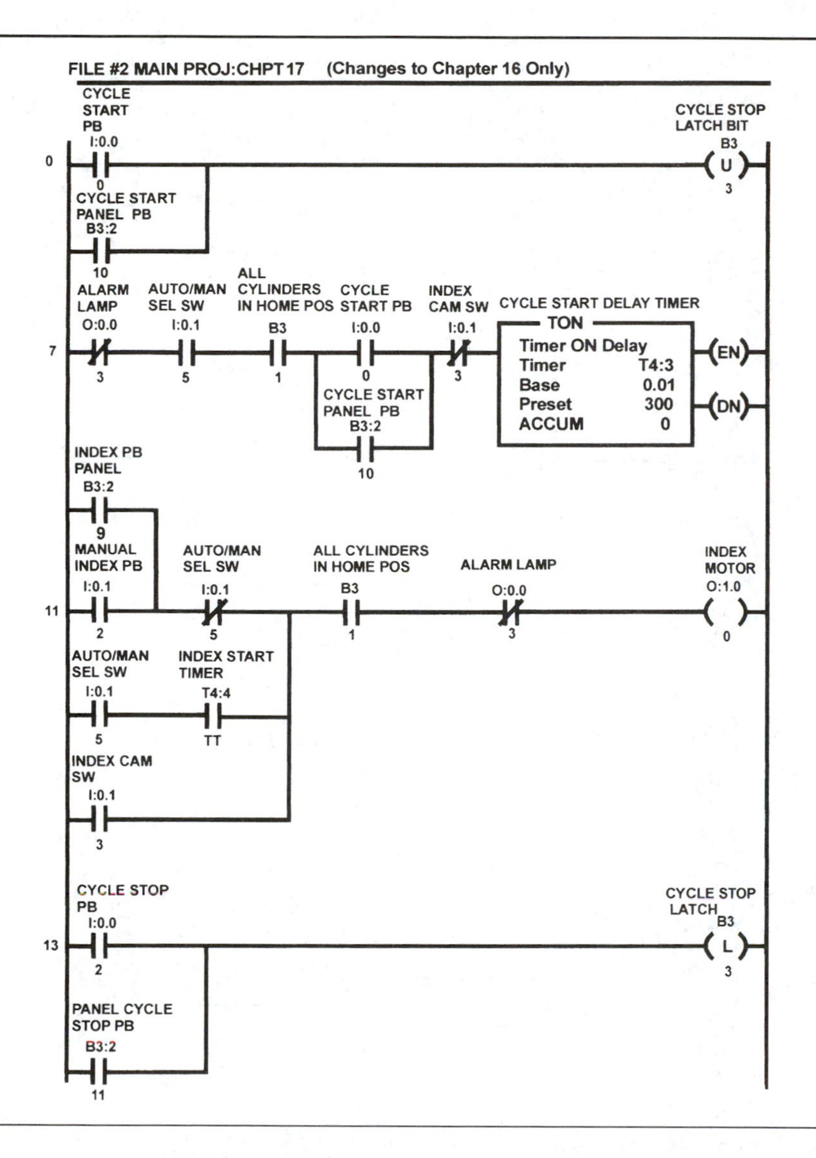

Figure 17–33
Changes to main program ladder file 2

added. The rest of the program, including the entire auto mode file, file 3, is unchanged from Chapter 16. Refer to Figures 17–33, 17–34, and 17–35 as these changes are discussed.

Main Program Changes

The first change occurs on rung 0, where we need to add our new panel cycle start button B3:2/10 to the reset for the stop latch. Next we need to add the panel cycle start B3:2/10 bit in parallel with the existing hardwired I/O cycle start input in rung 7.

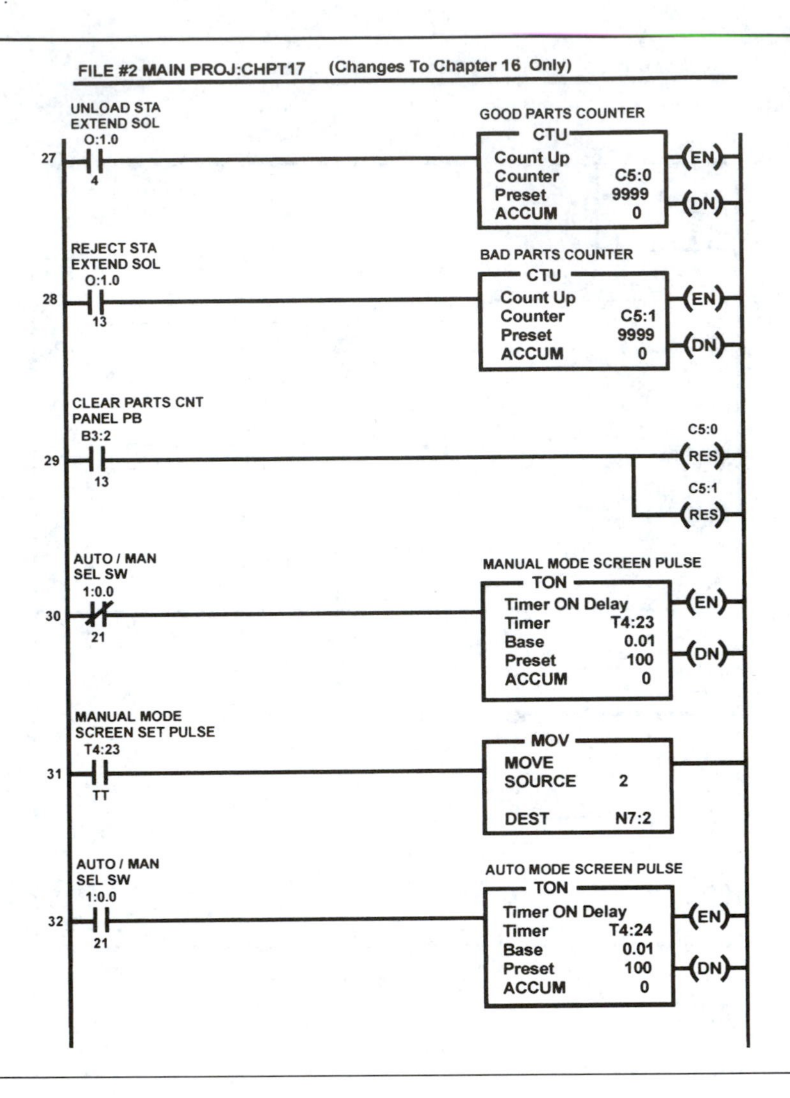

Figure 17–34 Changes to main program ladder file 2, *continued*

We next have to add the panel index PB B3:2/9 in parallel with the existing hard-wired I/O for manual index PB in rung 11. In rung 13 we need to add the panel cycle stop PB B3:2/11 in parallel with the existing hard-wired cycle stop PB. That concludes the rung changes, but we still have some new rungs to add to the end of the program.

Beginning with rung 27, we add the two counters for good and bad parts. These counters are driven off the unload and reject solenoids. Next we insert the two counter reset instructions being driven by the panel reset PB. Our next task is to send a number to the display's screen_call tag N7:2 whenever we change program states between the auto and manual modes. In order for Panel View's screen Goto buttons to work, we must not hold the

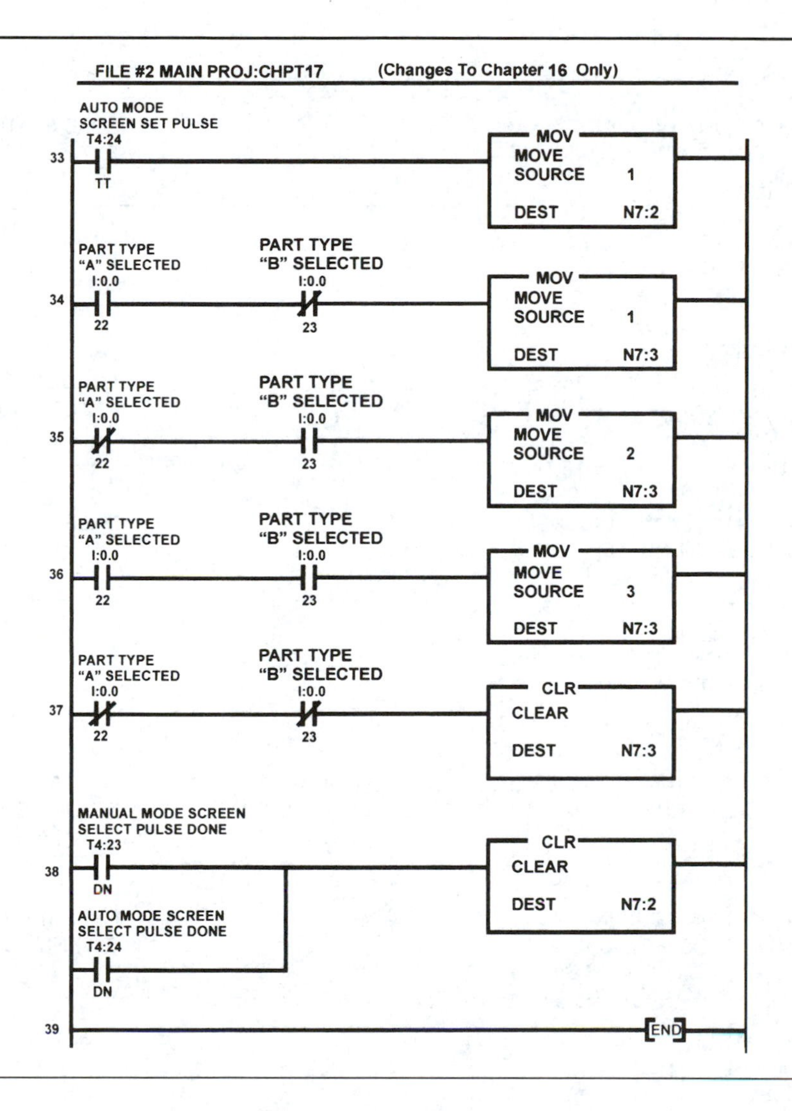

Figure 17–35
Changes to main program ladder file 2, *continued*

screen call value at a number above zero, so we generate a 1-second pulse and then return the value to zero. Rungs 30 and 31 send the manual screen call number of 2. Rungs 32 and 33 send the auto mode screen call of 1. Rung 38 clears the screen_call tag address N7:2 back to "0," otherwise screen Goto buttons local to the panel view terminal would not work. Rungs 34 through 37 send the part type value to the screen address for tag name Type.

Manual Mode Changes

In the manual mode subroutine we have to parallel the existing manual extend push buttons for the five air cylinders with our new panel buttons. Refer to Figures 17–36 and 17–37. We start with rung 0 and add B3:2/0 in

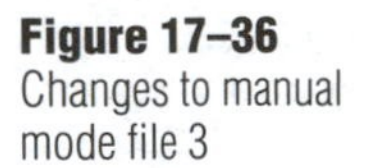

Figure 17–36 Changes to manual mode file 3

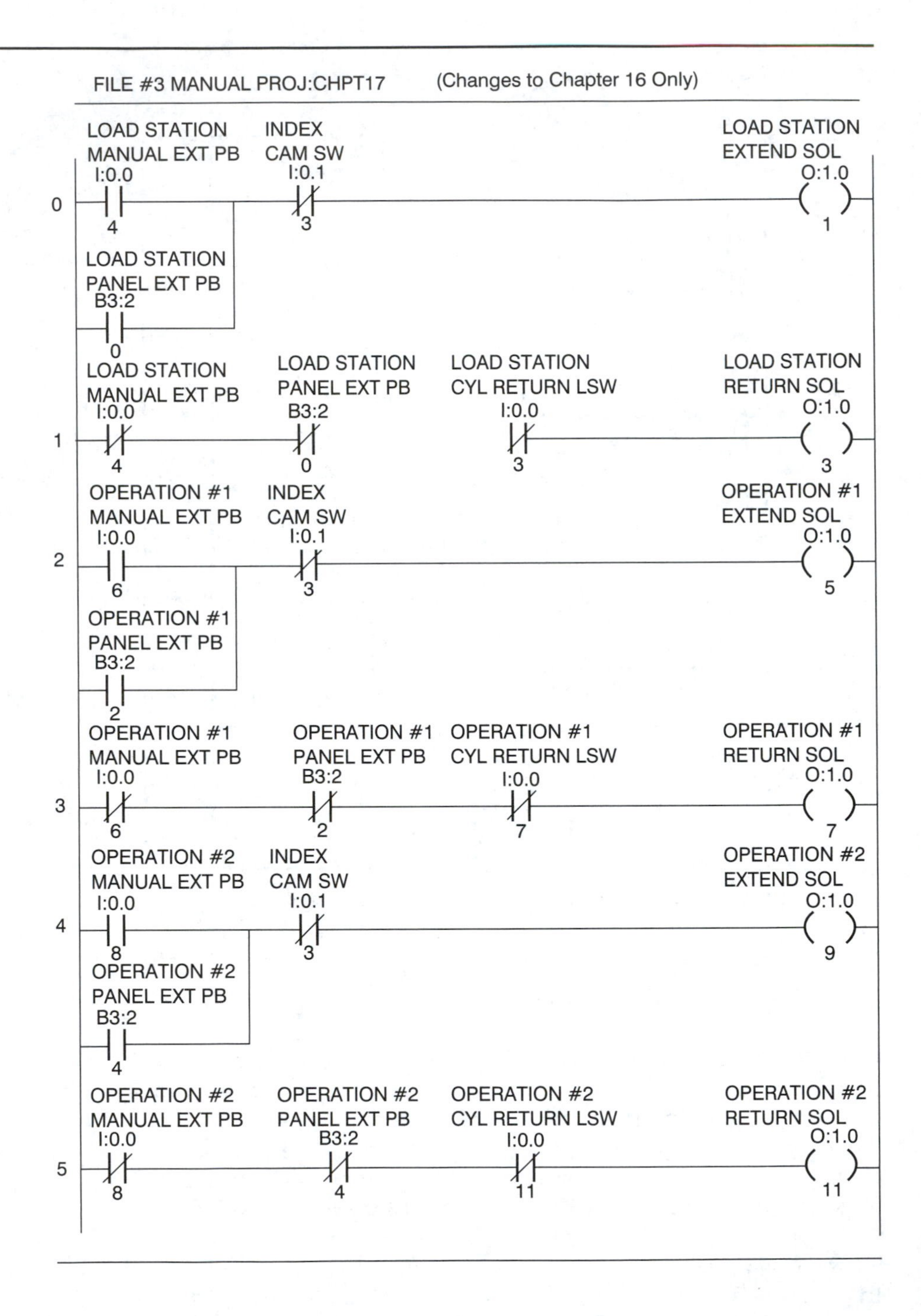

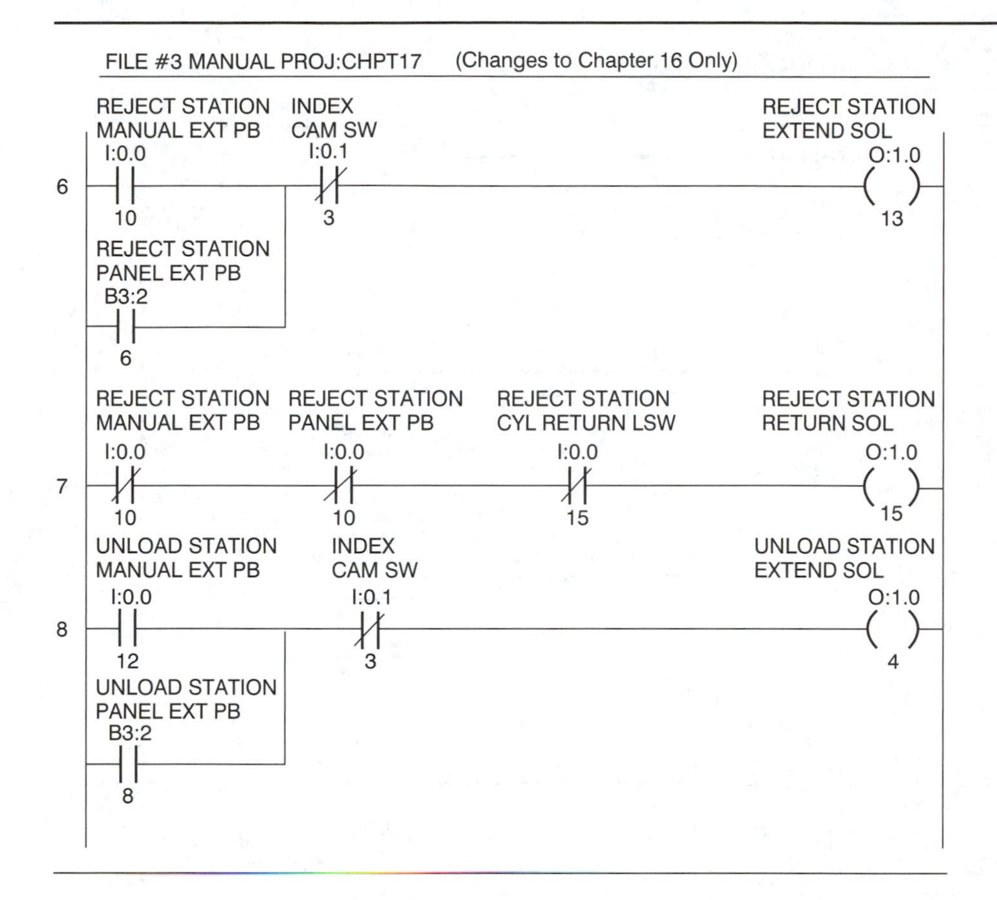

Figure 17–37 Changes to manual mode file 3, *continued*

parallel with the existing hard-wired button. In the next rung we add the NOT function of our panel button in series with the NOT function of the existing button. In this way if either button is pressed the load station cylinder will extend. If both buttons are released, then the cylinder will return.

Note that we have added the NOT contact of the return limit switch to each return solenoid rung. This way, when the air cylinder returns home the return solenoid will be shut off. Without these changes the return solenoid would stay energized most of the time and this may overheat the solenoid. We then repeat the exact same additions to the other four solenoid control rungs.

This ends the changes to the manual section. Note that there are no changes to the auto mode section.

Diagnostic Subroutine Changes

The first additions to the diagnostic subroutine are on rung 43. We stated earlier that we needed to time each major station for our display to show

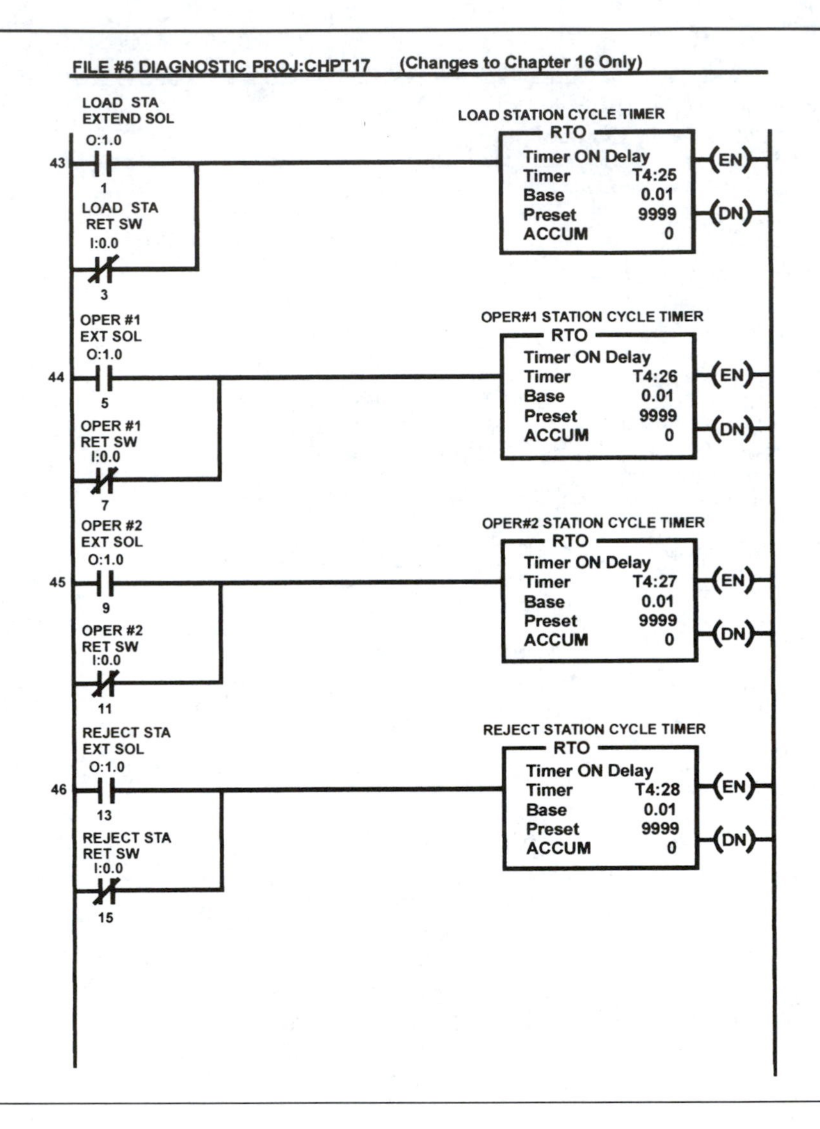

Figure 17–38 Changes to diagnostic ladder file 5

the last cycle time. This will be used when we have a timing problem and need to compare the last time to the normal cycle time. Refer to Figures 17–38, 17–39, and 17–40 as you read about the changes.

Note that we will use the RTO timer for this section. We must time the station and then hold that time until we can transfer the value to the display tag location in memory. Each timer is driven by the solenoid output bit and the cylinder home switch. Note that the preset

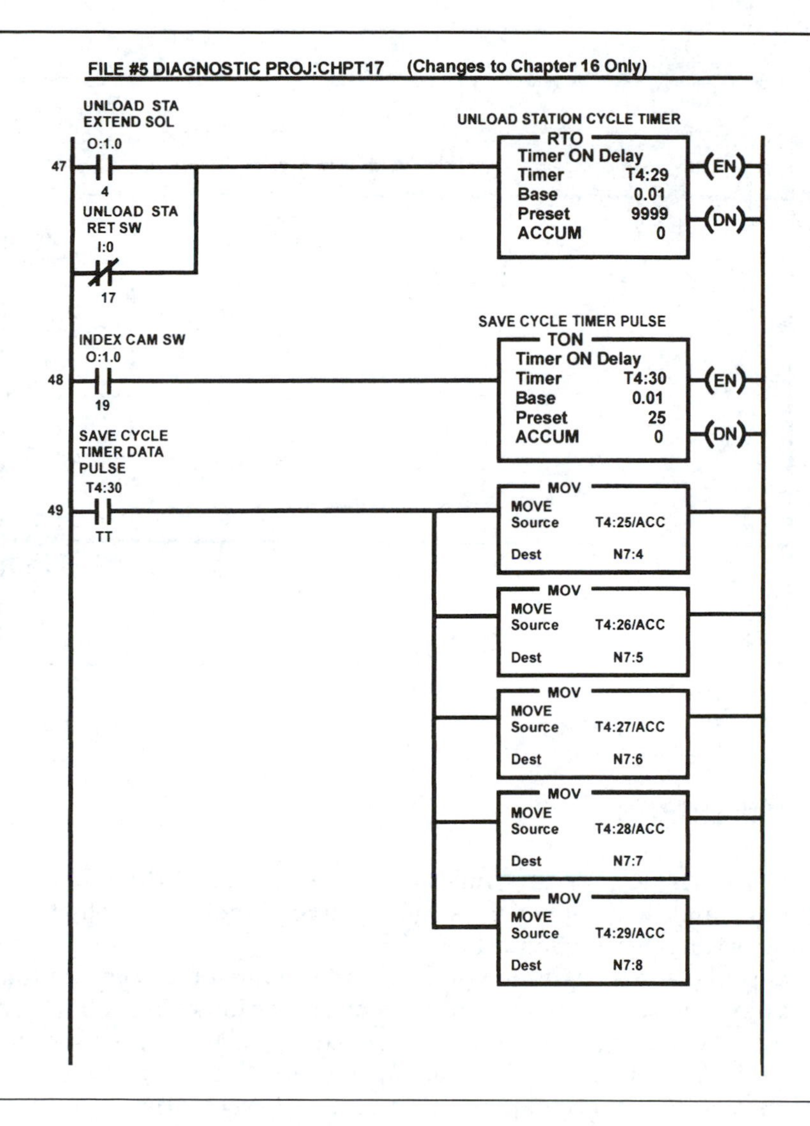

Figure 17–39 Changes to diagnostic ladder file 5, *continued*

is set for 9999 because we will not be looking for a done bit in this application.

In rung 48 we add a timer to give us a pulse of 250 milliseconds each time the index cam switch makes contact. In rung 49 we use the TT bit from this pulse timer to move the timer accumulated values to the tag memory locations. In rung 50 we use the DN bit to clear all of the timers to get ready for the next cycle.

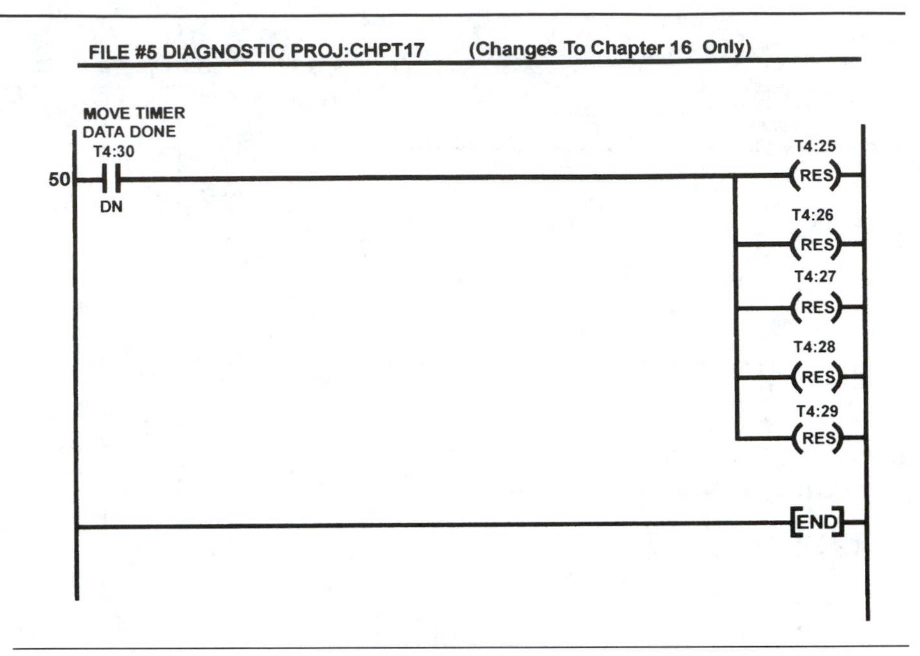

Figure 17–40
Changes to diagnostic ladder file 5, *continued*

CONCLUSIONS

This concludes our operator interface design project. While this project was very simple in concept it does contain most of the elements that go into making up an operator control interface.

As stated at the beginning of the chapter, a great number of different display systems are available on the market. Various Panel View terminals are shown in Figure 17–41. When picking an HMI display system for your project, you must consider the major characteristics and capabilities that are required and then pick the lowest cost system that can fill these requirements.

Be sure you also consider the environmental conditions in which the HMI display must live. Touch screens are a great option unless the operator works in an environment that contains sticky and tacky materials or inks that will contaminate the screen. Touch screens are also very sensitive to sharp objects like long fingernails, pens, or tools. Even the Mylar keypads can be damaged by sharp objects if they are used repeatedly to press on the key area.

If used in a high static electricity area or an area containing strong magnetic fields, be sure the HMI display device is properly grounded according to the manufacturer's specifications. In many applications

Figure 17–41 Several different Panel View terminals. *Courtesy*: Rockwell Automation, West Allis, Wisconsin.

the display can be mounted on a swivel arm that allows the operator to position it to meet her changing needs. On some large machines or assembly lines, the display can be a portable device used by the maintenance group for manual control. In this case we may have data highway plugs at various points along the line and in this way be able to plug into the closest point to the mechanism on which we need to work. This alleviates the need for dedicated manual panels and buttons on the line.

REVIEW QUESTIONS

1. Why is it important to get the machine operator's input before starting a new operator interface design?
2. What are *tags* and how do we assign them in a PLC system?
3. Why is it a good idea to make a list of all the I/Os that will be used to pass information between the PLC and display systems?
4. When creating push buttons using the Panel Builder software, what is the difference between the write tag and the indicator tag?
5. Why can we use only two of the states in multistate indicators if our tag is a bit type?
6. How does the Panel View terminal know if an alarm condition has occurred?
7. How can we call up a particular screen on the Panel View terminal from the PLC system?
8. When assigning alarm messages why do we always have to include the alarm word tag name?

9. What types of environmental conditions may make a touch screen a bad choice for a project display device?

10. Why is it not a good idea to use a normal office-style IBM PC on the factory floor as an interface and display system to a PLC system?

11. Why did we use RTO timers to record the cycle times of each station in our design?

LAB ASSIGNMENT

Modify the lab program you wrote in the Chapter 16 lab to include a Panel View terminal. Be sure to have at least a main run screen, a manual screen, and a screen to display data. All the error conditions you programmed to detect in Chapter 16 must have error messages in this new program.

APPENDIX A

PLC TEACHING KIT I/O LIST AND SCHEMATICS

Outputs:

O:0.0/0	LAMP #1
O:0.0/1	LAMP #2
O:0.0/2	LAMP #3
O:0.0/3	LAMP #4
O:0.0/4	LAMP #5
O:0.0/5	LAMP #6
O:0.0/6	LAMP #7
O:0.0/7	LAMP #8
O:0.0/8	LAMP #9
O:0.0/9	LAMP #10
O:0.0/10	LAMP #11
O:0.0/11	LAMP #12
O:0.0/12	LAMP #13
O:0.0/13	LAMP #14
O:0.0/14	LAMP #15
O:0.0/15	LAMP #16
O:1.0/0	INDEX MOTOR
O:1.0/1	CYL 1 EXT SOLENOID
O:1.0/2	NOT USED
O:1.0/3	CYL 1 RET SOLENOID
O:1.0/4	CYL 5 EXT SOLENOID
O:1.0/5	CYL 2 EXT SOLENOID
O:1.0/6	NOT USED
O:1.0/7	CYL 2 RET SOLENOID
O:1.0/8	NOT USED
O:1.0/9	CYL 3 EXT SOLENOID
O:1.0/10	NOT USED
O:1.0/11	CYL 3 RET SOLENOID
O:1.0/12	NOT USED
O:1.0/13	CYL 4 EXT SOLENOID
O:1.0/14	NOT USED
O:1.0/15	CYL 4 RET SOLENOID

Inputs:

I:0.0/0	PB #1
I:0.0/1	CYL 1 EXTEND PROX SW
I:0.0/2	PB #2
I:0.0/3	CYL 1 RETURN PROX SW
I:0.0/4	PB #3
I:0.0/5	CYL 2 EXTEND PROX SW
I:0.0/6	PB #4
I:0.0/7	CYL 2 RETURN PROX SW
I:0.0/8	PB #5
I:0.0/9	CYL 3 EXTEND PROX SW
I:0.0/10	PB #6
I:0.0/11	CYL 3 RETURN PROX SW
I:0.0/12	PB #7
I:0.0/13	CYL 4 EXTEND PROX SW
I:0.0/14	PB #8
I:0.0/15	CYL 4 RETURN PROX SW
I:0.1/0	PB #9
I:0.1/1	CYL 5 RETURN PROX SW
I:0.1/2	PB #10
I:0.1/3	INDEX CAM SWITCH
I:0.1/4	SEL SW #1
I:0.1/5	SEL SW #2
I:0.1/6	SEL SW #3
I:0.1/7	SEL SW #4

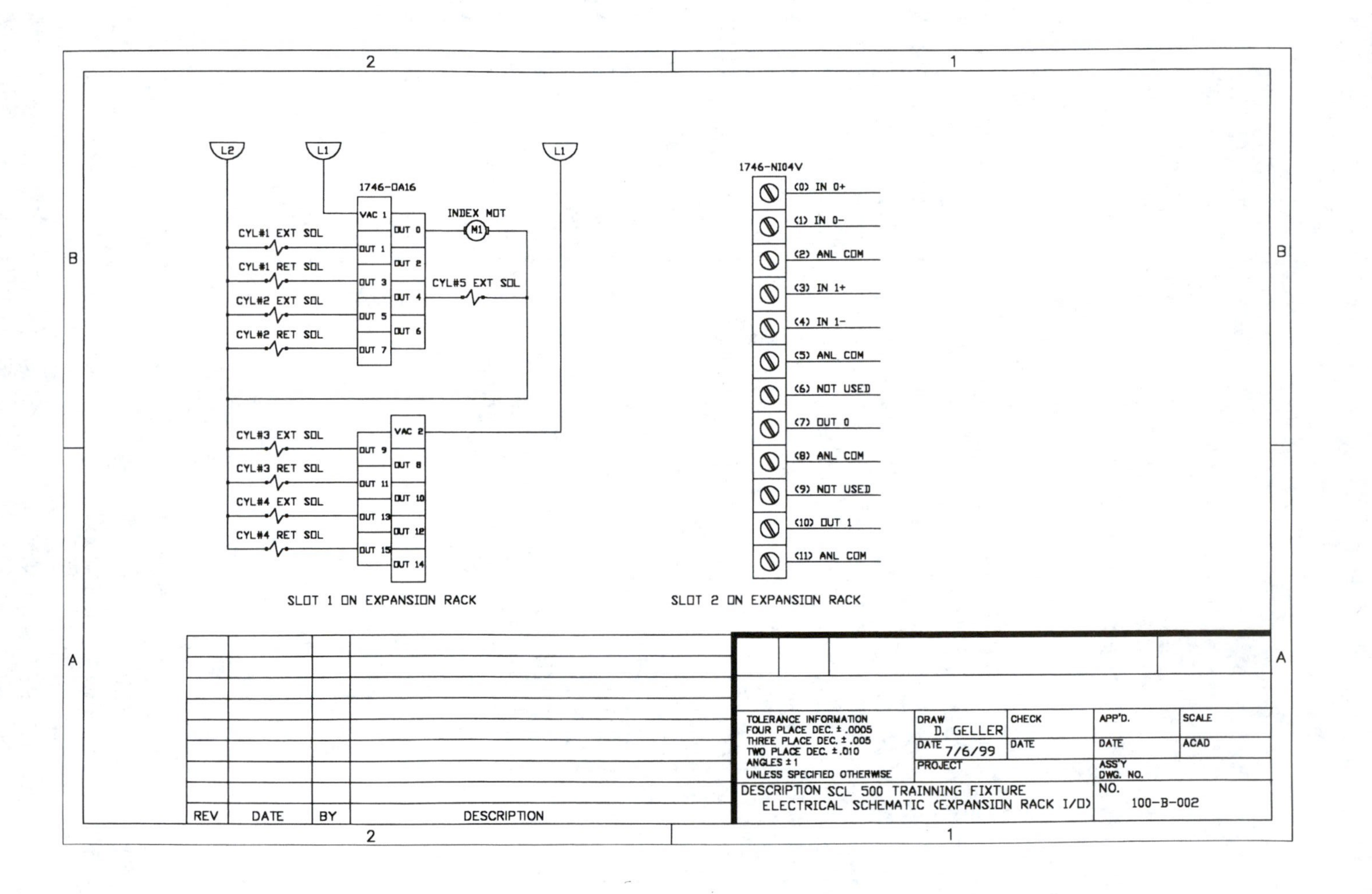

L2
L1
L1
1746-OA16
VAC 1
OUT 0
OUT 1
OUT 2
OUT 3
OUT 4
OUT 5
OUT 6
OUT 7
INDEX MOT
M1
CYL#5 EXT SOL
CYL#1 EXT SOL
CYL#1 RET SOL
CYL#2 EXT SOL
CYL#2 RET SOL
CYL#3 EXT SOL
CYL#3 RET SOL
CYL#4 EXT SOL
CYL#4 RET SOL
VAC 2
OUT 8
OUT 9
OUT 10
OUT 11
OUT 12
OUT 13
OUT 14
OUT 15
SLOT 1 ON EXPANSION RACK
1746-NI04V
(0) IN 0+
(1) IN 0-
(2) ANL COM
(3) IN 1+
(4) IN 1-
(5) ANL COM
(6) NOT USED
(7) OUT 0
(8) ANL COM
(9) NOT USED
(10) OUT 1
(11) ANL COM
SLOT 2 ON EXPANSION RACK
REV
DATE
BY
DESCRIPTION
TOLERANCE INFORMATION
FOUR PLACE DEC. ±.0005
THREE PLACE DEC. ±.005
TWO PLACE DEC. ±.010
ANGLES ±1
UNLESS SPECIFIED OTHERWISE
DRAW
D. GELLER
DATE 7/6/99
PROJECT
CHECK
DATE
APP'D.
DATE
ASS'Y DWG. NO.
SCALE
ACAD
DESCRIPTION SCL 500 TRAINNING FIXTURE
ELECTRICAL SCHEMATIC (EXPANSION RACK I/O)
NO. 100-B-002

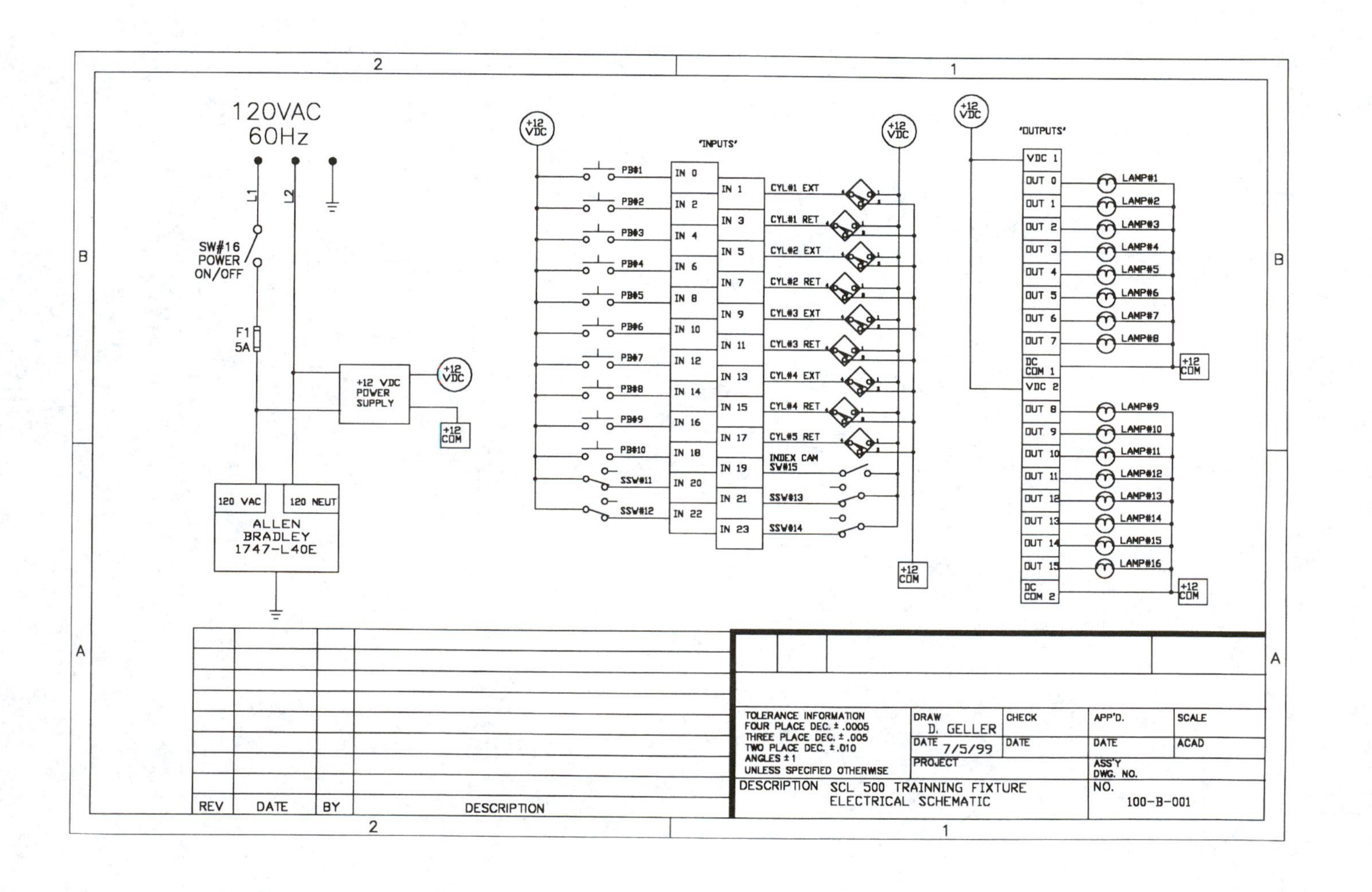

120VAC
60Hz
L1
L2
SW#16
POWER
ON/OFF
F1
5A
+12 VDC
POWER
SUPPLY
+12
VDC
+12
COM
120 VAC
120 NEUT
ALLEN
BRADLEY
1747-L40E
'INPUTS'
PB#1
PB#2
PB#3
PB#4
PB#5
PB#6
PB#7
PB#8
PB#9
PB#10
SSW#11
SSW#12
IN 0
IN 2
IN 4
IN 6
IN 8
IN 10
IN 12
IN 14
IN 16
IN 18
IN 20
IN 22
IN 1
IN 3
IN 5
IN 7
IN 9
IN 11
IN 13
IN 15
IN 17
IN 19
IN 21
IN 23
CYL#1 EXT
CYL#1 RET
CYL#2 EXT
CYL#2 RET
CYL#3 EXT
CYL#3 RET
CYL#4 EXT
CYL#4 RET
CYL#5 RET
INDEX CAM
SW#15
SSW#13
SSW#14
'OUTPUTS'
VDC 1
OUT 0
OUT 1
OUT 2
OUT 3
OUT 4
OUT 5
OUT 6
OUT 7
DC
COM 1
VDC 2
OUT 8
OUT 9
OUT 10
OUT 11
OUT 12
OUT 13
OUT 14
OUT 15
DC
COM 2
LAMP#1
LAMP#2
LAMP#3
LAMP#4
LAMP#5
LAMP#6
LAMP#7
LAMP#8
LAMP#9
LAMP#10
LAMP#11
LAMP#12
LAMP#13
LAMP#14
LAMP#15
LAMP#16
REV
DATE
BY
DESCRIPTION
TOLERANCE INFORMATION
FOUR PLACE DEC. ± .0005
THREE PLACE DEC. ± .005
TWO PLACE DEC. ± .010
ANGLES ± 1
UNLESS SPECIFIED OTHERWISE
DRAW
D. GELLER
DATE 7/5/99
CHECK
DATE
PROJECT
APP'D.
DATE
SCALE
ACAD
ASS'Y
DWG. NO.
DESCRIPTION SCL 500 TRAINNING FIXTURE
ELECTRICAL SCHEMATIC
NO.
100-B-001

INDEX